Mobile Communications

Mobile Communications

**Edited by
Adam Houle**

WILLFORD PRESS

www.willfordpress.com

Published by Willford Press,
118-35 Queens Blvd., Suite 400,
Forest Hills, NY 11375, USA

ISBN: 978-1-68285-343-6

Cataloging-in-Publication Data

Mobile communications / edited by Adam Houle.
   p. cm.
Includes bibliographical references and index.
ISBN 978-1-68285-343-6
1. Mobile communication systems. 2. Wireless communication systems. 3. Cognitive radio networks.
4. Communication--Network analysis. I. Houle, Adam.
TK5103.2 .M66 2017
621.384--dc23

For information on all Willford Press publications
visit our website at www.willfordpress.com

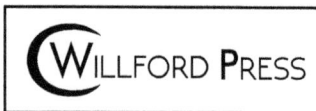

WILLFORD PRESS

Printed in the United States of America.

Contents

Preface

Every book is a source of knowledge and this one is no exception. The idea that led to the conceptualization of this book was the fact that the world is advancing rapidly; which makes it crucial to document the progress in every field. I am aware that a lot of data is already available, yet, there is a lot more to learn. Hence, I accepted the responsibility of editing this book and contributing my knowledge to the community.

Mobile communications refer to the complex networks and devices that enable digital transmission of voice, data or fax between locations and parties that are separate from each other. Mobile operation networks develop further as more and more users utilize mobile technologies. This book seeks to facilitate on-going research in the field of mobile communications and digital networks. This book, with its detailed analyses and data, will prove immensely beneficial to professionals and students involved in this area at various levels. The aim of this text is to present researchers that have transformed this discipline and aided its advancements. It aims to serve as a resource guide for students and experts alike and contribute to the growth of the discipline of mobile communications.

While editing this book, I had multiple visions for it. Then I finally narrowed down to make every chapter a sole standing text explaining a particular topic, so that they can be used independently. However, the umbrella subject sinews them into a common theme. This makes the book a unique platform of knowledge.

I would like to give the major credit of this book to the experts from every corner of the world, who took the time to share their expertise with us. Also, I owe the completion of this book to the never-ending support of my family, who supported me throughout the project.

Editor

Towards autonomous vehicular clouds

Stephan Olariu[1,*], Mohamed Eltoweissy[2], Mohamed Younis[3]

[1]Department of Computer Science, Old Dominion University, Norfolk, Virginia, USA; [2]Pacific Northwest National Laboratory, Richland, Washington DC, USA; [3]Department of Computer Science and Electrical Engineering, University of Maryland, College Park, MD 20742, USA

Abstract

The dawn of the 21st century has seen a growing interest in vehicular networking and its myriad potential applications. The initial view of practitioners and researchers was that radio-equipped vehicles could keep the drivers informed about potential safety risks and increase their awareness of road conditions. The view then expanded to include access to the Internet and associated services. This position paper proposes and promotes a novel and more comprehensive vision namely, that advances in vehicular networks, embedded devices and cloud computing will enable the formation of autonomous clouds of vehicular computing, communication, sensing, power and physical resources. Hence, we coin the term, autonomous vehicular clouds (AVCs). A key feature distinguishing AVCs from conventional cloud computing is that mobile AVC resources can be pooled dynamically to serve authorized users and to enable autonomy in real-time service sharing and management on terrestrial, aerial, or aquatic pathways or theaters of operations. In addition to general-purpose AVCs, we also envision the emergence of specialized AVCs such as mobile analytics laboratories. Furthermore, we envision that the integration of AVCs with ubiquitous smart infrastructures including intelligent transportation systems, smart cities and smart electric power grids will have an enormous societal impact enabling ubiquitous utility cyber-physical services at the right place, right time and with right-sized resources.

Keywords: autonomous systems, cloud computing, cyber-physical systems, resource management, vehicular networks

1. The vehicular model

The past 20 years have seen an unmistakable trend to make the vehicles on our roads smarter and the driving experience safer and more enjoyable. A typical car or truck today is likely to contain at least some of the following devices: an on-board computer, a GPS device, a radio transceiver, a short-range rear collision radar device, a camera, supplemented in high-end models with a variety of sophisticated sensing devices. Some high-end vehicles already offer the convenience of an event data recorder (EDR) that collects transactional data from most of the vehicle sub-assemblies. It is not widely known that some GM vehicles as old as model year 1994 were equipped with an EDR-like device able to store retrievable data. In general, the EDRs are intended to be tamper-proof, very much like the well-known black boxes on board commercial and military aircraft. Among other things, the EDRs are designed to optimize the 'up time' of vehicles by sophisticated self-checks and by scheduling vehicles for maintenance on a per-need basis as opposed to a fixed calendar date. In 2006 the National Highway Traffic Safety Administration (NHTSA) declared its intent to standardize the various EDR devices provided, on a voluntary basis, by car and truck manufacturers [1]. As it turns out, the EDRs are already having a societal impact, as they are finding their way into the courtrooms where insurance companies use EDR logs in their litigations.

*Corresponding author. Email: olariu@cs.odu.edu

As technology is moving closer and closer to packing sophisticated resources in individual vehicles, many manufacturers are turning their attention to making the vehicles on our roads more fuel- and energy-efficient than ever. It is sufficient to recall that the past decade has seen the emergence of hybrid vehicles from the automotive engineer's drawing board into production, to the point where today half a dozen car and truck manufacturers offer hybrid vehicles on the American market.

In addition to their sophisticated array of sensing and computation capabilities, the availability of virtually unlimited power supply and growing Internet presence will make our vehicles perfect candidates for housing powerful on-board computers augmented with huge storage devices that, collectively, may act as networked computing centers on wheels.

2. Why vehicular networks?

Wireless technology was available for the past 60 years yet, with few exceptions, it did not find its way into the arena of vehicular communications until very recently. In order to understand the sea change that we have witnessed in the past decade or so, it helps to recall that the US Department of Transportation (US-DOT) estimates that in a single year, congested highways due to various traffic events cost over $75 billion in lost worker productivity and over 8.4 billion gallons of fuel [2]. The US-DOT also notes that over half of all congestion events are caused by highway incidents rather than by rush-hour traffic in big cities [1]. Further, the NHTSA indicates that congested roads are one of the leading causes of traffic accidents, projecting data extrapolated from January to September 2009 statistics, that, for 2009, an estimated 25576 fatalities are directly attributable to traffic-related incidents [3].

Unfortunately, on most US highways, congestion is a daily event and with rare exceptions, advance notification of imminent congestion is unavailable [4–6]. It is worth mentioning that in the transportation science community several solutions for reducing the effects of congestion were contemplated over the years [7, 8]. One of the proposed solutions involves adding more traffic lanes to our roadways and streets. While at first sight this seems to be a reasonable course of action, a recent study has pointed out that this strategy is futile in the long run as it is likely to lead to more congestion and to increased levels of pollution [8]. On the other hand, it has been argued that given sufficient advance notification, drivers could make educated decisions about taking alternate routes; in turn, this would improve traffic safety by reducing the severity of congestion and, at the same time, save time and fuel [7, 9].

Under present-day technology traffic monitoring and incident reporting systems employ inductive loop detectors (ILDs), video cameras, acoustic tracking systems and microwave radar sensors [7]. By far the most prevalent among these devices are the ILDs embedded in highways every mile (or half-mile) [10–12]. ILDs measure traffic flow by registering a signal each time a vehicle passes over them. Each ILD (including hardware and controllers) costs around $8200; in addition, the ILDs are connected by optical fiber that costs $300000 per mile [9, 13]. Interestingly, official statistics shows that over 50% of the installed ILD base and 30% of the video cameras are defective [7, 12]. Not surprisingly, transportation departments worldwide are looking for less expensive, more reliable and more effective methods for traffic monitoring and incident detection.

To be effective, innovative traffic-event detection systems must enlist the help of the most recent technological advances. This has motivated extending the idea of Mobile Ad-hoc Networks (MANET) to roadway and street communications. The new type of networks, referred to as Vehicular Ad-hoc Networks (VANET) that employ a combination of Vehicle-to-Vehicle (V2V) and Vehicle-to-Infrastructure (V2I) communications, have been proposed to give drivers advance notification of traffic events. In V2V systems, each vehicle is responsible for inferring the presence of an incident based on reports from other vehicles.

As we just saw, the original impetus for the interest in VANET was provided by the need to inform fellow drivers of actual or imminent road conditions, delays, congestion, hazardous driving conditions and other similar traffic-related concerns. Therefore, most VANET applications focus on traffic status reports, collision avoidance, emergency alerts, cooperative driving and other similar concerns [14–16, 40, 42]. Almost across the board, the community of researchers and practitioners anticipate that advances in VANET, or other emerging vehicle-based computing and communications technology, are poised to have a huge societal impact. Because of this envisioned societal impact, numerous vehicle manufacturers, government agencies and standardization bodies around the world have spawned national and international consortia devoted exclusively to VANET. Examples include Networks-on-Wheels, the Car-2-Car Communication Consortium, the Vehicle Safety Communications Consortium, Honda's Advanced Safety Vehicle Program, among many others. We refer the interested readers to the survey articles [17, 18] where many US and European initiatives and standards are discussed in detail.

The past few years have witnessed a rapid convergence of Intelligent Transportation Systems (ITS) and VANET leading to the emergence of Intelligent Vehicular Networks with the expectation to revolutionize the way we drive by creating a safe, secure and robust ubiquitous computing environment that will eventually pervade our highways and city streets. In support of traffic-related communications, the US Federal Communications Commission (FCC) has allocated 75 MHz of spectrum in the

5.850–5.925 GHz band specially allocated by the FCC for Dedicated Short-Range Communications (DSRC) [5, 19]. It was recently noticed that the DSRC spectrum set aside by the FCC, by far exceeds the needs of traffic-related safety applications. This observation has motivated the emergence of a host of other applications that can take advantage of the allocated spectrum. Not surprisingly, we see more and more third-party providers offering non-safety-related applications ranging from location-specific services, to on-the-road peer-to-peer communications, to Internet access, to on-line gaming and other forms of mobile entertainment. In due time, we will see the emergence of commercial applications targeted at the traveling public and distributed via the excess bandwidth in DSRC. As a pleasant side benefit, the unsightly billboards that flank our highways will disappear and will be replaced by in-vehicle advertising that the driver can filter according to their wants and needs.

3. Looking into the crystal ball

Recently, there has been a good deal of commercial and research interests in utilizing broadband communications and wireless technologies to provide Internet connectivity to the public on the road [15, 20–23]. One such system that has received quite a bit of attention in the recent literature is known as Drive-thru Internet [24, 25] and relies on dedicated road-side access points (AP) on roadways and city streets to enable the driving public to connect to the Internet, at least while they are within the coverage of the infrastructure. An important feature of Drive-thru Internet systems is the multi-access sharing of the same AP's bandwidth by the vehicles that are simultaneously under its coverage [26]. Since, as a rule, the vehicles move very fast, and the coverage limitations are dictated by present-day technology, the amount of data that a passing vehicle can download from any given AP is rather limited, a state of affairs that is promoting vehicular interchanges and the blossoming of vehicular peer-to-peer connections that are key to mitigating the effects of the limited Internet coverage conferred by Drive-thru Internet and other similar schemes [14, 27–29].

We also fully expect third-party infrastructure providers to deploy various forms of road-side infrastructure as well as advanced in-vehicle resources such as embedded powerful computing and storage devices, cognitive radios and cognitive radio networks, and multi-modal programmable sensor nodes. As a result, in the near future, vehicles equipped with computing, communication and sensing capabilities will be organized into ubiquitous and pervasive networks with virtually unlimited Internet access while on the move. This will revolutionize the driving experience making it safer and more enjoyable [14, 20, 21].

The huge array of on-board capabilities is likely to remain underutilized by safety applications alone. The realization of this fact has already motivated the investigation of offering value-added services including on-line gaming, mobile infotainment, along with various location-specific services. We conjecture that the potential is even far beyond that. Specifically, we propose to 'take vehicular networks to the clouds' so that our prized transportation means can take their natural place integrated with our productivity, comfort, safety and economic prosperity.

4. Cloud computing

The notion cloud computing started from the realization of the fact that instead of investing in infrastructure, businesses may find it useful to rent the infrastructure and sometimes the needed software to run their applications [30, 31]. This powerful idea has been suggested, at least in part, by ubiquitous and relatively low-cost high-speed Internet, virtualization and advances in parallel and distributed computing and distributed databases. One of the key benefits of cloud computing is that it provides scalable access to computing resources and information technology (IT) services [23].

Cloud computing is a paradigm shift adopted by a large number of infrastructure providers who have a large installed infrastructure that often goes underutilized. Hand in hand with cloud computing go 'cloud IT services' where not only computational resources and storage are rented, but specialized services are also provided on demand. In this context, a user may purchase the amount of services they need at the moment. As their IT needs grow and as their services and customer base expand, the users will be in the market for more and more cloud services and more diversified computational and storage resources. In general there are three delivery models for cloud services: Software as a Service (SaaS): customers rent software hosted by the vendor; Platform as a Service (PaaS): customers rent infrastructure and programming tools hosted by the vendor to create their own applications; and Infrastructure as a Service (IaaS): customers rent processing, storage, networking and other fundamental computing resources for all purposes [17, 31–33].

5. Autonomous vehicular clouds

The huge fleet of energy-sufficient vehicles that crisscross our roadways, airways, and waterways, most of them with a permanent Internet presence, featuring substantial on-board computational, storage and sensing capabilities can be thought of as a huge farm of computers on the move. These attributes make vehicles ideal candidates for nodes in a cloud as described above. Indeed, the owner of a vehicle may decide to rent out their in-vehicle capabilities on demand, or a per-instance, or a per-day, per-week or per-month basis, just as owners of large computing facilities find it economically appealing to rent out excess capacity to seek pecuniary advantages.

More significantly, we postulate that vehicles will autonomously self-organize into clouds utilizing their corporate resources on demand and largely in real time in resolving critical problems that may occur unexpectedly. The new vehicular clouds will also contribute to unraveling some technical challenges of the increasingly complex transportation systems with their emergent behavior and uncertainty.

We believe it is only a matter of time before the huge vehicular fleets on our roadways, streets and parking lots will be recognized as an abundant and underutilized computational resource that can be tapped into for the purpose of providing third-party or community services. However, what distinguishes vehicles from standard nodes in a conventional cloud is autonomy and mobility. Indeed, large numbers of vehicles spend substantial amounts of time on the road and may be involved in dynamically changing situations; we argue that in such situations, the vehicles have the potential to cooperatively solve problems that would take a centralized system an inordinate amount of time, rendering the solution useless.

In [34] Olariu et al. argued convincingly that it is only a matter of time before the huge vehicular fleets on our roadways, streets and parking lots will be recognized as an abundant and under-utilized computational resource that can be tapped into for the purpose of providing third-party or community services. However, what distinguishes vehicles from standard nodes in a conventional cloud is autonomy and mobility. Indeed, large numbers of vehicles spend substantial amounts of time on the road and may be involved in dynamically changing situations; Olariu et al. [34] argued that in such situations, the vehicles have the potential to cooperatively solve problems that would take a centralized system an inordinate amount of time, rendering the solution useless.

In this work on continue to explore various aspects of the vehicular cloud (VC) concept proposed by Olariu et al. in their seminal paper [34].

With this in mind, we have coined the term autonomous vehicular cloud (AVC) to refer to: *a group of largely autonomous vehicles whose corporate computing, sensing, communication and physical resources can be coordinated and dynamically allocated to authorized users.*

In our view, the AVC concept is the next natural step in meeting the computational and situational awareness needs not only of the driving public but also of a much larger segment of the population. A primary goal of the AVC is to provide on-demand solutions to events that have occurred but cannot be met reasonably with pre-assigned assets or in a proactive fashion.

It is important to delineate the structural, functional and behavioral characteristics of AVCs. As a step in this direction, in this position paper, we identify autonomous

cooperation among vehicular resources as a distinguishing characteristic of AVCs. Another important characteristic of AVCs is the ability to offer a seamless integration and decentralized management of cyber-physical resources; an AVC can dynamically adapt its managed vehicular resources allocated to applications according to the applications' changing requirements and environmental and systems conditions.

As far as a simple taxonomy goes, AVCs can be public, private or various hybrids thereof. The public AVC will provide (typically short-term) services on the Internet, whereas a private AVC is proprietary and provides (typically long-term) services to a limited set of users and would belong to specific vehicle fleets such as FedEx, UPS, Costco or Wal-Mart. As an example of a hybrid AVC, one may consider inter-AVC cooperation as discussed in Section 6.

It is not too far-fetched to imagine, in the not-so-distant future, a large-scale federation of AVCs will be established *ad hoc* in support of mitigating a large-scale emergency. One of these large-scale emergencies could be a planned evacuation in the face of a potentially deadly hurricane or tsunami that is expected to make landfall in a coastal region [35, 36]. Yet another such emergency would be a natural or man-made disaster apt to destroy the existing infrastructure and to play havoc with cellular communications. In such a scenario, a federation of AVCs could provide a short-term replacement for the infrastructure and also provide a decision-support system.

6. Application scenarios

The main goal of this section is to illustrate the power of the AVC concept. We touch upon several important scenarios illustrating various aspects AVCs that are extremely important and that, under present-day technology are unlikely to see a satisfactory resolution. The outlined scenarios are representative of two application categories: (i) traffic management and (ii) asset management. Other categories include situational awareness, for example tipping and cueing for threat analysis and mitigation, and nomadic social and business spaces, for example mobile marketplace (Sue may want to sell perfumes on her way back from work).

6.1. Traffic management scenarios

The goal of this subsection is to discuss a number of traffic management scenarios where AVCs can be useful. While only a small number of scenarios are presented here, it is easy to extrapolate from them and to contemplate many variations of these scenarios where AVCs are likely to make a difference.

Scenario I: synchronizing traffic lights after clearing an accident. Consider for example, a city block where a traffic-related event (e.g. an accident) has occurred and where, as a consequence, a large number of vehicles are co-located. Once the traffic event has been cleared, relying on the existing scheduling of the traffic lights will not help dissipate the huge traffic backlog in an efficient way. We envision a solution to this problem where the vehicles themselves will pool their computational resources together creating the effect of a powerful super-computer that will recommend to a higher authority a way of rescheduling the traffic lights that will serve the purpose of de-congesting the afflicted area as fast as possible. We note that, in general, the solution cannot involve a handful of traffic lights but may require rescheduling the traffic lights in a large geographic area.

As mentioned before, the ability of vehicles to pool their resources, in a dynamic way, in support of the common good will have a huge societal impact alleviating, among others recurring congestion events that plague our cities around the morning or afternoon rush hour. Also, and very importantly, while congestion is a daily phenomenon, proactively solving the problem is infeasible because of the dynamic nature of the problem, and of the huge computational effort its resolution requires. The problem is best solved if and when it occurs in an on-demand fashion dedicating the right amount of resources rather than conservatively pre-allocating abundant resources based on the worst case, which is becoming increasingly infeasible. The key concept that allows the problem to be solved efficiently and economically is the engagement of the necessary resources from the available vehicles participating in the traffic event and their involvement in finding a solution autonomously without waiting for an authority to react to the complicated situation on the ground.

Scenario II: autonomous mitigation of recurring congestion. In face of traffic congestion some drivers often pursue detours and alternate routes that often involve local roads. Making the decision behind the steering wheel is often challenging. The driver does not know whether the congestion is about to ease or is worsening. In addition, when many vehicles decide to execute the same travel plan, local roads become flooded with traffic that exceeds its capacity and sometimes deadlocks take place. Contemporary ITS and traffic advisory schemes are both slow to report traffic problems and usually do not provide any mitigation plan. An AVC-based solution will be the most appropriate and effective choice. Basically, vehicles in the vicinity will be able to query the plan of each other and estimate the impact on local roads. In addition, an accurate assessment of the cause of the congestion and traffic flow can be made by contacting vehicles close to where the bottleneck is. In addition, appropriate safety precautions can be applied to cope with the incident, e.g. poor air quality due to the smoke of a burned vehicle.

Interestingly, this approach can be applied not to drivers on the road but also to those who are about to leave home. Delayed start and telecommuting may be considered as an alternative in order to increase productivity and avoid wasted energy and time.

Scenario III: sharing on-road safety messages. The trend in the car manufacturing industry is to equip new vehicles with major sensing capabilities in order to achieve efficient and safe operation. For example, Honda is already installing cameras on their Civic models in Japan. The cameras track the lines on the road and help the driver stay in lane. A vehicle would thus be a mobile sensor node and an AVC can be envisioned as a huge wireless sensor network with very dynamic membership. It would be beneficial for a vehicle to query the sensors of other vehicle in the vicinity in order to increase the fidelity of its own sensed data, get an assessment of the road conditions and the existence of potential hazard ahead. For example when the tire pressure sensor on a vehicle reports the loss of air, vehicles that are coming behind on the same lane should suspect the existence of nails on the road and may consider changing the lane. The same happens when a vehicle changes lane frequently and significantly exceeds the speed limit; vehicles that come behind, and which cannot see this vehicle, can suspect the presence of aggressive drivers on the road and consider staying away from the lanes and/or keeping a distance from the potentially dangerous driver. The same applies when detecting holes, unmarked speed breakers, black ice, etc. Contemporary VANET design cannot pull together the required solution and foster the level of coordination needed for providing these safety measures.

Scenario IV: dynamic management of HOV lanes. As pointed out by the US-DOT in its 2008 report and guidelines, *'the primary purpose of an HOV lane is to increase the total number of people moved through a congested corridor by offering two kinds of incentives: a savings in travel time and a reliable and predictable travel time. Because HOV lanes carry vehicles with a higher number of occupants, they may move significantly more people during congested periods, even when the number of vehicles that use the HOV lane is lower than on the adjoining general-purpose lanes. In general, carpoolers, vanpoolers, and transit users are the primary beneficiaries of HOV lanes'* [26].

It is, thus, plainly obvious that the main goal of HOV lanes is to promote traffic fluidity and to prevent traffic slowdowns and congestion. Due to insufficient cyber-physical means (appropriate signs being one key shortcoming) most cities in the USA only use HOV lanes at rush hour. However, AVC could make recommendations for setting up HOV lanes dynamically in the best interest of promoting traffic fluidity and of minimizing travel times for people using the designated HOV lanes. AVCs can enable such a dynamic solution by factoring data from sensors on board the individual vehicles, e.g. occupancy sensors, and local traffic intensity in order to optimally configure the HOV lanes. Such a solution is infeasible under present-day technology.

The same idea applies to the strategy of marking certain streets and thoroughfares as 'one-way' in support of improving the fluidity of traffic. Again, currently such an approach is infeasible mostly because of insufficient signaling means. This, however, should not be a problem in AVC since the drivers will be alerted in real time to road occlusions and other dynamic changes.

6.2. Asset management scenarios

The goal of this subsection is to present a number of plausible scenarios that illustrate the potential use of AVCs for managing existing assets.

Scenario V: mobile experimental and analytics laboratory in support of homeland security.   Sensor networks are expected to evolve into long-lived, open, ubiquitous, multi-purpose networked systems. Recently, the authors have proposed ANSWER, an autonomous networked sensor system whose mission is to provide in situ users with real-time, secure information that enhances their situational and location awareness [37]. ANSWER finds immediate applications in homeland security. The architectural model of ANSWER is composed of a large number of sensors and of a set of (mobile) aggregation-and-forwarding nodes, possibly AVC nodes, that organize and manage their sensors (if any) and the sensors in their vicinity. As argued in [37], ANSWER can provide secure, QoS-aware information and analysis services to *in situ* mobile users in support of application-level tasks and queries, while hiding network-level details. We anticipate that an AVC can naturally interface in a symbiotic relationship with ANSWER creating a powerful mission-oriented system.

Scenario VI: augmenting the capabilities of small businesses.   Consider a small business employing about 250 people and specializing in offering IT support and services. It is not hard to imagine that, even if we allow for car-pooling, there will be up to 150 vehicles parked in the company's parking lot. Day in and day out, the computational resources in those vehicles are sitting idle. We envision harvesting the corporate computational and storage resources in the vehicles sitting in the parking lot for the purpose of creating a computer cluster and a huge distributed data storage facility that, with proposer security safeguards in place, will turn out to be an important asset that the company cannot afford to waste.

Scenario VII: efficient tasking of law enforcement officers.   Law enforcement officers play a crucial role in keeping the road safe for motorists. Even if a police vehicle is so visible on the road, it serves as a deterrent for aggressive drivers and vehicle safety violators. An AVC can be used as an effective resource-planning tool for the police squad. Moving vehicles form an AVC and report to the police so that decisions can be made efficiently about deploying troopers in certain spots and/or employing surveillance cameras and aircraft to identify and video tape violators for further assigning fines. That will allow effective usage of officers' time and enable them to allocate resources for other vital tasks such as criminal investigation and prevention. Implementing this idea through today's technology is resource-prohibitive and requires major infrastructure investment.

Scenario VIII: dynamic management of parking facilities.   Anyone who has attempted to find a convenient parking spot in the downtown area of a big city or close to a university campus where the need for parking by far outstrips the supply would certainly be interested to enlist the help of an automated parking management facility. The problem of managing parking availability is a ubiquitous and a pervasive one, and several solutions were reported recently [38, 39, 41, 43]. However, most of the existing solutions rely on a centralized solution where reports from individual parking garages and parking meters are aggregated at a central (city-wide) location and then disseminated to the public. The difficulty is with the real-time management of parking availability since the information that reaches the public is often stale and outdated. This, in turn, may worsen the situations especially when a large number of drivers are trying to park, say, to attend a down-town event.

We envision that by real-time pooling the information about the availability of parking at various locations inside the city, an AVC consisting of the vehicles that happen to be in a certain neighborhood will be able to maintain real-time information about the availability of parking and direct the drivers to the most promising location where parking is (still) available.

Scenario IX: dynamic asset management in planned evacuations.   In cases of predicted disasters, such as hurricanes, massive evacuations are often necessary in order to minimize the impact of the disaster on human lives. However, there are several issues involved in a large-scale evacuation. For example, once an evacuation is underway, finding available resources, such as gasoline, drinking water, medical facilities and shelter, quickly becomes an issue [12]. In its recent report on hurricane evacuations [9], the US-DOT found that emergency evacuation plans often do not even consider availability of such resources. The US-DOT also determined that emergency managers need a method for communicating with evacuees during the evacuation in order to provide updated information. The report suggested that traffic monitoring equipment should be deployed to provide real-time traffic information along evacuation routes.

We now point out natural ways in which AVC can work with the emergency management center overseeing the evacuation in order to provide travel time estimates, notification of available resources, such as gasoline, food, and shelter, and notification of contra-flow roadways to the evacuees. We anticipate that the vehicles involved in the evacuation will self-organize into one or several inter-operating vehicular clouds that will work hand in hand with the emergency management center. In the course of this interaction, the emergency managers can upload information about open shelters to the central server.

It is important to note that this system would be used to facilitate an evacuation before disaster strikes, so we assume that electricity and network connections are available. In addition to having state authorities send information to the AVCs about evacuations or contra-flow lanes, using role-based communication as described earlier, the AVCs themselves could determine the direction and speed that traffic is flowing. The evacuees entering entrance ramps onto contra-flow roadways (these ramps would likely have been used as exit ramps previously) will be alerted to the direction that traffic is moving. The AVCs could also alert drivers to upcoming entrance ramps that were previously used as exit ramps during non-contra-flow travel. Since the AVC system can easily monitor traffic flow, it could offer recommendations to the emergency center about which roadways are good contra-flow candidates.

Scenario X: AVCs in developing counties. We conjecture that the usefulness and practicality of the AVC concept will become even more apparent in developing countries lacking a sophisticated centralized decision-support infrastructure. We further conjecture that, in such contexts, AVCs will play an essential role in bringing together a huge number of relatively modest computational resources available in the vehicular network into one or several foci of computing and communications that will find and/or recommend solutions to problems arising dynamically and that cannot possibly be resolved with the existing infrastructure. We have seen a similar phenomenon happening with the penetration of cell phones in developing countries where they were adopted rapidly and unhesitatingly by a population that had access to a modest landline telephony system.

7. AVC research issues

The application scenarios discussed above require better V2V and V2I collaboration in order to reach critical and mutually beneficial decisions, effective and unconventional management to cope with the highly dynamic nature of the computing, communication, sensing and physical resources, and well-defined operation structures that enable autonomy and authority in adjusting local settings with the potential of making wide impact.

Currently, our group is initiating several research projects in AVC engineering. We now present some of the research issues along the three systems engineering dimensions, namely structure, function and behavior (operation and policy).

7.1. Architectural challenges

- Elastic mobile architecture
  - The AVC networking and associated protocol architecture must be developed to accommodate changing application demands and resource availability on the move.

- Resilient AVC architecture in the wild
  - AVC basic structural and composed building blocks must be designed and engineered to withstand structural stresses induced by the inherent instability in the operating environment. Research is needed on architectures enabling vehicle visualization and migration of virtual vehicles.

- Service-oriented network architecture
  - Contemporary layered network architectures, for example the TCP/IP stack, have proven limited in face of evolving applications and technologies. We envision the adoption of service-oriented component-based network architectures with intrinsic monitoring and learning capabilities.

7.2. Functional challenges

- Enabling AVC autonomy
  - Research is needed on developing a trustworthy base, negotiation and strategy formulation methodology (game theory, etc.), efficient communication protocols, data processing and decision-support systems, etc.

- Managing highly dynamic cloud membership
  - There is a critical need to efficiently manage mobility, resource heterogeneity (including sensing, computation and communication), trust and vehicle membership (change in interest, change in location, resource denial and/or failure).

- Cyber-physical control
  - AVCs can be defined by their aggregated cyber and physical resources. Their aggregation, coordination and control are non-trivial research issues.

- Cooperation between AVCs
  - To motivate the need for AVCs to cooperate, imagine that in adjacent areas of a municipality there is a sporting event as well as a rock concert downtown. Both these events are very likely to draw a huge crowd. Now, assuming that due to bad planning by the municipality, the two events end at the same time, creating two distinct zones of massive conges-

tion. In such a scenario, each congestion event will trigger the formation of an ad hoc AVC. These two AVCs will have to coordinate and to solve the congestion problem collectively, since they cannot proceed to selfishly reschedule traffic lights in a way that benefits each of them individually.

7.3. Operational and policy challenges

- Trust and trust assurance
  - In order for the vision outlined above to become reality, the problems of assuring emergent trust and security in AVC communication and information need to be addressed. The establishment of trust relationships between the various players is a key component of trustworthy computation and communication. We argue that since typically most, if not all the vehicles involved, must have met before, the task of establishing proactively, a basic trust relationship between vehicles is possible and may be even desirable (think in terms of vehicles that meet day after day in a parking garage).
  - Also, in order to be effective at cooperative problem solving, an AVC may need to have delegated authority to take local action in lieu of a central authority. Referring to our motivating scenarios, it is clearly useless for the vehicles involved in a traffic jam to produce a workable schedule of the traffic lights that will best promote the rapid dissipation of congestion if they do not have the authority to implement such a schedule. Clearly, the resolution of this problem resides in some form of a trust relationship that needs to be forged between the municipal or county authority and the AVC.

- Contract-driven versus *ad hoc* AVC
  - We anticipate that AVCs will be largely contract-driven, where the owner of the vehicle or fleet consents to renting out some form of excess computational or storage capacity. At the same time, mobility concerns dictate that in addition to the contract-based form of AVC, it should be possible to form an AVC in an *ad hoc* manner as necessitated by dynamically changing situations like those discussed in the scenarios above.

- Effective operational policies
  - In order for the AVCs to operate and inter-operate seamlessly, issues related to authority establishment and management, decision support and control structure, the establishment of accountability metrics, assessment and intervention strategies, rules and regulations, standardization, etc. must all be addressed. Dealing with these will require a broad participation and must involve local, state or even federal decision makers.

- AVC utility computing
  - There is a need for economic models and metrics to determine reasonable pricing and billing for AVC services.

8. Architectures for AVCs

Although our ultimate goal is to produce a *unified* architectural framework for the AVC, the main goal of this section is to review several possible architectures, of increasing complexity that suit various particular manifestations of AVCs.

8.1. A static architecture

In some cases, an AVC may behave just as a conventional cloud. This is, no doubt, the case in static environments as the one we contemplate below. Indeed, consider a small business employing about 250 people and specializing in offering IT support and services. It is not hard to imagine that, even if we allow for car-pooling, there will be up to 150 vehicles parked in the company's parking lot. Day in and day out, the computational resources in those vehicles are sitting idle.

The company may proactively seek the formation of a static AVC by providing appropriate incentives to its employees who will rent the resources of their vehicles to the company on a per-day, per-week or per-month basis. The resulting (more or less) static AVC will harvest the corporate computational and storage resources of the participating vehicles sitting in the parking lot for the purpose of creating a computer cluster and a huge distributed data storage facility that, with proper security safeguards in place, will turn out to be an important asset that the company cannot afford to waste.

In the scenario above, the architecture of the AVC will be almost identical to the architecture of a conventional cloud, with the additional twist of, perhaps, limiting the interaction to week-days.

8.2. Interfacing with a static infrastructure

It is often the case that an AVC is created and evolves in an area instrumented by the deployment of some form of a static infrastructure supportive of the management of various activities. In an urban setting, such an infrastructure includes traffic lights, cameras as well as the utility or street lighting poles. On our roadways, the static infrastructure includes ILDs, the road-side units and other ITS hardware deployed in support of traffic monitoring and management.

We note here that in a not-so-distant future, a pre-deployed set of tiny sensors, even if not organized in a permanent sensor network, may play the role of the static infrastructure that the AVC may find it beneficial to interact with. In fact, this view is consistent with our NSF-funded ANSWER project [6] where the place of the PSAR is taken by the AVC that is constantly interacting with the static infrastructure.

It is self-evident that the AVC benefits from the interaction with the existing static infrastructure. Consider for example, a city block where a minor traffic-related event has occurred and where, as a consequence, a number of vehicles are co-located. Once the traffic event has been cleared, relying on the existing scheduling of the traffic lights will not help dissipate the traffic backlog in an efficient way. We envision a solution to this problem where the vehicles themselves will pool their computational resources together creating the effect of a powerful super-computer that will recommend to a higher authority a way of rescheduling the traffic lights that will serve the purpose of de-congesting the afflicted area as fast as possible.

It is worth noting that in this particular instance, the scope of the traffic lights to reschedule is relatively modest and does not require the federation of several AVCs.

8.3. A simple dynamic architecture

Consider for example, a city block where a minor traffic-related event has occurred and where, as a consequence, a number of vehicles are co-located. Once the traffic event has been cleared, relying on the existing scheduling of the traffic lights will not help dissipate the traffic backlog in an efficient way. We envision a solution to this problem where the vehicles themselves will pool their computational resources together creating the effect of a powerful super-computer that will recommend to a higher authority a way of rescheduling the traffic lights that will serve the purpose of de-congesting the afflicted area as fast as possible.

It is worth noting that in this particular instance, the scope of the traffic lights to reschedule is relatively modest and does not require the federation of several AVCs. The architecture that will support the formation of this AVC will involve the following elements:

- A broker elected spontaneously among the vehicles that will attempt to spontaneously form an AVC.

- The broker will then secure a preliminary authorization from a higher (city) forum for the formation of an AVC. If several brokers attempt to secure such an authorization simultaneously, one will succeed and the others will possibly form a team that will coordinate the formation of the AVC. In the sequel we assume that there is a unique broker.

- The broker will inform the vehicles in the area of the received authorization and will invite participation in the AVC.

- The cars will/or will not respond to the invitation on a purely autonomous basis.

- The broker decides if a sufficient number of vehicles have volunteered and will then announce the formation of the AVC.

- The AVC will pool their computational resources to form a powerful super-computer that, using a digital map of the area, will produce a proposal schedule to the higher (city) forum for approval and implementation.

- Once the proposal has been accepted and implemented, the AVC is dissolved.

While the scenario above and the resulting architecture are slightly more complex than those of a conventional cloud, we note that, in general, the solution cannot involve only a handful of traffic lights but will require rescheduling the traffic lights in a large area. This motivates the collaboration of several AVCs.

9. Concluding remarks

The main goal of this paper was to put forth a novel concept whose time has come: namely that of AVCs. AVCs are emerging from the convergence of advances in mobility, powerful embedded in-vehicle resources, ubiquitous sensing and cloud computing. When fully realized and deployed, AVCs would yield significant enhancements in safety, security and economic vitality of our modern society. They would enable non-conventional applications that go far beyond what people are expecting from today's VANET and ITS. Not surprisingly, the practical realization of our vision and the production of AVC standards will require tackling numerous novel technical challenges, whose resolution will certainly involve adopting a clean-slate approach. New research and development programs are needed to build AVC reference models, architectures and protocols, to address emergent trust and trust assurance issues, to provide AVC-driven cyber-physical resource coupling and coordination, to realize broader benefits through AVC federations, among many others. In addition to terrestrial vehicles the AVC concept also applies to aerial and aquatic vehicles or any hybrid combination thereof.

Acknowledgements. This work was supported, in part, by the following NSF Grants CNS 0721523, CNS 0721563, CNS 0721586 and CNS 0721644. A preliminary version of this paper has appeared in the Proceedings of Second International ICST Conference on Ad Hoc Networks, 18–20 August 2010, Victoria, British Columbia, Canada.

References

[1] NATIONAL HIGHWAY TRAFFIC SAFETY ADMINISTRATION (2006) *Traffic Safety Facts*, http://www-nrd.nhtsa.dot.gov.

[2] US Department of Transportation (2008) *National Transportation Statistics*.

[3] National Highway Traffic Safety Administration (2010, March) *Traffic Safety Facts—Preliminary 2009 Report*, http://www-nrd.nhtsa.dot.gov/Pubs/811255.pdf.

[4] ElBatt, T., Goel, S., Holland, G., Krishnan, H. and Parikhan, J. (2006) Cooperative collision warning using dedicated short range wireless communications. In *Proceedings of ACM VANET* (ACM Press).

[5] Sengupta, R., Rezaei, S., Shlavoder, S.-E., Cody, D., Dickey, S. and Krishnan, H. (2006, May) Cooperative collision warning systems: concept definition and experimental implementation. *California PATH Technical Report UCB-ITS-PRR-2006-6*.

[6] Virginia Department of Transportation (2006) *Commonwealth of Virginia's Strategic Highway Safety Plan, 2006–2010*, http://virginiadot.org/info/resources/Strat_Hway_Safety_Plan_FREPT.pdf.

[7] Fontaine, M. (2009) Traffic monitoring. In Olariu, S. and Weigle, M.C. [eds.] *Vehicular Networks: From Theory to Practice* (Boca Raton, FL: Taylor and Francis), 1.1–1.28.

[8] Sightline (2009), http://www.sightline.org/research/energy/res_pubs/analysis-ghg-roads.

[9] Sreedevi, I. and Black, J. (2001, February) *Loop Detectors*, California Center for Innovative Transportation, http://www.calccit.org/itsdecision/serv_and_tech/Traffic_Surveillance/road-based/in-road/loop_report.html.

[10] Roess, R.-P., Prassas, E.-S. and McShane, W.-R. (2004) *Traffic Engineering* (Erewhon, NC: Pearson Prentice Hall), 3rd ed.

[11] US Department of Transportation (2008, August) *Federal-Aid Highway Program Guidance on High Occupancy Vehicle (HOV) Lanes*, http://ops.fhwa.dot.gov/freewaymgmt/hovguidance/index.htm.

[12] Varaiya, P., Lu, X.-Y. and Horowitz, R. (2006, October) *Deliver a Set of Tools for Resolving Bad Inductive Loops and Correcting Bad Data*, http://path.berkeley.edu/~xylu/TO6327/TO6327_SEMP.pdf.

[13] University of Virginia Center for Transportation Studies (2005, November) *Virginia Transportation Research Council, Probe-Based Traffic Monitoring State-of-the-Practice Report*.

[14] Lee, U., Cheung, R. and Gerla, M. (2009) Emerging vehicular applications. In Olariu, S. and Weigle, M.C. [eds.] *Vehicular Networks: From Theory to Practice* (Boca Raton, FL: Taylor and Francis), 6.1–6.30.

[15] Lochert, C., Scheuermann, B., Wewetzer, C., Luebke, A. and Mauve, M. (2008, September) Data aggregation and roadside unit placement for a VANET traffic information system. In *Proceedings of ACM VANET* (ACM Press).

[16] Yang, Y. and Bagrodia, R. (2009, September) Evaluation of VANET-based advanced intelligent transportation systems. In *Proceedings of ACM VANET*, Beijing, China (ACM).

[17] Le, L., Festag, A., Baldesari, R. and Zhang, W. (2009) CAR-2-X communications in Europe. In Olariu, S. and Weigle, M.C. [eds.] *Vehicular Networks: From Theory to Practice* (Boca Raton, FL: Taylor and Francis), 4.1–8.32.

[18] Misener, J.A., Dickey, S., VanderWerf, J. and Sengupta, R. (2009) Vehicle-infrastructure cooperation. In Olariu, S. and Weigle, M.C. [eds.] *Vehicular Networks: From Theory to Practice* (Boca Raton, FL: Taylor and Francis, CRC Press), 3.1–8.35.

[19] US Federal Communications Commission (FCC) (2003, September) *Standard Specification for Telecommunications and Information Exchange Between Roadside and Vehicle Systems—5 GHz Band Dedicated Short Range Communications (DSRC) Medium Access Control (MAC) and Physical Layer (PHY) Specifications* (Washington, DC).

[20] Anda, J., LeBrun, J., Ghosal, D., Chuah, C.-N. and Zhang, M. (2005, May) VGrid: vehicular ad hoc networking and computing grid for intelligent traffic control. In *Proceedings of IEEE Vehicular Technology Conference—Spring* (IEEE Press), 2905–2909.

[21] Czajkowski, K., Fitzgerald, S., Foster, I. and Kesselman, C. (2001) Grid information services for distributed resource sharing. In *Proceedings of 10th IEEE International Symposium on High Performance Distributed Computing* (New York: IEEE), 181–184.

[22] Eriksson, J., Balakrishnan, H. and Madden, S. (2008, September) Cabernet: vehicular content delivery using WiFi. In *Proceedings of 14th ACM International Conference on Mobile Computing and Networking* (San Francisco, CA: MobiCom'2008).

[23] Lochert, C., Scheuermann, B., Caliskan, M. and Mauve, M. (2007, January) The feasibility of information dissemination in vehicular ad-hoc networks. In *Proceedings of 4th Annual Conference on Wireless On-demand Network Systems and Services (WONS '07)* (ACM Press), 92–99.

[24] Ott, J. and Kutscher, D. (2004) Drive-thru Internet: IEEE 802.11b for automobile users. In *Proceedings of IEEE INFOCOM* (IEEE Press).

[25] Ott, J. and Kutscher, D. (2005) A disconnection-tolerant transport for drive-thru Internet environments. In *Proceedings of IEEE INFOCOM* (IEEE Press).

[26] Tan, W.-L. Lau, W.C. and Yue, O.-C. (2009, September) Modeling resource sharing for a road-side access point supporting drive-thru Internet. In *Proceedings of ACM VANET*, Beijing, China (ACM).

[27] Abuelela, M. and Olariu, S. (2009) Content delivery in zero-infrastructure VANET. In Olariu, S. and Weigle, M.C. [eds.] *Vehicular Networks: From Theory to Practice* (Boca Raton, FL: Taylor and Francis), 8.1–8.15.

[28] Rybicki, J., Scheuermann, B., Kiess, W., Lochert, C. Fallahi, P. and Mauve M. (2007, September) Challenge: peers on wheels—a road to new traffic information systems. In *Proceedings of 13th Annual ACM International Conference on Mobile Computing and Networking* (Montreal: ACM).

[29] Rybicki, J., Scheuermann, B., Koegel, M. and Mauve, M. (2009, September) PeerTIS—a peer-to-peer traffic information system. In *Proceedings of ACM VANET* (Beijing, China: ACM).

[30] Foley, J. (2008, August) Private clouds take shape. In *Information Week*.

[31] Hodson, S. (2008, May) *What is Cloud Computing?* http://www.winextra.com/2008/05/02/what-is-cloud-computing.

[32] HOOVER, J.N. and MARTIN, R. (2008, June) Demystifying the cloud. In *Information Week Research & Reports*, 30–37.

[33] KIM, W. (2009, January–February) Cloud computing: today and tomorrow. *J. Object Technol.* **8**(1): 65–72, http://www.jot.fm/issues/issue_2009_01/column4/.

[34] OLARIU, S., KHALIL, I. and ABUELELA, M. (2011) Taking VANET to the Clouds. *Pervasive Comput. Commun.* **7**(1): 7–21.

[35] FELDSTEIN, D. and STILES, M. (2005, September). Too many people and no way out. In *The Houston Chronicle*.

[36] US DEPARTMENT OF TRANSPORTATION (2006, June) *Catastrophic Hurricane Evacuation Plan Evaluation: A Report to Congress*, http://www.fhwa.dot.gov/reports/hurricane-evacuation/.

[37] ELTOWEISSY, M., OLARIU, S. and YOUNIS, M. (2007). ANSWER: autonomous networked sensor system. *J. Parallel Distrib. Comput.* **67**(1): 111–124.

[38] XU, Q., MAK, T., KO, J. and SENGUPTA, R. (2004, October) Vehicle-to-vehicle safety messaging in DSRC. In *Proceedings of 1st ACM International Workshop on Vehicular Ad Hoc Networks* (ACM Press).

[39] YAN, G. OLARIU, S. and WEIGLE, M.C. (2008) Providing VANET security through active position detection. *Comput. Commun.* **31**(12): 2883–2897.

[40] AIJAZ, A., BOCHOW, B., DÖTZER, F., FESTAG, A., GERLACH, M., KROH, R. and LEINMÜLLER, T. (2006, March) Attacks on inter-vehicle communication systems: an analysis. In *Proceedings of International Workshop on Intelligent Transportation* (Hamburg, Germany: WIT'2006).

[41] TROPOS NETWORKS (2010) http://www.tropos.com/pdf/solutions/Parking-Final.pdf.

[42] YAN, G. OLARIU, S. and WEIGLE, M.C. (2009) Providing location security in vehicular ad-hoc networks. *IEEE Wireless Commun.* **16**(6): 48–55.

[43] Automated Parking Management System at New Hyderabad International Airport, http://www.inrnews.com/realestateproperty/india/hyderabad/automated_parking_management_s.html.

Optimal Scanning Bandwidth Strategy Incorporating Uncertainty about Adversary's Characteristics

Andrey Garnaev[1,*], Wade Trappe[2]

[1]WINLAB, Rutgers University, North Brunswick, USA
[2]WINLAB, Rutgers University, North Brunswick, USA

Abstract

In this paper, we investigate the problem of designing a spectrum scanning strategy to detect an intelligent Invader who wants to utilize spectrum undetected for his/her unapproved purposes. To deal with this problem we model the situation as two games, between a Scanner and an Invader, and solve them sequentially. The first game is formulated to design the optimal (in maxmin sense) scanning algorithm, while the second one allows one to find the optimal values of the parameters for the algorithm depending on the parameters of the network. These games provide solutions for two dilemmas that the rivals face. The Invader's dilemma consists of the following: the more bandwidth the Invader attempts to use leads to a larger payoff if he is not detected, but at the same time also increases the probability of being detected and thus fined. Similarly, the Scanner faces a dilemma: the wider the bandwidth scanned, the higher the probability of detecting the Invader, but at the expense of increasing the cost of building the scanning system. The equilibrium strategies are found explicitly and reveal interesting properties. In particular, we have found a discontinuous dependence of the equilibrium strategies on the network parameters, fine and the type of the Invader's award. This discontinuity of the fine means that the network provider has to take into account a human/social factor since some threshold values of fine could be very sensible for the Invader, while in other situations simply increasing the fine has a minimal deterrence impact. Also we show how incomplete information about the Invader's technical characteristics and reward (e.g. motivated by using different type of application, say, video-streaming or downloading files) can be incorporated into the scanning strategy to increase its efficiency.

Keywords: Scanning, Bandwidth, Detection, Bayesian game

1. Introduction

Over the last few decades, the increasing demand for wireless communications has motivated the exploration for more efficient usage of spectral resources ([1, 2]). In particular, it has been noticed that there are large portions of spectrum that are severely underutilized [3]. Recently, cognitive radio technologies (CR) have been proposed as a means to intelligently use such spectrum opportunities by sensing the radio environment and exploiting available spectrum holes for secondary usage [4]. In CR systems, secondary users are allowed to "borrow (or lease)" the usage of spectrum from primary users (licensed users), as long as they do not hinder in the proper operation of the primary users' communications. Unfortunately, as we move to make the CR technologies commercial, which will allow secondary users to access spectrum owned by primary users, we will face the inevitable risk

that adversaries will be tempted to use CR technology for illicit and selfish purposes [5]. If we imagine an unauthorized user (Invader) attempting to sneak usage of spectrum without obeying proper regulations or leasing the usage of the spectrum, the result will be that both legitimate secondary users and primary users will face unexpected interference, resulting in significant performance degradation across the system.

The challenge of enforcing the proper usage of spectrum requires the notion of a "spectrum policing agent", whose primary job is to ensure the proper usage of spectrum and identify anomalous activities occurring within the spectrum[5]. As a starting point to being able to police the usage of spectrum, we must have the ability to scan spectrum and effectively identify anomalous activities. Towards this objective, there have been several research efforts in signal processing techniques that can be applied to the spectrum scanning problem. For example, in [6, 7], the authors presented methods for detecting a desired signal contained within interference. Similarly,

---
*Corresponding author. Email: garnaev@yahoo.com

detection of unknown signals in noise without prior knowledge of authorized users was studied in [8, 9]. As another example, in [5], the authors proposed a method to detect anomalous transmission by making use of radio propagation characteristics. In [10], the authors investigated what impact on spectrum scanning can have information about the over-arching application that a spectrum thief might try to run, while, in [11], a stationary bandwidth scanning strategy in a discounted repeated game was suggested.

However, these pieces of work tend to not examine the important "interplay" between the two participants inherent in the problem– the Invader, who is smart and will attempt to use the spectrum in a manner to minimize the chance of being detected and fined, while also striving to maximize the benefit he/she receives from illicit usage of this spectrum; and the Scanner, who must be smart and employ a strategy that strategically maximizes the chance of detecting and fining the smart Invader, with minimal cost. This challenge is made more difficult by the complexity of the underlying scanning problem itself: there will be large swaths of bandwidth to scan, and the system costs (e.g., analog-to-digital conversion, and the computation associated with running signal classifiers) associated with scanning very wide bandwidth makes it impossible to scan the full range of spectrum in a single instance. Consequently, it is important to understand the strategic dynamics that exist between the Scanner and the Invader, while also taking into account the underlying costs and benefits that exist for each participant as well as information or its lack on the technical characteristics of the Invader and his object to intrude into the bandwidth. This paper[1] focuses on finding the optimal scanning strategy by selecting the scanned (and, similarly, the invaded) bandwidth that should be employed in spectrum scanning and examining how incorporating information or the lack of information about the technical characteristics of the Invader and his object can improve the scanning strategy. In order to solve this problem we will apply a Bayesian approach. Note that Bayesian approaches have been widely employed in dealing with different problems in networks, for example, intrusion detection [13–15], scanning bandwidth [10] and transmission under incomplete information [16–21]. Finally note that the optimal scanning problem also relates the problem of designing security systems. Note that an extensive literature exists on the construction and modeling of different aspects of such security systems for communication and network security [13, 22–29],

security in wireless networks [30, 31] and cyber-security [14, 16, 32]. In [33], the readers can find a structured and comprehensive survey of the research contributions that analyze and solve security and privacy problems in computer networks via game-theoretic approaches.

The organization of this paper is as follows: in Section 2, we first define the problem by formulating two games, which will be solved sequentially in terms of payoff and cost functions. In the first game, the Scanner looks for the maxmin scanning algorithm, if parameters (widths of used bandwidths) of scanning and intrusion are fixed and known. In the second game each player, using the first game's result, which supplies detection probability, looks for the optimal values of these parameters. To gain insight into the problem, in Section 4.1, we outline a linearized model for detection probability and arrive at the corresponding best response strategies for each player in Section 4.2. We then explicitly obtain the equilibrium strategies, in Section 4.3 and Section 4.4, for cases involving complete and incomplete knowledge of the Invader's technical characteristics (radio's capabilities). In Section 5, numerical illustrations are supplied. Finally, in Section 6, discussions and conclusions are supplied, and, in Appendix, the proofs of the announced results are offered to close the paper.

2. Formulation of the scanning problem

In this section, we set up our problem formulation. Our formulation of the spectrum scanning problem involves two players: the Scanner and the Invader. The Scanner, who is always present in the system, scans a part of the band of frequencies that are to be monitored, in order to prevent illegal usage by a potential Invader of the primary (Scanner) network's ownership of this band. We assume that the amount of bandwidth that needs to be scanned is much larger than is possible using a single scan by the Scanner, and hence the Scanner faces a dilemma: the more bandwidth that is scanned, the higher the probability of detecting the Invader, but at the expense of increasing the cost of the RF scanning system.

We assume that if the Scanner scans a particular frequency band I_S and the Invader uses the band I_I then the invasion will be detected with certainty if $I_S \cap I_I \neq \emptyset$, and it will not be detected otherwise. Without loss of generality, we can assume that the size of the protected frequency band is normalized to 1. The Invader wants to use spectrum undetected for some illicit purpose. We consider two scenarios: (a) The reward for the Invader is related to the width of the frequency band he uses if he is undetected. If he is detected he will be fined. Thus, the Invader faces a dilemma: the more bandwidth he tries to use yields a larger payoff if he is not detected but also it increases the probability of being detected

---

[1]The authors note that a shortened version of this research was presented at Crowncom 2013 [12], and this paper extends the idea presented at Crowncom.

and thus to be fined, (b) The reward for the Invader is unknown to the Scanner: he only knows whether it is related to the width of the frequency band the Invader uses, or not. We formulate this problem as two games, which will be solved separately in the following two subsections.

2.1. Formulation of the first game – the scanning algorithm

In the first game, where we look for a maxmin scanning algorithm, the Scanner selects the band $B_S = [t_S, t_S + x] \subseteq [0, 1]$ with a fixed upper bound of frequency width x to scan i.e. $t_S \leq 1 - x$. The Invader selects the band $B_I = [t_I, t_I + y] \subseteq [0, 1]$ with a fixed upper bound frequency width y to intrude, i.e., $t_I \leq 1 - y$. Thus, B_S and B_I are pure strategies for the Scanner and the Invader. The Scanner's payoff $v(B_S, B_I)$ is 1 if the Invader is detected (i.e. $[t_S, t_S + x] \cap [t_I, t_I + y] \neq \emptyset$) and his payoff is zero otherwise. The goal of the Scanner is to maximize his payoff, while the Invader wants to minimize it. Thus, the Scanner and the Invader play a zero-sum game. The saddle point (equilibrium) of the game is a couple of strategies (B_{S*}, B_{I*}) such that for each couple of strategies (B_S, B_I) the following inequalities hold [34]:

$$v(B_S, B_{I*}) \leq v := v(B_{S*}, B_{I*}) \leq v(B_{S*}, B_I),$$

where v is the value of the game. It is clear that the game does not have a saddle point in the pure strategy if $x + y < 1$. To find the saddle point we have to extend the game by mixed strategies, where we assign a probability distribution over pure strategies. Then instead of the payoff v we have its expected value. The game has a saddle point in mixed strategies, and let $P(x, y)$ be the value of the game. Then $P(x, y)$ is the maximal detection probability of the Invader under worst conditions.

2.2. Formulation of the second game – the optimal parameters of the scanning algorithm

In the second game the rivals knowing their equilibrium strategies from the first game as well as detection probability $P(x, y)$, want to find the equilibrium frequency widths x and y. We here consider three sub-scenarios: (a) the Invader's type is known: namely, it is known how the reward for the Invader is related to the width of the frequency band he uses if he is undetected, (b) the technical characteristics of the Invader are known: namely, it is known which frequency band is available for him to use, (c) the Invader's type is unknown: instead, there is only a chance that the Invader reward is related to the width in use. Otherwise, it is not related. Different types of rewards can be motivated by using different types of applications (say, file-download or streaming video).

Invader reward is related to the bandwidth used. A strategy for the Scanner is to scan a width of frequency of size $x \in [a, b]$, and a strategy for the Invader is to employ a width of frequency of size $y \in [a, c]$, where $c \leq b < 1/2$. Thus, we assume that the Invader's technical characteristics (e.g., radio's capabilities) are not better than the Scanner's ones.

If the Scanner and the Invader use the strategies x and y, then the payoff to the Invader is the expected reward (which is a function $U(y)$ of bandwidth y illegally used by the Invader) minus intrusion expenses (which is a function $C_I(y)$ of bandwidth y) and expected fine F to pay, i.e.,

$$v_I(x, y) = (1 - P(x, y))U(y) - FP(x, y) - C_I(y). \quad (1)$$

The Scanner wants to detect intrusion taking into account scanning the expenses and damage caused by the illegal use of the bandwidth by the Invader. For detection, he is rewarded by a fine F imposed on the Invader. Thus, the payoff to the Scanner is the difference between the expected reward for detection, and the damage from intrusion into the bandwidth (which is a function $V(y)$ of bandwidth y illegally used by the Invader) with the scanning expenses (which is a function $C_S(x)$ of scanned bandwidth x),

$$v_S(x, y) = FP(x, y) - V(y)(1 - P(x, y)) - C_S(x). \quad (2)$$

Note that introducing transmission costs in such a formulation is common for CDMA [31, 35] and ALOHA networks ([36, 37]).

Incomplete information of the Invader's reward and technical characteristics. In this section we assume that the Invader's reward is defined by the reason he intruded into the bandwidth illegally for, and we consider two such reasons:

(a) With probability $1 - q_0$ for the Invader it is just important to work in the network without being detected. Thus, if he is not detected his reward is U which does not depend on the width of bandwidth employed for the intrusion. Then, of course, to minimize the probability of detection he will employ the minimal bandwidth allowed, thus, his strategy is $y = a$.

(b) With probability q_0 for the Invader the bandwidth he uses is important. Thus, his reward is the same as in Section 2.2. We assume that his technical characteristics can be different, the Invader knows his characteristics, but the Scanner does not know them. Under Invader's technical characteristics we assume an upper bound on the spectrum width he can employ. The Scanner knows only this upper bound on bandwidth as c with a conditional probability $q(c) \geq 0$ for $c \in [a, b]$, i.e., $\int_a^b q(c) \, dc = 1$.

To deal with this situation we are going to apply a Bayesian approach, namely, we introduce type $c \in [a, b]$ for the Invader related to the corresponding upper bounds (thus, we here deal with a continuum of Invader's types). The Invader knows his type, while the Scanner knows only its distribution. Denote by $y(c) \in [a, c]$ the strategy of the Invader of type c. Then his payoff is given as follows:

$$v_I^c(x, \boldsymbol{y}(c)) = (1 - P(x, \boldsymbol{y}(c)))U(\boldsymbol{y}(c)) \\ - FP(x, \boldsymbol{y}(c)) - C_I(\boldsymbol{y}(c)). \quad (3)$$

The payoff to the Scanner is the expected payoff taking into account the *type* of Invader:

$$v_S^E(x, \boldsymbol{y}) = (1 - q_0)v_S(x, a) + q_0 \int_a^b q(c)v_S(x, \boldsymbol{y}(c)) \, dc \quad (4)$$

with $v_S(x, \boldsymbol{y}(c))$ given by Eq. (2).

Here we look for Bayesian equilibrium [34], i.e., for such couple of strategies $(x_*, \boldsymbol{y}_*)$ that for any $(x, \boldsymbol{y})$ the following inequalities hold:

$$v_S^E(x, \boldsymbol{y}_*) \le v_S^E(x_*, \boldsymbol{y}_*), \\ v_I^c(x_*, \boldsymbol{y}(c)) \le v_I^c(x_*, \boldsymbol{y}_*(c)), c \in \text{supp}(q), \quad (5)$$

with $\text{supp}(q) = \{c \in [a, b] : q(c) > 0\}$.

We assume that the Scanner and the Invader know (as in the case with complete information) the parameters F, C_I, C_S, V, U, a, b as well as the probabilities $q(c)$ ($c \in [a, b]$) and q_0.

3. Equilibrium strategies for the first game

In the following theorem we give the equilibrium strategies for the first game (thus, maxmin scanning algorithm) for fixed bound width of the rivals.

Theorem 1. In the first game with fixed width to scan x and to invade y, the rivals employ a uniform tiling behavior. Namely,

(a) Let $1 - (x + y)M \le y$ with

$$M = \lfloor 1/(x + y) \rfloor, \quad (6)$$

where $\lfloor \xi \rfloor$ is the greatest integer less or equal to ξ. Then the Scanner and the Invader will, with equal probability $1/M$, employ a band of the set A_{-S} and A_{-I} correspondingly.

(b) Let $1 - (x + y)M > y$. Then the Scanner and the Invader will, with equal probability $1/(M + 1)$, employ a band of the set A_{+S} and A_{+I} correspondingly, where

$$A_{-S} = \{[k(x+y) - x, k(x+y)], k = 1, ..., M\},$$
$$A_{-I} = \{[k(x+y) - y - \epsilon(M + 1 - k), k(x+y) - \epsilon(M - k)],$$
$$k = 1, ..., M\}, \quad 0 < \epsilon < x/M,$$
$$A_{+S} = A_{-S} \cup [1 - x, 1],$$
$$A_{+I} = \{[(k - 1)(x + y + \epsilon), (k - 1)(x + y + \epsilon) + y],$$
$$k = 1, ..., M\} \cup [1 - y, 1], \quad 0 < \epsilon < \frac{1 - y - M(x + y)}{M - 1}.$$

The value of the game (detection probability) $P(x, y)$ is given as follows:

$$P(x, y) = \begin{cases} 1/M, & 1 - (x + y)M \le y, \\ 1/(M + 1), & 1 - (x + y)M > y. \end{cases} \quad (7)$$

4. Equilibrium strategy for the second game

In this section, which is split into five subsections, we find the equilibrium strategies for the second game explicitly. First, in Subsection 4.1 we linearize our model to get an explicit solution, then in Subsection 4.2 the best response strategies are given, and they are employed in Subsections 4.3 and 4.4 to construct equilibrium strategies for known and unknown Invader's technical characteristics correspondingly.

4.1. Linearized model

In order to get an insight into the problem, we consider a situation where the detection's probability $P(x, y)$ for $x, y \in [a, b]$ is approximated by a linear function as follows:

$$P(x, y) = x + y. \quad (8)$$

Thus, Eq. (7) and Eq. (8) coincide for $x + y = 1/n, n = 2, 3, ...$ We assume that the scanning and intrusion cost as well as the Invader's and Scanner's utilities are linear in the bandwidth involved, i.e., $C_S(x) = C_S x$, $C_I(y) = C_I y$, $U(y) = Uy$, $V(y) = Vy$ where $C_S, C_I, U, V > 0$. Then the payoffs to the Invader and the Scanner, if they use strategies $x \in [a, b]$ and $y \in [a, c]$ ($\boldsymbol{y}(c) \in [a, c]$) respectively, become:

(i) For the known Invader's reward:

$$v_I(x, y) = U(1 - x - y)y - F(x + y) - C_I y, \\ v_S(x, y) = F(x + y) - Vy(1 - x - y) - C_S x,$$

(ii) For the unknown Invader's reward and technical characteristics:

$$v_I^c(x, \boldsymbol{y}(c)) = U(1 - x - \boldsymbol{y}(c))\boldsymbol{y}(c) \\ - F(x + \boldsymbol{y}(c)) - C_I \boldsymbol{y}(c), \text{ for } c \in \text{supp}(q),$$

$$v_S^E(x, \boldsymbol{y}) = q_0 \int_a^b \Big[F(x + \boldsymbol{y}(\xi)) \\ - V\boldsymbol{y}(\xi)(1 - x - \boldsymbol{y}(\xi))\Big]q(\xi) \, d\xi \\ + (1 - q_0)\Big(F(x + a) - Va(1 - x - a)\Big) - C_S x.$$

Note that linearized payoffs have found extensive usage for a wide array of problems in wireless networks [36, 38–41]. Of course, such an approach simplifies the original problem and only gives an approximated solution. Meanwhile, it can also be very useful: sometimes it allows one to obtain a solution explicitly, and allows one to look inside of the structure of the solution as well as the correlation between parameters of the system.

4.2. The best response strategies

In this section, we give the best response strategies for the Scanner and the Invader when the Invader's reward and technical characteristics are unknown, i.e., such strategies that $BR_S^E(y) = \arg\max_x v_S^E(x, y)$ and $BR_I^c(x) = \arg\max_{y(c)} v_I^c(x, y(c))$.

Theorem 2. In the second step of the considered game with unknown Invader's reward and technical characteristics the Scanner and the Invader have the best response strategies $BR_S^E(y)$ and $BR_I^c(x)$ given as follows:

$$BR_S^E(y) = \begin{cases} a, & \bar{y} < R_{q_0}, \\ \text{any from } [a, b], & \bar{y} = R_{q_0}, \\ b, & \bar{y} > R_{q_0}, \end{cases} \quad (9)$$

$$BR_I^c(x) = \begin{cases} c, & c \le L(x), \\ L(x), & a < L(x) < c, \\ a, & L(x) \le a \end{cases} \quad (10)$$

with

$$
\begin{aligned}
L(x) &= \frac{T - x}{2}, \\
T &= \frac{U - F - C_I}{U}, \\
R &= \frac{C_S - F}{V}, \\
R_{q_0} &= \frac{C_S - F - (1 - q_0)Va}{q_0 V} = \frac{R - a(1 - q_0)}{q_0}
\end{aligned}
\quad (11)
$$

and

$$\bar{y} = \int_a^b q(\xi)y(\xi)\, d\xi.$$

4.3. Equilibrium strategies: the unknown Invader's reward and technical characteristics

The equilibrium for the game exists since the payoff to the Scanner is linear in x and the payoff to the Invader of type c is concave in $y(c)$. The equilibrium can be found by Eq. (5) as a couple of strategies (x, y) which are the best response to each other, i.e., $x = BR_S^E(y)$ and $y(c) = BR_I^c(x)$, $c \in [a, b]$ and such a solution always exists and is unique as shown in the following theorem.

Theorem 3. The considered second game with unknown Invader's reward and technical characteristics has unique Bayesian equilibrium (x, y), and it is given as follows:

$$x = \begin{cases} b, & R_{q_0} \le \overline{\mathbf{BR}}_I(b), \\ \overline{\mathbf{BR}}_I^{-1}(R_{q_0}), & \overline{\mathbf{BR}}_I(b) < R_{q_0} < \overline{\mathbf{BR}}_I(a), \\ a, & \overline{\mathbf{BR}}_I(a) \le R_{q_0}, \end{cases}$$

$$y(c) = \begin{cases} BR_I^c(b), & R_{q_0} \le \overline{\mathbf{BR}}_I(b), \\ BR_I^c\left(\overline{\mathbf{BR}}_I^{-1}(R_{q_0})\right), & \overline{\mathbf{BR}}_I(b) < R_{q_0} < \overline{\mathbf{BR}}_I(a), \\ BR_I^c(a), & \overline{\mathbf{BR}}_I(a) \le R_{q_0}, \end{cases} \quad (12)$$

where $c \in \text{supp}(q)$ with

$$\overline{\mathbf{BR}}_I(x) = \int_a^b q(\xi) BR_I^\xi(x)\, d\xi \quad (13)$$

and $\overline{\mathbf{BR}}_I^{-1}(x)$ is inverse function to $\overline{\mathbf{BR}}_I(x)$, i.e., $\overline{\mathbf{BR}}_I^{-1}\left(\overline{\mathbf{BR}}_I(x)\right) = x$.

4.4. Equilibrium strategies: the known Invader's technical characteristics and unknown reward

The equilibrium for the second game with complete information about the technical characteristics of the Invader and unknown reward can be presented explicitly as follows:

Theorem 4. Let the Invader's technical characteristics be known but his reward can be unknown. This second game has unique Nash equilibrium, and it is given by Table 1.

Note that the Scanner's and Invader's equilibrium strategies can have sudden jumps (discontinuities) as one continuously varies the fine F and probability q_0 that the Invader's reward related bandwidth used. It is caused by the fact that R_{q_0} depends on these parameters, while L depends only on F. For example, (i_1)-(i_6) implies that the Invader's equilibrium strategy can jump while probability q_0 varies, and (i_2) and (i_6) yield about the possibility of such a jump by fine F. The possibility of jumps for the Scanner's equilibrium strategy follows from (i_2) and (i_5).

5. Numerical illustrations

As a numerical illustration of the scenario with complete information on the Invader's technical characteristics, we consider $U = V = 1$, $a = 0.01$, $b = 0.3$, $C_S = 0.4$, $C_I = 0.1$ and q is the uniform distribution in $[a_0, b_0] = [a + (b - a)/10, b] = [0.039, 0.3]$. Figure 1 demonstrates the Scanner's equilibrium strategy and payoff as functions of the fine $F \in [0.1, 0.4]$ and the

Case	Condition	Condition	x	y	P_R	P_U
i_1	$R_{q_0} < a$	$L(b) < a$	b	a	$a+b$	$2a$
i_2	$R_{q_0} < a$	$a \le L(b) \le c$	b	$L(b)$	$b + L(b)$	$b + a$
i_3	$R_{q_0} < a$	$c < L(b)$	b	c	$b + c$	$b + a$
i_4	$c < R_{q_0}$	$L(a) < a$	a	a	$2a$	$2a$
i_5	$c < R_{q_0}$	$a \le L(a) \le c$	a	$L(a)$	$a + L(a)$	$2a$
i_6	$c < R_{q_0}$	$c < L(a)$	a	c	$a + c$	$2a$
i_7	$a \le R_{q_0} \le c$	$L(b) \le R_{q_0} \le L(a)$	$L^{-1}(R_{q_0})$	R_{q_0}	$L^{-1}(R_{q_0}) + R_{q_0}$	$L^{-1}(R_{q_0}) + a$
i_8	$a \le R_{q_0} \le c$	$L(a) \le a$	a	a	$2a$	$2a$
i_9	$a \le R_{q_0} \le c$	$a < L(a) < R_{q_0}$	a	$L(a)$	$a + L(a)$	$2a$
i_{10}	$a \le R_{q_0} \le c$	$c < L(b)$	b	c	$b + c$	$b + a$
i_{11}	$a \le R_{q_0} \le c$	$R_{q_0} < L(b) < c$	b	$L(b)$	$b + L(b)$	$b + a$

Table 1. The equilibrium strategies (x, y) with $L^{-1}(R_{q_0}) = T - 2(C_S - F - (1 - q_0)V)/(q_0 V)$ and P_R and P_U are detection probabilities of the Invader with reward related and un-related to the bandwidth used.

probability $q_0 \in [0.01, 0.99]$ that the Invader's reward related to bandwidth used. Increasing fine F and probability q_0 makes the Scanner employ a larger band and impacts the Scanner's payoff in a multi-directional way, namely, it increases F and decreases q_0. This is caused by the fact that the Invader, who wants to minimize his detection probability, causes less damage to the network than the one who benefits from using a larger bandwidth.

Figures 2 and 3 illustrate the Invader's equilibrium strategy and payoff if his reward is related to the bandwidth used for $c = a_0$ and $c = b_0$ respectively. Figure 4 demonstrates corresponding detection probabilities. The Invader of type $c = a_0$ employs a constant strategy $y(c) = a_0$ independent of the fine F and probability q_0. The Invader's payoff and detection probability vary in opposite directions while fine F and probability q_0 are increasing, namely, the Invader's payoff is decreasing, while the detection probability is increasing, since it also makes the Scanner to employ a larger bandwidth. What is interesting is that the Invader's payoff experiences a sudden drop and the detection probability experiences a sudden jump due to the Scanner's behaviour, who alters his strategy by a sudden jump at threshold values. For the Invader with a reward un-related to the bandwidth used, the payoff and detection behave similarly but with some shift since such an Invader also employs a constant strategy $y = a$ (Figure 5). The Invader of type $c = b_0$ uses a strategy depending on fine F and probability q_0. Increasing fine F and probability q_0 makes the Invader employ a smaller bandwidth and reduces his payoff. What is interesting that his detection probability is not monotonous by fine F and probability q_0 and increasing fine F and probability q_0 could even reduce the detection probability. It can be explained that at the threshold values of fine F and probability q_0 the Scanner already gets the upper band, while the Invader still does not get to the lower band, and

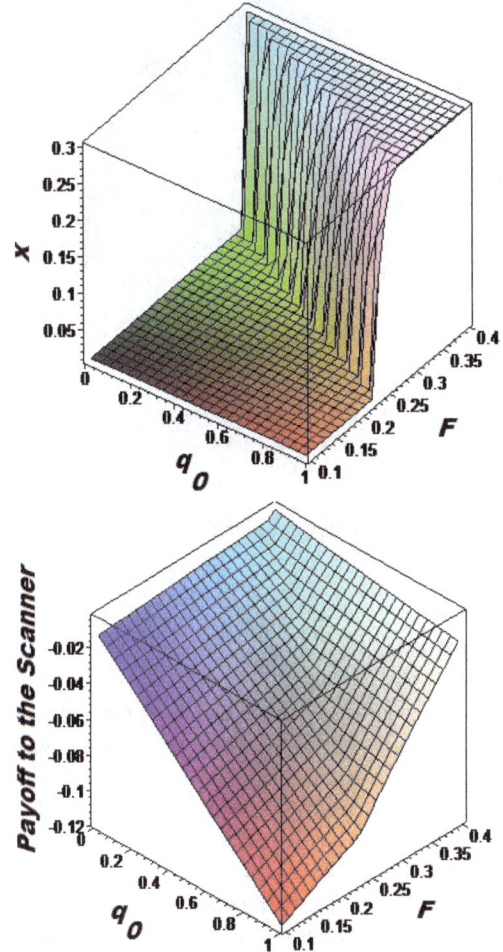

Figure 1. Equilibrium strategy x (upper) and payoff (bottom) to the Scanner.

further increasing of the fine and probability leads to continuous decreasing of the detection probability due to the smaller bandwidth employed by the Invader.

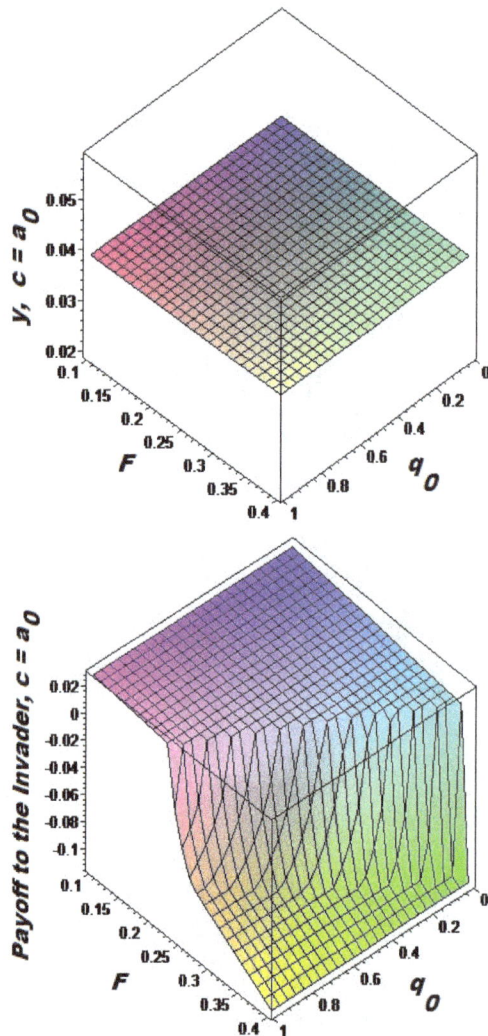

Figure 2. Equilibrium strategy y (upper) and payoff (bottom) to the Invader for $c = a_0$.

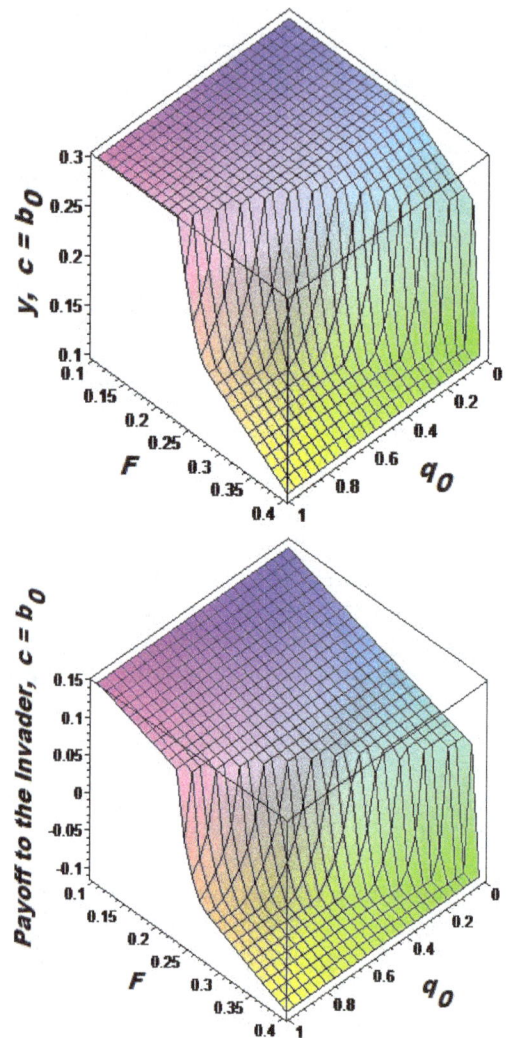

Figure 3. Equilibrium strategy y (upper) and payoff (bottom) to the Invader for $c = b_0$.

6. Discussion

In this paper, we suggest a simple model for designing a maxmin scanning algorithm for detection of an Invader with incomplete information about the Invader's reward and technical characteristics and we find the optimal parameters (width of bandwidth to scan) for this algorithm. We have shown that this optimal width essentially depends on the network's and agent's characteristics and under some conditions a small variation of network parameters and fine could lead to jump changes in the optimal strategies, as well as in the payoffs of the rivals. This mixture between continuous and discontinuous behavior of the Invader under the influence of fine implies that the network provider has to carefully make a value judgement: some threshold values of fine could have a huge impact on the Invader, while in the other situations a small increase will have a minimal impact on the strategies used. A goal for our

future investigation is to investigate the non-linearized detection probability. Also, we intend to extend our model to the case of multi-step scanning algorithms with learning.

References

[1] Haykin, S. (2005) Cognitive radio: brain-empowered wireless communications. *IEEE Journal on Selected Areas in Communications* **23**: 201–220.

[2] Mitola, J. (1999) Cognitive radio: brain-empowered wireless communications. In *Proceedings of IEEE International Workshop on Mobile Multimedia Communications (MoMuC '99)*: 3–10.

[3] Akyildiz, I.F., Lee, W.Y., Vuran, M.C. and Mohanty, S. (2006) Next generation /dynamic spectrum access/ cognitive radio wireless networks: a survey. *Computer Networks* **50**: 2127–2159.

[4] Fette, B.A. (2009) *Cognitive radio technology* (Burlington, MA: Academic Press).

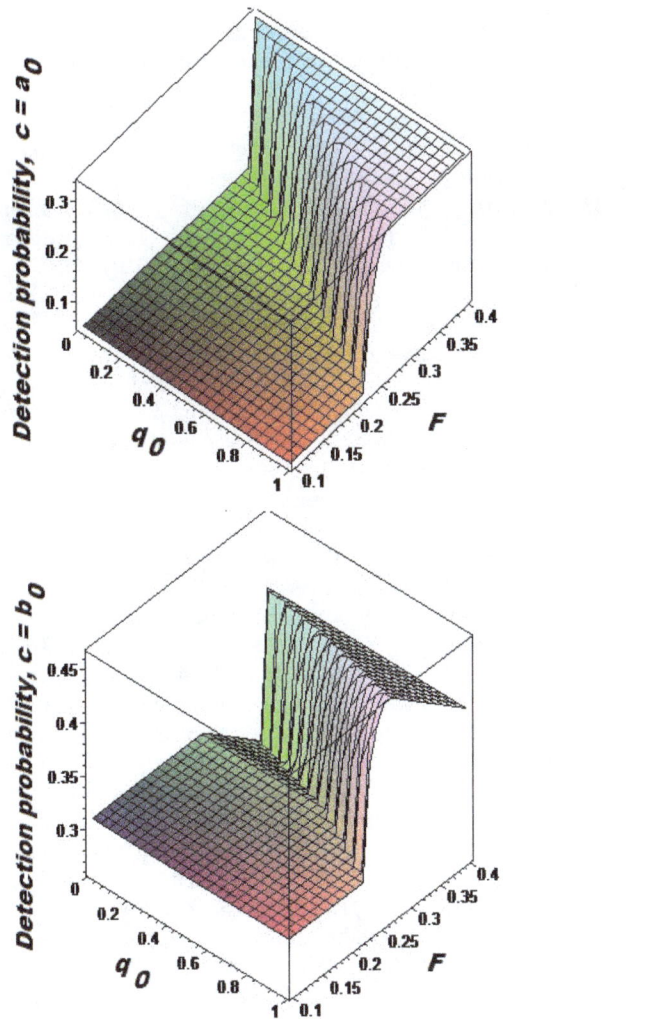

Figure 4. Detection probability of the Invader with $c = a_0$ (upper) and $c = b_0$ (bottom).

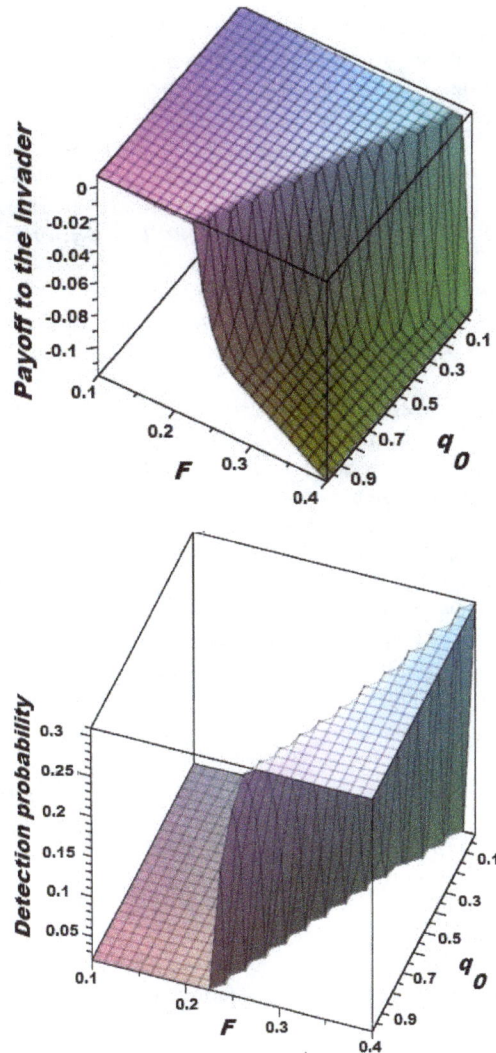

Figure 5. Payoff (upper) and detection probability (bottom) of the Invader with reward un-related to the bandwidth used.

[5] Liu, S., Chen, Y., Trappe, W. and Greenstein, L.J. (2009) ALD: An anomaly detection framework for dynamic spectrum access networks. In *Proceedings of IEEE INFOCOM 2009*: 675–683.

[6] Verdu, S. (1998) *Multiuser detection* (Boston, MA: Cambridge University Press).

[7] Trees, H.V. (2004) *Detection, Estimation, and Modulation Theory, Radar-Sonar Signal Processing and Gaussian Signals in Noise* (Wiley-Interscience).

[8] Digham, F.F., Alouini, M.S. and Simon, M.K. (2007) On the energy detection of unknown signals over fading channels. *IEEE Transactions on Communications* **55**: 21–24.

[9] Urkowitz, H. (1967) Energy detection of unknown deterministic signals. *Proceedings of the IEEE* **55**: 523–531.

[10] Garnaev, A., Trappe, W. and Kung, C.-T. (2012) Dependence of optimal monitoring strategy on the application to be protected. In *Proceedings of 2012 IEEE Global Communications Conference (GLOBECOM)*: 1054–1059.

[11] Garnaev, A. and Trappe, W. (2014) To eavesdrop or jam, that is the question. In Sherif, M. and et al. [eds.] *Proceedings of ADHOCNETS 2013* (Springer), *LNICST* **129**: 146–161.

[12] Garnaev, A., Trappe, W. and Kung, C.-T. (2013) Optimizing scanning strategies: Selecting scanning bandwidth in adversarial RF environments. In *Proceedings of the 8th International Conference on Cognitive Radio Oriented Wireless Networks (CROWNCOM)*: 148–153.

[13] Liu, Y., Comaniciu, C. and Man, H. (2006) A Bayesian game approach for intrusion detection in wireless Ad Hoc networks. In *Proceedings of the 2006 Workshop on Game Theory for Communications and Networks*: 1–12.

[14] Agah, A., Das, S.K., Basu, K. and Asadi, M. (2004) A Bayesian game approach for intrusion detection in wireless Ad Hoc networks. In *Proceedings of the 3rd IEEE International Symposium on Network Computing and Applications (NCA 2004)*: 243–346.

[15] GARNAEV, A., BAYKAL-GURSOY, M. and POOR, H.V. (2014) Incorporating attack-type uncertainty into network protection. *IEEE Transactions on Information Forensics and Security* **9**: 1278–1287.

[16] HAN, Z., MARINA, N., DEBBAH, M. and HJRUNGNES, A. (2009) Physical layer security game: Interaction between source, eavesdropper, and friendly jammer. *EURASIP Journal on Wireless Communications and Networking* Article ID 452907.

[17] ALTMAN, E., AVRACHENKOV, K. and GARNAEV, A. (2010) Fair resource allocation in wireless networks in the presence of a jammer. *Performance Evaluation* **67**: 338–349.

[18] HE, G., DEBBAH, M. and ALTMAN, E. (2009) *k*-player Bayesian waterfilling game for fading multiple access channels. In *Proceedings of the 3rd IEEE International Workshop on Computational Advances in Multi-Sensor Adaptive Processing (CAMSAP 2009)*: 17–20.

[19] HEIKKINEN, T. (1999) *A Minmax Game of Power Control in a Wireless Network under Incomplete Information*. Tech. Rep. 99–43, DIMACS, New Brunswick, NJ.

[20] JEAN, S. and JABBARI, B. (2004) Bayesian game-theoretic modeling of transmit power determination in a self-organizing CDMA wireless network. In *Proceedings of 60th IEEE Vehicular Technology Conference (VTC)*, **5**: 3496–3500.

[21] ADLAKHA, S., JOHARI, R. and GOLDSMITH, A. (2007), Competition in wireless systems via bayesian inference games, http://arxiv.org/abs/0709.0516.

[22] ALPCAN, T. and BASAR, T. (2003) A game theoretic approach to decision and analysis in network intrusion detection. In *Proceedings of IEEE CDC 2003*: 2595–2600.

[23] NGUYEN, K.C., ALPCAN, T. and BASAR, T. (2009) Security games with incomplete information. In *Proceedings of IEEE ICC 2009*: 1–6.

[24] ALTMAN, E., AVRACHENKOV, K. and GARNAEV, A. (2009) Jamming in wireless networks: The case of several jammers. In *Proceedings of International Conference on Game Theory for Networks (GameNets '09)*: 585–592.

[25] HAMILTON, S.N., MILLER, W.L., OTTAND, A. and SAYDJARI, O.S. (2002) Challenges in applying game theory to the domain of information warfare. In *Proceedings of ISW 2002*: 1–6.

[26] ROY, S., ELLIS, C., SHIVA, S., DASGUPTA, D., SHANDILYA, V. and WU, Q. (2010) A survey of game theory as applied to network security. In *Proceedings of HICSS 2010*: 1–10.

[27] GARNAEV, A., GARNAEVA, G. and GOUTAL, P. (1997) On the infiltration game. *International journal of game theory* **26**: 215–221.

[28] COMANICIU, C., MANDAYAM, N.B. and POOR, H.V. (2005) *Wireless Networks Multiuser Detection in Cross-Layer Design* (New York: Springer).

[29] BASTON, V.J. and GARNAEV, A.Y. (2000) A search game with a protector. *Naval Research Logistics* **47**(2): 85–96.

[30] MUKHERJEE, A. and SWINDLEHURST, A.L. (2010) Optimal strategies for countering dual-threat jamming/eavesdropping-capable adversaries in MIMO channels. In *Proceedings of MILCOM 2010*: 1695–1700.

[31] ZHU, Q., SAAD, W., HAN, Z., POOR, H.V. and BASAR, T. (2011) Eavesdropping and jamming in next-generation wireless networks: a game-theoretic approach. In *Proceedings of Military communications conference (MILCOM)*: 119–124.

[32] KONG-WEI, L. and WING, J. (2005) Game strategies in network security. *International journal of information security* **4**: 71–86.

[33] MANSHAEI, M.H., ZHU, Q., ALPCAN, T., BASAR, T. and HUBAUX, J.-P. (2013) Game theory meets network security and privacy. *ACM Computing Survey* **45**(3).

[34] FUDENBERG, D. and TIROLE, J. (1991) *Game theory* (Boston, MA: MIT Press).

[35] ALTMAN, E., AVRACHENKOV, K. and GARNAEV, A. (2010) Taxation for green communication. In *Proceedings of the 8th International Symposium on Modeling and Optimization in Mobile, Ad Hoc and Wireless Networks (WiOpt 2010)*: 108–112.

[36] SAGDUYU, Y.E. and EPHREMIDESS, A. (2009) A game-theoretic analysis of denial of service attacks in wireless random access. *Journal of Wireless Networks* **15**: 651–666.

[37] GARNAEV, A., HAYEL, Y., ALTMAN, E. and AVRACHENKOV, K. (2012) Jamming game in a dynamic slotted ALOHA network. In JAIN, R. and KANNAN, R. [eds.] *Game Theory for Networks* (Springer), *LNICST* **75**: 429–443.

[38] ALTMAN, E., AVRACHENKOV, K. and GARNAEV, A. (2011) Jamming in wireless networks under uncertainty. *Mobile Networks and Applications* **16**: 246–254.

[39] KIM, S.L., ROSBERG, Z. and ZANDER, J. (1999) Combined power control and transmission rate selection in cellular networks. In *Proceedings of 50th IEEE Vehicular Technology Conference (VTC)*, **3**: 3 –10.

[40] KOO, I., AHN, J., LEE, H.A. and KIM, K. (1999) Analysis of Erlang capacity for the multimedia DS-CDMA systems. *IEICE Trans. Fundamentals* **E82-A**(5): 849–855.

[41] GARNAEV, A. and TRAPPE, W. (2013) The eavesdropping and jamming dilemma in multi-channel communications. In *Proceedings of 2013 IEEE International Conference on Communications (ICC)*: 753–757.

Appendix A. Proof of Theorem 1

Suppose that the Invader uses a band B_I with width y and the Scanner with equal probability employ a band from the set A_{-S} (A_{+S}) for $1 - (x + y)M \leq y$ (for $1 - (x + y)M > y$). The intervals composing A_{-S} and A_{+S} are separated from each other by at most y. Thus, at least one band from A_{-S} for $1 - (x + y)M \leq y$ and from A_{+S} for $1 - (x + y)M > y$ intersects with B_I. Thus, detection probability is greater or equal to $1/M$ for $1 - (x + y)M \leq y$ and it is is greater or equal to $1/(M + 1)$ for $1 - (x + y)M > y$.

Suppose that the Scanner uses a band B_S with width x and the Invader with equal probability employ a band from the set A_{-I} (A_{+I}) for $1 - (x + y)M \leq y$ (for $1 - (x + y)M > y$). The intervals composing A_{-I} and A_{+I} are separated from each other by more that x. Thus, at most one band from A_{-I} for $1 - (x + y)M \leq y$ and from A_{+I} for $1 - (x + y)M > y$ intersects with B_S. Thus, detection probability is less or equal to $1/M$ for $1 - (x + y)M \leq y$

and it is is less or equal to $1/(M + 1)$ for $1 - (x + y)M > y$ and the result follows. ∎

Appendix B. Proof of Theorem 2

Note that

$$v_S^E(x, y) = \left(F - C_S + q_0 V \bar{y} + (1 - q_0)Va\right)x$$
$$+ q_0 \left[(F - V)\bar{y} + V \int_a^b y^2(\xi)q(\xi)\,d\xi\right]$$
$$+ (1 - q_0)(F - V + Va)a.$$

Thus, for a fixed y the payoff $v_S^E(x, y)$ is linear on x. Thus, $\mathrm{BR}_S^E(y) = \arg\max_x v_S^E(x, y)$ is defined by sign of $F - C_S + q_0 V \bar{y} + (1 - q_0)Va$ as it is given by Eq. (9).

Note that, the Invader's payoff has the following form:

$$v_I^c(x, y(c)) = (U(1 - x) - F - C_I)y(c) - Uy^2(c) - xF.$$

Thus, for a fixed x the payoff $v_I^c(x, y(c))$ is a concave quadratic polynomial on $y(c)$ getting its absolute maximum at $y(c) = (U(1 - x) - F - C_I)/(2U)$. Thus, the maximum of $v_I^c(x, y(c))$ by $y(c)$ within $[a, c]$ is reached either on its bounds $y(c) = a$ and $y(c) = c$ or at $y(c) = (U(1 - x) - F - C_I)/(2U)$ if it belongs to $[a, c]$ as it is given by Eq. (10). ∎

Appendix C. Proof of Theorem 3

First note that (x, y) is a Nash equilibrium if and only if it is a solution of equations $x = \mathrm{BR}_S^E(y)$ and $y(c) = \mathrm{BR}_I^c(x)$, $c \in [a, b]$ with $\mathrm{BR}_S^E(y)$ and $\mathrm{BR}_I^c(x)$ given by Theorem 2.

By Eq. (11) we have that Eq. (9) in equilibrium point is equivalent to

$$x = \begin{cases} a, & \overline{\mathbf{BR}}_I(x) < R_{q_0}, \\ \text{any from } [a, b], & \overline{\mathbf{BR}}_I(x) = R_{q_0}, \\ b, & \overline{\mathbf{BR}}_I(x) > R_{q_0} \end{cases} \quad \text{(C.1)}$$

with $\overline{\mathbf{BR}}_I(x)$ given by Eq. (13).

Note that $\overline{\mathbf{BR}}_I(x)$ is non-increasing on x. Thus, if $\overline{\mathbf{BR}}_I(a) < R_{q_0}$, then $\overline{\mathbf{BR}}_I(x) < R_{q_0}$ for any x and Eq. (C.1) yields that x has to be equal to a. If $\overline{\mathbf{BR}}_I(b) > R_{q_0}$, then $\overline{\mathbf{BR}}_I(x) > R_{q_0}$ for any x and Eq. (C.1) yields that x has to be equal to b. If $\overline{\mathbf{BR}}_I(b) \leq R_{q_0} \leq \overline{\mathbf{BR}}_I(a)$ then $x = \overline{\mathbf{BR}}_I^{-1}\left(R_{q_0}\right)$ and the result follows. ∎

Appendix D. Proof of Theorem 4

For the situation with complete information of the Invader's technical characteristics the best response strategies turn into

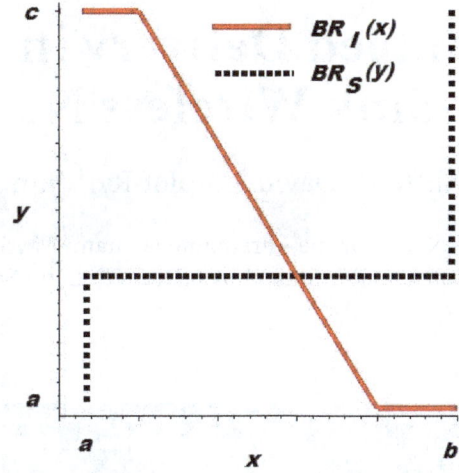

Figure D.1. The Nash equilibrium as an intersection of the best response curves

$$\mathrm{BR}_S(y) = \begin{cases} a, & y < R_{q_0}, \\ \text{any from } [a, b], & y = R_{q_0}, \\ b, & y > R_{q_0}, \end{cases} \quad \text{(D.1)}$$

$$\mathrm{BR}_I(x) = \begin{cases} a, & L(x) \leq a, \\ L(x), & a < L(x) < c, \\ c, & c \leq L(x) \end{cases}$$
$$= \begin{cases} a, & x \leq T - 2c, \\ L(x), & T - 2c < x < T - 2a, \\ c, & T - 2a \leq x. \end{cases} \quad \text{(D.2)}$$

Thus, the equilibrium can be described as an intersection of the best response curves (Figure D.1). Such intersection always exists.

Let $a > R_{q_0}$. By Eq. (D.1), $\mathrm{BR}_S(y) \equiv b$. This, jointly with Eq. (D.2), implies (i_1)-(i_3).

Let $R_{q_0} > c$. By Eq. (D.1), $\mathrm{BR}_S(y) \equiv a$. Then, Eq. (D.2) implies (i_4)-(i_6).

Let $a \leq R_{q_0} \leq c$. First note $L(x)$ is linear decreasing function from $L(a)$ for $x = a$ to $L(b)$ for $x = b$.

(a) Let $L(b) \leq R_{q_0} \leq L(a)$. Then the equation $L(x) = R_{q_0}$ has the unique root within $[a, b]$. Thus, Eq. (D.1) and Eq. (D.2) yield (i_7).

(b) Let $L(a) \leq R_{q_0}$. Then, $L(x) < R_{q_0}$ for $x \in (a, b]$. Thus, by Eq. (D.2), $\mathrm{BR}_I(x) < c$ for $x \in [a, b]$. Besides, by the assumption, the equation $L(x) = R_{q_0}$ does not has root in $[a, b]$. Thus, by Eq. (D.1), $\mathrm{BR}_S(y) \equiv a$. Thus, Eq. (D.2) implies (i_8) and (i_9).

(c) Let $R_{q_0} < L(b)$. Then $L(x) > R_{q_0}$ for $x \in [a, b)$. Thus, by Eq. (D.2), $\mathrm{BR}_I(x) > a$ for $x \in [a, b]$. Besides, by the assumption, the equation $L(x) = R_{q_0}$ does not has root in $[a, b]$. Thus, by Eq. (D.1), $\mathrm{BR}_S(y) = b$, and, Eq. (D.2) implies (i_{10}) and (i_{11}). ∎

3

Guaranteed Delivery in k-Anycast Routing in Multi-Sink Wireless Networks

Nathalie Mitton[1], David Simplot-Ryl[1], Jun Zheng[2]

[1]Inria Lille - Nord Europe - firstname.lastname@inria.fr
[2]National Mobile Communications Research Lab - Southeast University, China

Abstract

In k-anycasting, a sensor wants to report event information to any k sinks in the network. In this paper, we describe KanGuRou, the first position-based energy efficient k-anycast routing which guarantees the packet delivery to k sinks as long as the connected component that contains s also contains at least k sinks. A node s running KanGuRou first computes a tree including k sinks with weight as low as possible. If this tree has $m \geq 1$ edges originated at node s, s duplicates the message m times and runs m times KanGuRou over a subset of defined sinks. We present two variants of KanGuRou, each of them being more efficient than the other depending of application settings. Simulation results show that KanGuRou allows up to 62% of energy saving compared to plain anycasting.

1. Introduction

Wireless sensor networks have been receiving a lot of attention in recent years due to their potential applications in various areas such as monitoring and data gathering. Sensor measurements from the environment may be sent to a base station (sink) in order to be analyzed. Other sensors may serve as routers on a path established to deliver the report. In large sensor networks, there may exist a bottleneck (around sink) if a single sink collects reports from all sensors. Scenarios with multiple sinks are then being considered, where each sensor reports to at least one sink, usually the nearest one. In wireless multi-sink sensor networks, anycasting is performed when several sinks are available, each offering same services. Then any of sinks may receive the report from sensors, and meet application demands. However, the cost of anycasting may depend on the distance between the receiving sinks and the reporting sensor. It is therefore desirable that selected algorithm reaches one of sinks close to the event. For reliability, load-balancing and security purposes, it is then useful to ensure that at least k sinks receive the messages (where the overall number of sinks is greater than k) whatever the k sinks. Although many anycasting protocols have been deployed in wired networks [14], developing an efficient anycast routing protocol for wireless networks is challenging. Energy consumption and scalability are two challenging issues in designing protocols for sensor networks since they operate on limited capacity batteries while the number of deployed sensors could be very large. To the best of our knowledge, only few protocols have been designed for anycasting (when $k = 1$) in wireless networks. Most of them are based on an adaptation of an anycast routing for wired networks [1] and need flooding techniques that do not scale. Other ones [3, 7, 8] need a costly tree structure that is not robust and does not scale well in dynamic networks. Position based anycasting algorithm [8] is greedy and localized but optimizes neither hop count nor power consumption. Only algorithms proposed in [9] are geographic localized anycast routing protocols that guaranteed delivery (therefore loop-less), are memory-less, and scalable. In case of localized position based anycasting problem considered in this article, sensor nodes are merely aware of their positions, positions of their neighbors, and positions of all actors/sinks. But to the best of our knowledge, so far, there is no efficient position-based k-anycasting. To date, there is no so much work in the literature. Most of works are adaptation of wired solutions [13] and are thus centralized. Others use flooding [15] and not suitable for high dynamic networks (such as wireless sensor networks). A distributed k-anycast routing protocol based on mobile agents is proposed in [16] but requires a regular update of routing tables which also have to maintain paths towards every sink.

In this paper, we introduce KanGuRou (k-ANycast GUaranteed delivery ROUting protocol), a **position-based**, **energy-efficient** localized k-anycast routing protocol that **guarantees delivery** (therefore loop-less), is **memory-less**, and **scalable**. Unlike [16], it does not maintain any routing table and does not need to add any information neither on nodes nor in the message, which makes it scalable regardless of the number of sinks/nodes. It inspires from energy-efficient anycast EEGDA algorithm [9] and the splitting techniques of MSTEAM [5], proposing a new tree construction to ensure reaching k sinks. At each step, the current node s computes a spanning tree over k sinks with minimal cost. A message replication occurs when the tree spanning s and the set of sinks has multiple edges (later called branches) originated at the current node. Since there may be more sinks than the k to be reached, all of them are not spanned by the tree. The number of sinks k' spanned by each branch determines the number of sinks to be reached by each message. All sinks (not only the ones spanned by the tree) are distributed over every edge. The next hop is chosen in a cost-over-progress (COP) fashion, *i.e.* to the neighbor v which minimizes the ratio between the cost to reach v and the progress provided by v. The cost from s to v is the cost of the energy-weighted shortest path (ESP). The progress is computed as the difference between the weight of the trees computed by s and v resp. If s has no neighbor with positive progress, node s applies a EEGDA-face like routing, which is a face-based recovery mode. We prove that KanGuRou guarantees delivery to exactly k sinks. We present two variants which differ in the way the tree is computed. KanGuRou is evaluated through extensive simulations and results show that both variants of KanGuRou are energy efficient. Results show that KanGuRou allows up to 62% of energy saving and that every variant performs better regarding the percentage of sinks to reach. When this latter is less than 30% while the second one (KanGuRou-kPRIM) offers better energy saving when the percentage of sinks to reach is greater than 30%.

The remaining of the paper is organized as follows. Section 2 gives an overview of the literature about k-anycasting and present works on which KanGuRou is based. Section 3 introduces our notations. Section 4 presents KanGuRou. Section 5 presents simulation results. Finally Section 6 concludes the paper.

2. Related work

k-Anycast was first introduced in [13] for wired networks. Propositions in wireless networks firstly appeared in [12] proposing centralized solutions and thus does not really meet wireless networks requirements. [15] presents a reactive approach (flooding) and two advanced proactive approaches in which sinks have previously been gathered into components of at most k members and these components are then reached during the routing. To the best of our knowledge, the only distributed k-Anycast routing protocol is based on mobile agents and proposed in [16]. The protocol forms multiple components and each component has at least k members. Each component can be treated as a virtual server, so k-anycast service is distributed to each component. In this protocol, each routing node only needs to exchange routing information with its neighbors, so the protocol saves much communication cost and adapts to high dynamic networks. Nevertheless, although a first step toward, this algorithm needs to maintain routing tables at each node with as many entries as sinks and is not scalable.

Anycasting for wireless networks has first been modeled in [1]. Although many anycast protocols have been deployed in wired networks [14], there are very few for anycasting in wireless networks in the literature and only one of them [8] is geographical. Most of existing solutions are based on anycast for wired networks and need to build some structures. For example, in [7] a shortest path anycast tree rooted at each source is constructed for each event source. Sinks are the only leaves of the tree, and can dynamically join/leave the tree, which is updated accordingly. Data is delivered to the nearest sink on the tree. The algorithm thus simultaneously maintains paths to all sinks, and requires memorization of routing steps. Building a tree requires a lot of message exchanges. Tree-based protocols are not scalable, since the maintenance is costly when network has dynamic changes or when actors are moving.

In this paper, we introduce KanGuRou which is a position-based k-anycasting protocol. KanGuRou is an extension of the anycasting protocol proposed in [9] to the k anycasting. The only known position based anycasting algorithm is proposed in [8], where energy consumption needed to communicate at distance d is proportional to $u(d) = d^\alpha + c$. In the startup phase, each sensor node selects its next hop as follows. Let Q be a sensor, N be one of its neighboring nodes, and A be one of actors. Sensor Q selects neighbor N for which $u(|QN|) + u(|NA|)$ is minimized, over all neighbors and over all actors. This localized anycasting algorithm does not really optimize the power consumption (despite the claim), because it makes decision in the neighbor selection process based on long edges $|NA|$ which are not power optimal (an analytical proof of this fact was given in [5]). Further, it does not guarantee delivery in the presence of void areas. Initial routes are then used in a centralized data collection algorithm [8] as follows. All sensors within a region, when events occur, are reporting, each one to its actor selected by initial anycasting step. Authors [8] formulate integer linear program to construct data aggregation tree to minimize overall energy spent for reporting. In [9], authors describe EEGPA the first localized anycasting algorithms that guarantee delivery for connected multi-sink sensor networks based on a GFG approach. Three geographic localized anycast routing protocols are described loop-less, memory-less and scalable. They are generalizations of the EtE [4] protocol itself based on the well-known greedy-face-greedy (GFG) [2] unicast routing protocol to anycasting. Let $S(x)$ be the closest actor/sink to sensor x, and $|xS(x)|$ be distance between them. In greedy phase, a node s forwards the packet to its neighbor v that minimizes the ratio of cost of sending packet to v through an ESP over the reduction in distance ($|sS(s)| - |vS(v)|$) to the closest sink. EEGDA variant is to forward to the first neighbor on the shortest weighted path toward v. If none of neighbors reduces that distance then recovery mode is invoked. It is done by face traversal where edges are replaced by paths optimizing given cost.

KanGuRou also inspires from the multicast routing MSTEAM proposed in [5]. MSTEAM is a localized geographic multicast scheme based on the construction of local minimum spanning trees (MSTs), that requires information only on 1-hop neighbors. A message replication occurs when the MST spans the current node and the set of destinations has multiple edges originated at the current node. Destinations spanned by these edges are grouped together, and for each of these subsets the best neighbor is selected as the next hop. MSTEAM has been proved to be loop-free and to achieve delivery of the multicast message as long as a path to the destinations exists. To date, MSTEAM is the best known multicast algorithm.

3. Model and Notations

Network. We model the network as a graph $G = (V, E)$ where V is the set of sensor nodes and $uv \in E$ iff there exists a wireless link between u and $v \in V$. We suppose that nodes are equipped with a location service hardware such a GPS and are able to tune their range between 0 and R. We note $|uv|$ the Euclidean distance between nodes u and v. We note $N(u)$ the set of physical neighbors of node u, i.e. the set of nodes in communication range of node u ($N(u) = \{v \,|\, uv \in E\}$) and $V(G)$ the set V of vertices in G. $S = \{s_i\}_{i=0,1,..M}$ is the set of sinks, with M the number of sinks. Every node is aware of every sink and of its position. We note as $CT_S(s)$ the closest node in S to node s ($CT_S(s) = \{v \,|\, |sv| = \min_{w \in S} |sw|\}$). For a graph $G = (V, E)$ and a set $A \subseteq V$, we denote by $G|_A$ the subgraph of G which contains only nodes of A: $G|_A = (A, E \cap A^2)$.

Tree. Let $T = (V', E')$ be a tree and $a \in V'$ a vertex of T. $st(T, a)$ is the subtree of T with root a. T is an MST if its weight noted $\|T\|$ is minimal. The weight of the tree denotes the sum of the weight over all tree edges ($\|T\| = \sum_{uv \in E'} |uv|$). In an Euclidean MST, the weight of an edge is equal to its Euclidean length. A tree $T = (V', E') \subset G$ is a k-MST if $|V'| = k$ and that $\|T\|$ is the tree with minimum weight over all trees of k vertices from G.

Energy. We assume that every node is able to adapt its transmission range. We use the energy model defined in [11], i.e. the energy spent to send a message from nodes u to v is such that $cost(|uv|) = |uv|^\alpha + c$ if $|uv| \neq 0$. where c is signal processing overhead; α is a real constant (> 1) for signal attenuation. From this energy cost, we introduce the cost of the energy-weighted shortest path ($cost_{ESP}(s, d, t)$) from nodes s to d when aiming at target t. We compute the energy-weighted shortest path (ESP) only over nodes that are in the forwarding direction of the final target to avoid either creating routing loops or embedding the path in the message. Therefore, the shortest path computed from node s to node d is relative to the final target t. Let $x_0 x_1 ... x_i x_{i+1} .. x_n$, be the node IDs on the ESP from $s = x_0$ to $d = x_n$. We define the ESP cost as

$$cost_{ESP}(s, d, t) = \sum_{i=0}^{n-1} cost(|x_i x_{i+1}|) \qquad (1)$$

4. Contribution

4.1. General Idea

In this section, we present the main idea of KanGuRou which goal is to reach any k sinks among all available sinks in S. Nevertheless, given a source node s, the k closest sinks to s in Euclidean distance are not necessarily the k closest sinks in number of hops. Therefore, the routing messages in KanGuRou may change target sinks along the routing path. For instance, on Fig. 1, 5 closest sinks of s are S_1, S_2, S_5, S_6 and S_7. But S_1 is not reachable directly and the path to S_1 will get closer to S_4 which may be reached also. In addition, the source cannot determine the k sinks in advance and send k messages, one toward each sink because (i) several messages may follow the same path by sections which is useless and costly and (ii) since targets may change along the path, this cannot ensure that several messages will not reach the same sink.

KanGuRou (Algo. 1) proceeds as follows. Fig. 1 illustrates it.

1. Node s holding the message first checks whether it is a sink. If so, it removes itself from the set of available sinks and decrements the number of sinks k to reach. If $k = 0$, the algorithm stops. (Line 2).

2. Node s computes a tree $T(s)$ by running Algo. 3 (k-MST(s,S,k)) or Algo. 4 (k-Prim(s,S,k))) detailed later in Section 4.4, depending of the variant of KanGuRou (Line 7). $T(s)$ contains node s and exactly k sinks of S. If there are several edges/branches originated at s, a message duplication occurs. On Fig. 1, $T(s)$ appear in red and contains sinks S_1, S_3, S_5, S_6 and S_7. There are two branches originated at node s: one toward S_1 and one toward S_5.

3. s distributes the remaining sinks (Line 8), i.e. sinks that are not in $T(s)$ (Sinks S_2, S_4 and S_8 on Fig. 1) over every branch. Thus, for every successor a of s in $T(s)$ ($a \in \mathrm{succ}_{T(s)}$), a subset $S_a \subset S$ of the sinks is assigned to a as detailed in Section 4.5. On Fig. 1, branch of S_1 is assigned with Sinks S_1, S_3 and S_4 while Sinks S_2, S_5, S_6, S_7 and S_8 are associated to branch of S_5.

At this step, node s knows: (i) its successors $a \in \mathrm{succ}_{T(s)}$ in $T(s)$ (Sinks S_1 and S_5 on Fig. 1),

(ii) the number of sinks k_a to reach per successor a, i.e. the number of sinks in the subtree of a $st(T,a)$ (2 in branch of S_1 and 3 in branch of S_5 on Fig. 1),

(iii) the set of available sinks to reach per branch, i.e. S_a defined at the previous step.

Node s then sends as many packets as the number of its successors in $T(s)$. (Loop line 9)

Thus, for each branch of $T(s)$, i.e. $\forall a \in \mathrm{succ}_{T(s)}$, s selects a next hop based on a Greedy-Face-Greedy approach as follows. For every a, s computes the weight of the k_a-MST for each of its neighbors $u \in N(s)$ over S_a targets $\|k\text{-MST}(u, S_a, k_a)\|$. On Fig. 1, s will compute 3-MST over Sinks S_2, S_5, S_6, S_7 and S_8 to find the next hop for branch S_5 and 2-MST over Sinks S_1, S_3 and S_4 for branch S_1. If there exists no neighbor u for which the weight of tree over S_a $\|k\text{-MST}(u, S_a, k_a)\|$ is smaller than $\|sT(T,a)\| + |sa|$ (weight of the branch of $T(s)$ dedicated to a), node s switches to recovery mode (line 16) till reaching a node with positive progress towards a. If so, next hop v for branch toward a is determined through the greedy mode in a COP fashion (Line 18). Message is sent to node v with parameters k_a and S_a which will run KanGuRou again (Line 19) and so on till k_a sinks have been reached in this branch. As shown in [9], this ensures the packet delivery as soon as the network is connected.

Algorithm 1 KanGuRou(s, k, S) – Run at node s to reach k targets in S.

```
 1: if s ∈ S then
 2:     k ← k − 1; S ← S \ {s}
 3:     if k = 0 then
 4:         exit {All sinks of this branch have been reached}
 5:     end if
 6: end if
 7: T(s) ← k-MST(s, S, k) or k-Prim(s, S, k)  {k-MST of
        S ∪ {s} rooted in s}
 8: T'(s) ← AllocateMST(s, S, T(s)) {Allocate remaining
        targets to T(s)}
 9: for all a ∈ succ_{T(s)}(s) do
10:     S_a ← V(st(T', a)) {Nodes in sub-tree of T' rooted
            in a}
11:     k_a ← |T ∩ S_a| {Number of targets to be reached in
            S_a.}
12:     v ← CT_{S_a}(s)
13:     W ← ||sT(T, a)|| + |sa|
14:     A ← {v ∈ N(s) | ||k-MST(v, S_a, k_a)|| < W}
15:     if A = ∅ then
16:         RECOVERY(s, k_a, S_a, W)
17:     else
18:         v ← u ∈ A which minimizes (cost_{ESP}(s,u,a)) / (W − ||k-MST(u, S_a, k_a)||)
19:         KanGuRou(v, k_a, S_a)
20:     end if
21: end for
```

To sum up, let us assume that node s on Fig. 1 runs KanGuRou toward $k = 5$ sinks. First, s computes a 5-MST, $T(s)$ (red tree). $T(s)$ has two branches, so s duplicates the message. First message is sent toward branch of S_1 and has to reach 2 sinks among S_1, S_3 and S_4. s computes the COP and selects node a. To reach node a, message is sent to node f since path sfa is less energy consuming than following the direct edge sa. Node a runs KanGuRou and its tree has two branches. So node a duplicates again the message. First copy has to reach one sink among S_1 and S_3 while second copy has to reach S_4. S_4 is reached via path aeS_4 in a greedy way while other copy is sent along path boS_3. Second message sent by node s has to reach 3 sinks among S_2, S_5, S_6, S_7 and S_8. Greedy algorithm chooses node q. Tree computed on node q has 2 branches originated at q, so q duplicates the message. First copy is sent to node g which forwards it to Sink S_7. Second copy is sent to S_5. S_5 is a sink but the message still has to reach another sink so S_5 forwards it to its neighbor i which directly forwards the message to S_6. At last, 5 sinks have been reached: S_3, S_4, S_5, S_6 and S_7.

4.2. The greedy mode

Greedy mode is similar to the one used in [9]. When node s runs greedy algorithm toward Sink

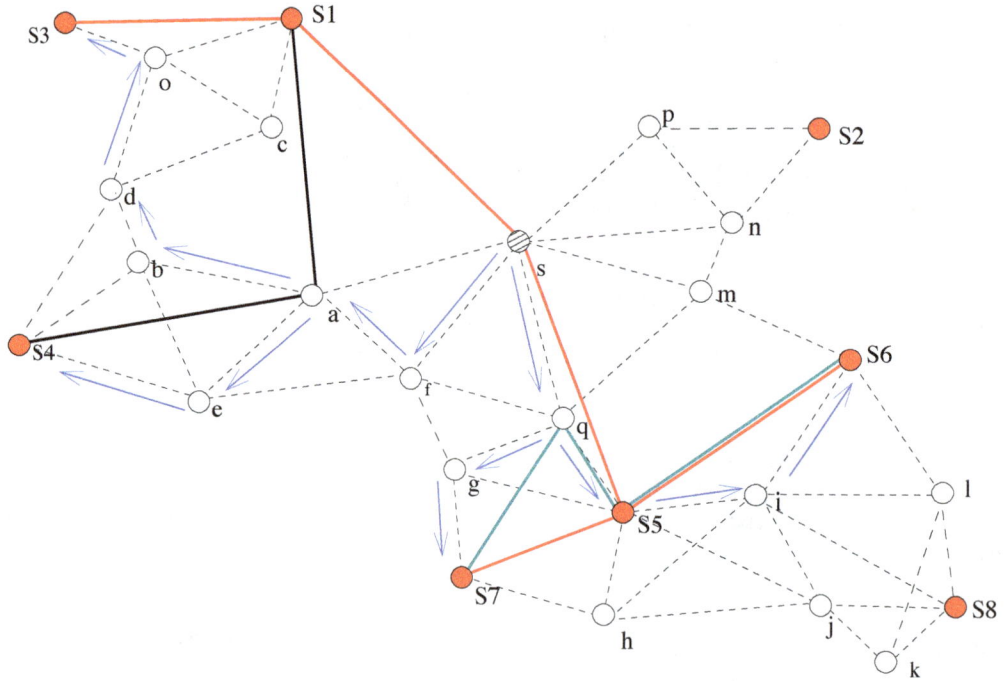

Figure 1. Sinks appear in red. Red links represent the 5–MST rooted in s, blue links the 2–MST rooted in a over S_1, S_3 and S_4, green links the 3–MST rooted in q over S_2, S_5, S_6, S_7 and S_8. Arrows show the message path.

a, it computes the subtree $sT(T(s), a)$ of $T(s)$ rooted in a. The weight W of the subtree issued from s toward a is thus the weight of $\|sT(T(s), a)\|$ plus the weight of the edge sa to reach it: $W = \|sT(T(s), a)\| + |sa|$. Then, to select the next hop, node s performs a COP approach in which (i) the cost considered is the cost of the energy weighted shorted path (Eq. 1) from node u to its neighbor v, (ii) the progress is the reduction of the weight of trees $W - \|k\text{-MST}(u, S_a, k_a)\|$. Only neighbors providing a positive progress are considered. If no such node exists, the greedy approach fails and s switches to recovery mode. If there exist neighbors u such that $W > \|k\text{-MST}(u, S_a, k_a)\|$, node u which minimizes $\frac{\text{cost}_{ESP}(s,v,a)}{W - \|k\text{-MST}(u, S_a, k_a)\|}$ is selected. Note that when computing $k\text{-MST}(u, S_a, k_a)$, all potential sinks are considered, not only the ones in $sT(T(s), a)$. Therefore, $k\text{-MST}(u, S_a, k_a)$ can include different sinks than $sT(T(s), a)$. For instance, on Fig. 1, 2-MST computed by node a (blue tree) over S_1, S_3 and S_4 includes S_1 and S_4 (while the one rooted in s includes S_1 and S_3).

4.3. The Recovery Mode

Recovery mode is detailed in Algo. 2. A node u enters the recovery mode while trying to reach k targets among the sinks in S if it has no neighbor which k-MST has a smaller weight than its own weight W toward the considered branch. u runs RECOVERY till reaching a sink or a node v for which $\|k\text{-MST}(v, k, S)\|$ is smaller

than W (Line 4 in Algo. 2). Unlike in anycasting, recovery in k-anycasting may reach a sink since the distance considered is not between a node and the closest sink but to the closest k sinks.. Yet, the weight of the tree issued in t may have a highest weight than the tree issued on the node which have launched the recovery step.

To determine what neighbor to reach, it applies an EtE-like Face routing [4]. EtE-like Face routing differs from the traditional Face [2] routing in the way that it does not run over the planar of the whole graph but on the planar of a connected dominated set (CDS) graph only (Lines 1-2). This allows considering longer edges. Face algorithm is applied to determine next hop v to reach over the faces on the CDS (Line 5). v is then reached by following an ESP (Line 7) and not necessarily by following the direct edge.

4.4. Computing the k–MST

Note that computing an exact k-MST is NP-complete. Also note that a k-MST is not necessarily included in the MST as example plotted on Fig. 2 shows. Thus, KanGuRou proposes to use two different tree constructions, both of them being an approximation of the k-MST algorithm. As we will see later, the choice of the variant used in the tree construction will depend on the number of sinks M available in the network and the number k of sinks that need to receive the information. It is important to highlight that this tree is computed on the complete graph of sinks $\varsigma = (S, E_\varsigma)$

Algorithm 2 RECOVERY(u,k,S,W) - Run at node u.

1: $(V', E') \leftarrow$ CDS$(V, E) \cup S \cup \{u\}$ {Extract a CDS graph from G}
2: $(V', E'') \leftarrow$ GG(V', E') {Build the Gabriel Graph of G'}
3: $u' \leftarrow u$, $T \leftarrow$ k-MST(u', S, k)
4: **while** $\|k - MST(v, k, S\| > W$ **do**
5:    $v \leftarrow$ FACE(u', T) {Compute the next node on the proper face}
6:    **while** $u' \neq v$ **do**
7:       $u' \leftarrow$ ESP$(u', v, CT_T(u'))$ {Compute the ESP from u' to v}
8:    **end while**
9: **end while**
10: KanGuRou(v, k, S)

Algorithm 3 k-MST(u, S, k) – Return a k-MST of $S \cup \{u\}$ rooted in u.

1: $T \leftarrow (\{u\}, \emptyset)$ {initialize the tree with root u}
2: $A \leftarrow S$ {set of nodes to be considered.}
3: **while** $k > 0$ **do**
4:    **for** $v \in A$ **do**
5:       $w \leftarrow x \in T$ which minimizes $|xv|$
6:       $P(v, 1) \leftarrow w$ {Path from v to T in 1 hop with minimum cost.}
7:       $l(v, 1) \leftarrow |vw|$ {Weight of the path from v to T in 1 hop with minimum cost.}
8:    **end for**
9:    **for** $i = 2$ to k **do**
10:       **for all** $v \in A$ **do**
11:          $y \leftarrow x \in T$ which minimizes $|vx|$
12:          $\forall w \in A$ $z \leftarrow x \in T$ which minimizes $|wx|$
13:          Select $w \in A$ such that $|wz| < |vy|$ which minimizes $(l(w, i - 1) + |vw|)/i$
14:          $p(v, i) \leftarrow p(w, i - 1).w$ {Path from v to T in i hops with minimum cost.}
15:          $l(v, i) \leftarrow l(w, i - 1) + |vw|${Weight of $p(v, i)$.}
16:       **end for**
17:    **end for**
18:    select $v \in A$ and $j \in [1 \ldots k]$ which minimizes $l(v, l)/j$
19:    **while** $p(v, j) \neq \emptyset$ **do**
20:       $(w, x) \leftarrow$ first edge in $p(v, j)$ {w is supposed to be in T while x is not in T}
21:       $T \leftarrow T \cup (\{x\}, \{(w, x)\})$; $A \leftarrow A \setminus \{x\}$; $k \leftarrow k - 1$
22:       $p(v, j) \leftarrow p(v, j) \setminus \{(w, x)\}$
23:    **end while**
24: **end while**
25: Return T.

---

with $E_c = \{uv \mid u, v \in S^2\}$. This is independent from the underlying topology.

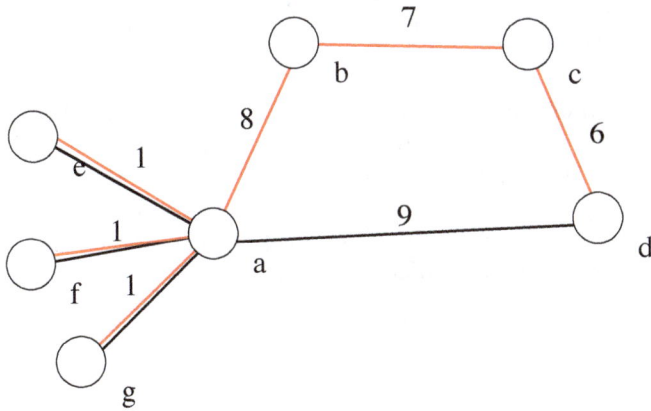

Figure 2. Illustration of MST and k-MST for $k = 4$. If root is node d, the optimal 4-MST (in blue) includes edges da, ae, af, ag while edge ad will not be included in the MST (in red). So, k-MST is not always included in the MST.

First variant: The first variant (later called KanGuRou) applies Algo. 3 and builds a tree with exactly $k + 1$ vertices (k sinks and the source) in an iterative way. It starts with a tree which only contains the root (Line 1), node s on Fig. 1. It then has to choose exactly k sinks in S to add in T. To do so, at each step, it computes the shortest path from any vertex to the tree in exactly i hops, for all i from 1 to $k - i$ for all vertices. On Fig. 1, for $i = 1$, s computes the distance from itself to every sink. For $i = 2$, s considers 2-hop paths from itself to every sink and keeps the shorter one as sS_1S_3 to reach S_3. To reduce the complexity of computing a path from a node u to T, it only considers nodes closer than u to T. On Fig. 1, node s will not compute any 2-hop path from s to S_2 since S_2 is the closest sink. Weight of every path is then normalized by the progress it provides, i.e.

the number of sinks on the path (Line 18) and the path with the lowest weight is then added to the tree. And so on till the final tree includes k sinks. In this way, note that S_2 is not included in path since step 1, path sS_5S_7 (weight 2) is chosen ($\frac{|sS_5| + |S_5S_7|}{2}$ is smaller than all other path ratios as $\frac{|sS_1| + |S_1S_3|}{2}$ or $\frac{sS_2}{1}$). Then at step 2, path sS_1S_3 is added ($\frac{|sS_1| + |S_1S_3|}{2} < \frac{|sS_1| + |S_1S_3| + |S_3S_6|}{3}$, etc) and at last, path S_5S_6 is added.

Second variant: Original Prim algorithm [10] consists in adding iteratively to the current tree (initialized with the root node) the edge with minimum weight which has exactly one extremity vertex in the tree, and so on till every vertex has been added to the tree. KanGuRou-kPrim (Algo. 4) performs similarly but stops when the tree includes and exactly k sinks.

To illustrate the difference between both variants, let us consider Fig. 2 and assume a tree construction rooted in node d with $k = 4$. Algo. 4 adds iteratively the edge (and corresponding nodes) with the lowest weight, i.e.

Algorithm 4 k-Prim(u, S, k) – Return a k-MST of $S \cup \{u\}$ rooted in u.

1: $T \leftarrow (\{u\}, \emptyset)$ {initialize the tree with root u}
2: $A \leftarrow S$ {set of nodes to be considered.}
3: **while** $k > 0$ **do**
4:    $w \leftarrow x \in A$ which minimizes $|xCT_T(x)|$
5:    $T \leftarrow T \cup (\{w\}, \{(w, CT_T(w))\})$
6:    $A \leftarrow A \setminus \{w\}; k \leftarrow k - 1$
7: **end while**
8: Return T.

---

nodes c, b, a and e (in the order). Resulting tree has a weight of 22. Algo. 3 does not consider edges one by one but multi-hop paths. It thus adds nodes a and e at once ($\frac{|da| + |ae|}{2}$ is the best ratio), then nodes f and g. Resulting tree has a weight of 12.

4.5. Distributing Sinks over Branches

Once the k-tree rooted in current node has been computed, the set of sinks has to be distributed over each branch. The number of sinks to be reached by branch is given by the number of sinks actually part of the branch. If s is the node in charge of the message, it computes its k-MST $T(s)$. If k_a is the number of sinks to be reached in the branch of $T(s)$ rooted in a, we have $\sum_{a \in succ_{T(s)}} k_a = k$. Nevertheless, the k_a sinks attached to branch of a are not necessarily the closest ones in number of hops while other sinks which are not in the tree can be closer (like S_1 on Fig. 1 which is the closest sink to s but not in hop count). The set of potential sinks to reach S_a is sent with the message over each branch a. S_a includes the k_a sinks included in the tree but also part of 'free' ones. S_a sinks have to be selected carefully in order to ensure that exactly k sinks will receive the message. They are such that: (i) $\bigcup_{a \in succ_{T(s)}} S_a = S$ since every sink is candidate and (ii) $S_a \cup S_b = \emptyset \ \forall \ a, b \in succ_{T(s)}$ in order to avoid that a message sent on 2 different branches reaches the same sink in which case, the overall number of sinks receiving the message will be less than k.

Algo. 5 details how the remaining sinks can be attached to every branch. In KanGuRou, each sink is assigned to the closest branch regardless of the size of the branches. However, we are aware that this solution is not necessarily the most adequate one since most of remaining sinks may be assigned to the same branch which might be the smallest one. Alternative solutions might be:

- Sinks may be distributed evenly between both branches, based on distance. For instance, on Fig. 3, $S_A = \{u_1, u_2, u_3, u_4, u_6, u_7\}$ and $S_B = \{u_5, u_8, u_9, u_10, u_11, u_12\}$.

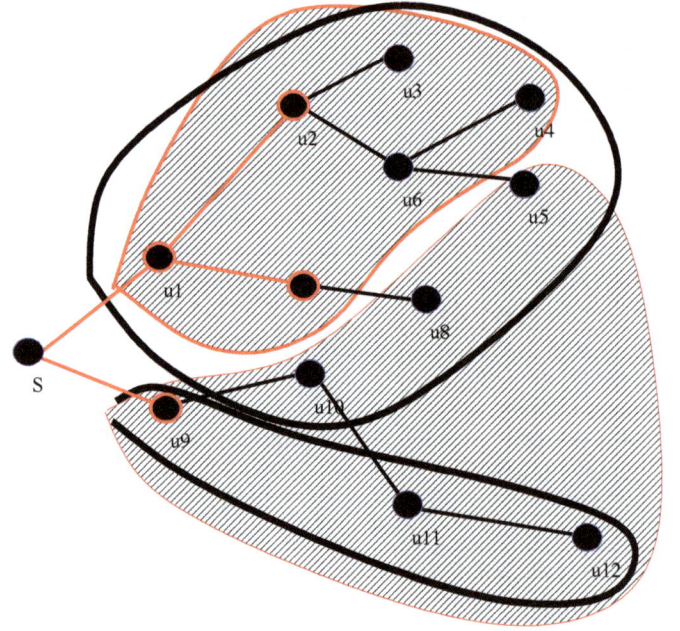

Figure 3. Illustration of sink allocation over branches.

- Sinks may be distributed proportionally to the number of sinks to reach per branch. For instance, on Fig. 3, since branch A is supposed to reach 3 sinks over 4, it will be assigned $\frac{3}{4} = 9$ sinks (sinks $u_1, u_2, u_3, u_4, u_5, u_6, u_7, u_8, u_10$) while branch B will receive 3 sinks (Sinks u_9, u_11, u_12).

However, setting in advance the number of sinks to assign to each branch will lead to some other issues. Indeed, issue will appear when sinks are at equal distance of several branches and when a sink p is closer to Branch A, but that Branch A has already been assigned enough sinks, all closer than p. We leave to further work a deeper study on this point.

Algorithm 5 AllocateMST(u, S, T). Allocate nodes in S not in T over T branches.

1: $A \leftarrow S \setminus T$
2: **while** $A \neq \emptyset$ **do**
3:    select edge uv in $V(T) \times A$ which minimizes $|uv|$
4:    $T \leftarrow (\{u\}, \{(u, v)\})$
5:    $A \leftarrow A \setminus \{v\}$
6: **end while**
7: Return T

4.6. Packet delivery to exactly k sinks guaranteed

We show that KanGuRou delivers a message to exactly k sinks as long as the underlying network is connected. To do so, we first need to introduce some lemma.

Lemma 1. Greedy step is loop free.

Proof. Greedy step reduces the distance between current node s and the closest set of k sinks $T(s)$. Loop cannot be created, since it is impossible to go back to a larger distance at a given node by repeated applications of greedy algorithm. □

Theorem 2. KanGuRou guarantees the packet delivery to exactly k sinks as long as the network is connected and that the number of sinks in the connected component including s is greater or equal to k.

Proof. We apply a mathematical induction demonstrating that Theorem 2 is true. **Initial step.** *Theorem 2 is true for $k = 1$.* When $k = 1$, the 1-MST computed by s running KanGuRou comes to finding $CT_S(s)$, *i.e.* the closest sink to s. The greedy step of KanGuRou thus computes the progress provided by neighbor u of s as $|sCTS(s) - uCTS(u)|$. Recovery step initiated in node v will switch back to greedy step as soon as it has reached a node w such that $|vCTS(v)| > |wCTS(w)|$. Yet, KanGuRou comes to EEGDA [9], been proven to guarantee packet delivery as long as the underlying network is connected. Yet, Theorem 2 is true for $k = 1$. **Induction step.** *Assuming that Theorem 2 is true for $k = i - 1$, $1 < i$, we have to prove that Theorem 2 is true for $k = i$.* When a node s runs KanGuRou, it may either duplicate and forward several times the message or just forward it once. Let us consider these two cases singly.

1. CASE 1: the message is sent on every branch originated at s in the k-MST, i.e. between 2 and k times. (On Fig. 1, the message is sent over 2 branches in s.) Since in KanGuRou:

   - *(i)* when a sink is reached, it is removed from the list of available sinks,
   - *(ii)* the set of available sinks is split over every branch such that an available sink is assigned to exactly one branch,
   - *(iii)* the sum of the number of sinks to reach per branch is equal to k,
   - *(iv)* a message forwarding stops if and only if it has reached k sinks among sinks it has been assigned,

   then node s runs independently KanGuRou with k such that $1 \leq k \leq i - 1$. for every branch. Thus, as Theorem 2 is true for $k < i$, the theorem is proven in this case.

2. CASE 2 (message is only forwarded), the message is forwarded in a repeated application of greedy and recovery phases. Greedy step is only applied if distance of current node u to the closest set of k actors/sinks $W = \|T(s)\|$ can be reduced (Lemma 1). The recovery step also has the same goal (reducing W) after following a face. Gabriel

graph preserves connectivity, and following very first face recover reduces distance W [6]. Distance W continues to decrease, and loop cannot be created until either delivery to a sink or a node on which a duplication of the message will be made. The delivery is guaranteed either to a sink in the set of available sinks assigned to the message or to a node which duplicates the message because its k-tree has several edges originating at itself since W, at each iteration, can always be reduced, until it eventually becomes 0.

   - CASE 2.1 (a sink is reached): the sink will remove itself from the list of available sinks and runs KanGuRou with $k = i - 1$, which guarantees delivery to exactly $i - 1$ sinks. Thus, in this case, KanGuRou eventually reaches exactly i sinks. Theorem 2 is true for $k = i$.
   - CASE 2.2 (a node which will split the message is reached): the current node u initiates between 2 and i duplications and thus runs between 2 and i times KanGuRou with $1 \leq k \leq i - 1$. As shown for Case 1 of this proof, every branch guarantees delivery to the number of sinks they have been asked. Since a sink may be assigned to at most one branch, a same sink cannot be reached by several branches (and counted as it), thus, the overall number of sinks reached is $k = i$. Therefore, in this case, KanGuRou reaches exactly i sinks.

Theorem 2 is true. □

5. Simulation Results

In this section, we evaluate the performances of KanGuRou under the WSNet[1] simulator with an IEEE 802.15.4 MAC layer. As there is no comparable algorithm in the literature since KanGuRou is the first position-based algorithm from the literature, we compare the two variants KanGuRou and KanGuRou-kPrim to running k times the plain EEGDA anycast routing protocol [9] to measure the gain provided by KanGuRou. We deploy N nodes (from 35 to 115) at random in a square of 100m × 100m, every node can adapt its range between 0 and 30m.

We first evaluate the behavior of different algorithms. Fig. 4 shows the number of times the message is split/duplicated for each algorithm. Obviously, the number of splits performed by EEGDA is equal to 1 whatever the parameters since EEGDA performs

---

[1]WSNet: http://wsnet.gforge.inria.fr/

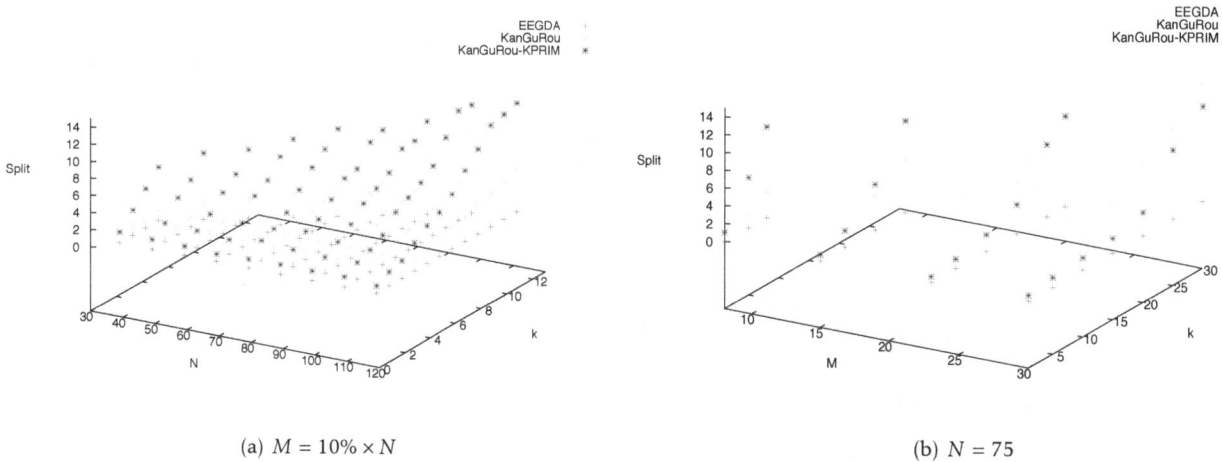

(a) $M = 10\% \times N$

(b) $N = 75$

Figure 4. Number of splits for each algorithm. M = number of sinks.

independent anycast routings. For both versions of KanGuRou, it is worth noting that when k increases for a given number of available sinks M and of nodes N, the number of splits also increases. This is expected since algorithms need to reach more sinks and respective trees are bigger and thus the message is more likely to be duplicated to reach sinks.

Also, for a fixed k, the number of splits increases when the number of nodes (and thus of available sinks) increases. This is due to the fact that more choices are given to the algorithm and thus more ramifications appear (Fig. 4(a)). We can also note (Fig. 4(b)) that the number of duplications is not really impacted by the overall number of available sinks M in the network (number of splits for a given k). At last, we can observe that the number of duplications increases when M increases (in proportion of N) more quickly for KanGuRou than for KanGuRou-kPrim. Yet, for a low value of M, KanGuRou-kPrim produces more duplications than KanGuRou while for high values of M, KanGuRou duplicates more often messages.

First, the number of sinks M is set to be 10% of the total deployed nodes N. We simulate the performance of three algorithms (EEGDA, KanGuRou, KanGuRou-kPrim) for 100 times in terms of N as well as k, and calculate the average values of results. We generate in random a new distribution of N nodes for each simulation. Fig.5 shows the energy consumption (computed based on Eq. 1) and the path length in terms of N and k (k varies from 1 to M). Note that for $k = 1$, results are the same for all three algorithms since KanGuRou comes to EEGDA independently of the tree construction. Simulation results show clearly that KanGuRou, KanGuRou-kPrim result in significant gains on the energy consumption (up to 62.51% (44.33% in average) and up to 74.22%

(53.84% in average) respectively) and path length (up to 62.17% (49.07% in average) and up to 56.61% (21.90% in average) respectively) compared to the traditional algorithm EEGDA. An amelioration was indeed expected since in KanGuRou, part of the path is mutualized. Nevertheless, the gain remains important. Globally, we can see that behavior of every algorithm is similar whatever the parameters. Regarding the energy consumption, results show that KanGuRou-kPrim consumes less energy compared to KanGuRou when k is important, and KanGuRou performs better for low k. This is due to the fact that when k increases (for a constant M), k-Prim algorithm gets closer and closer to the optimal k-MST construction. This is also linked to the number of message duplications illustrated by Fig. 4. A high number of splits implies shorter paths.

Figure 6 gives a closer look at the energy consumption and the path length in terms of k when the total deployed nodes N is a constant ($N = 75$) and M is set to be 8 sinks. We can see KanGuRou-kPrim performs better regarding energy consumption when k is greater than 3, and KanGuRou always has a gain of the path length compared to the other two algorithms in this case.

In the second scenario (Fig. 7), we fix the number of the total deployed nodes N to 75 and evaluate the performances of the three algorithms (EEGDA, KanGuRou, KanGuRou-kPrim) regarding the overall number of sinks M in the network. Obviously, when k increases for a given number of available sinks M, the path and the energy consumption increase since there are more sinks to reach. Similarly, when the number of sinks to reach k is fixed and that the number of available sinks M increases, the path and the energy consumption decrease since algorithms have more choice among

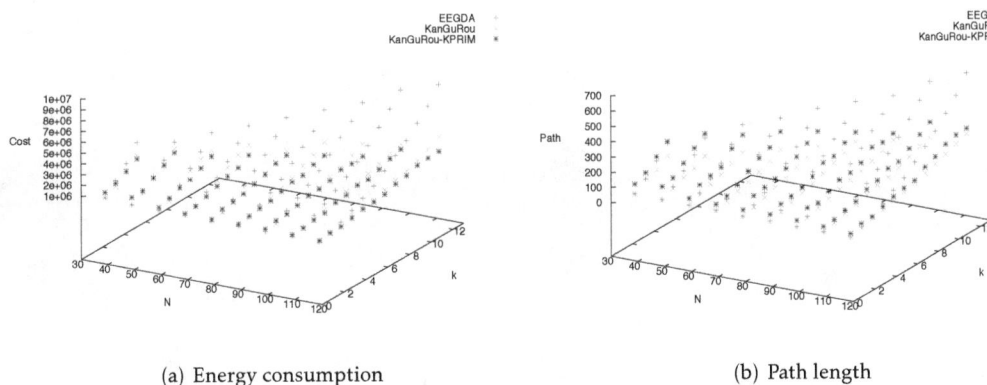

(a) Energy consumption   (b) Path length

Figure 5. Algorithms performances with regards to N and k for $M = 10\% \times N$.

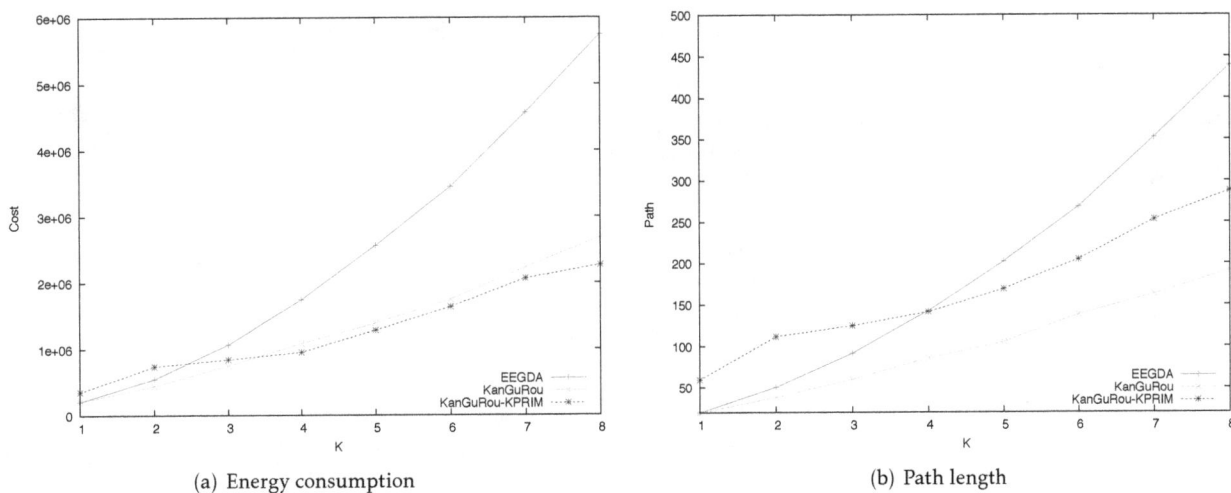

(a) Energy consumption   (b) Path length

Figure 6. Algorithms performance in terms of k over $M = 8$ sinks among $N = 75$ nodes.

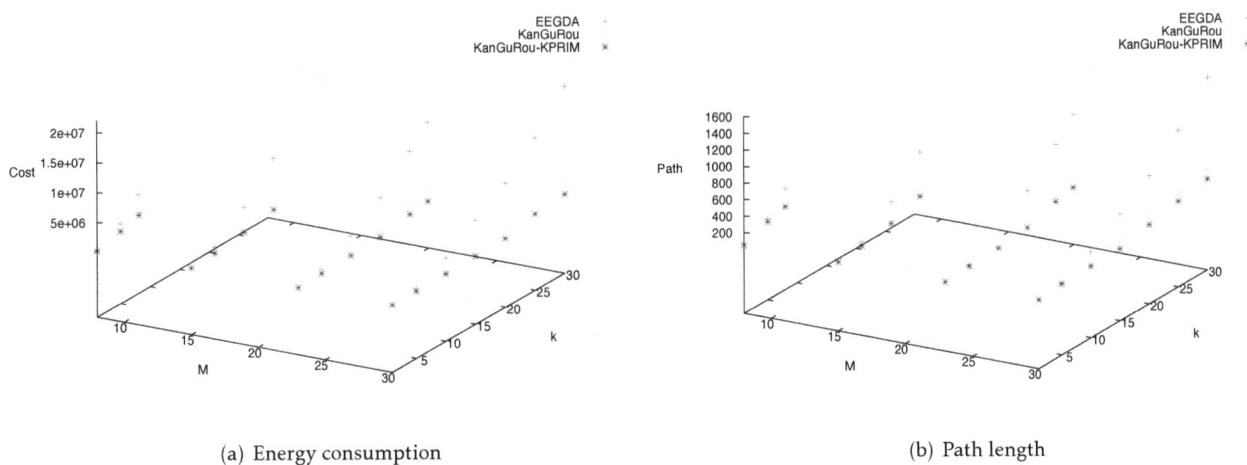

(a) Energy consumption   (b) Path length

Figure 7. Algorithms performances with regards to M and k for $N = 75$ nodes.

sinks and can join closer ones. An important feature   is that results show that KanGuRou-kPrim performs

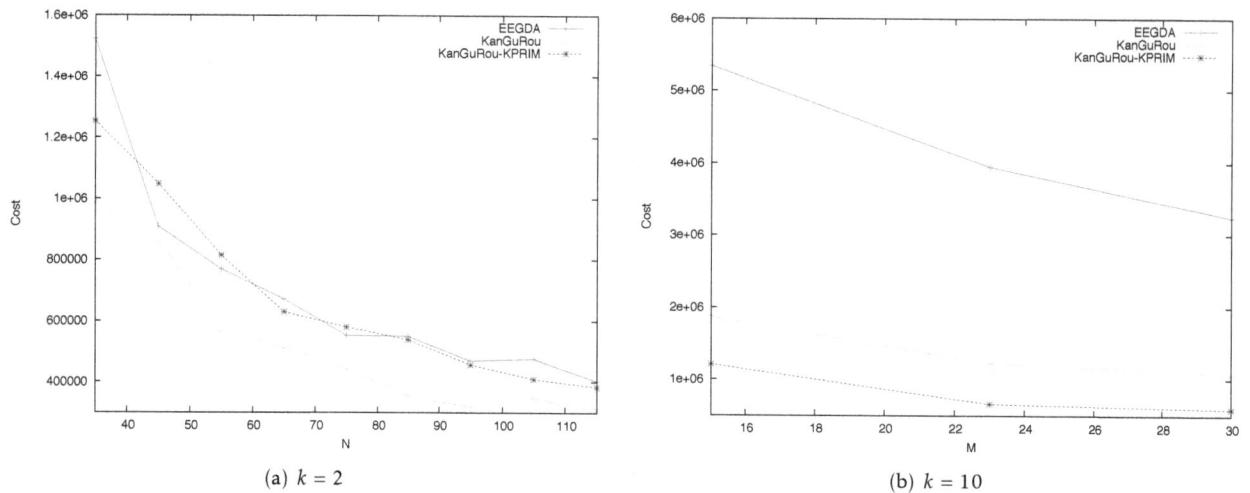

(a) $k = 2$

(b) $k = 10$

Figure 8. Algorithms performances for $M = 20$ sinks and $N = 75$ nodes.

better than KanGuRou for high values of M and k. Once again, this is linked to the number of path splitting and that the greater k, the closer to the optimal k-MST, k-Prim algorithm is.

Figure 8 shows the impact of the number of sinks to reach for a fixed number of nodes and a constant side of sink set. It shows that for a lo k, KanGuRou outperforms other solutions while it is not so efficient when k increases.

To sum up, the simulation results of different scenarios clearly show that *(i)* KanGuRou variants result in a significant gain of energy consumption and path length compared to the traditional algorithm EEGDA, *(ii)* depending of the percentage of sinks to be reached, one variant of KanGuRou performs better than the other one. When k is small (when $k \leq 30\% \times M$), KanGuRou always consumes less energy than KanGuRou-kPrim, *(iii)* when k is important (when $k > 30\% * M$), KanGuRou-kPrim brings a significant gain compared to KanGuRou especially when M is important. This is also highlighted by Fig. 9 which has a closer look at this feature. Figure clearly shows that up to a given number of available sinks, KanGuRou-kPrim performs better than KanGuRou ($M = 23$ on figure).

6. Conclusion and Future Work

In this paper, we have introduced KanGuRou, the very first position-based k-anycast routing protocol which is energy efficient and guarantees the packet delivery. Two variants are proposed for the construction of the tree. KanGuRou performs well when the number of sinks to reach is lower than 30% of the available sinks in the network while KanGuRou-kPrim performs better for higher values of k. In future work, we intend to claim theoretically how far KanGuRou is from the optimal centralized algorithm and provide some complexity

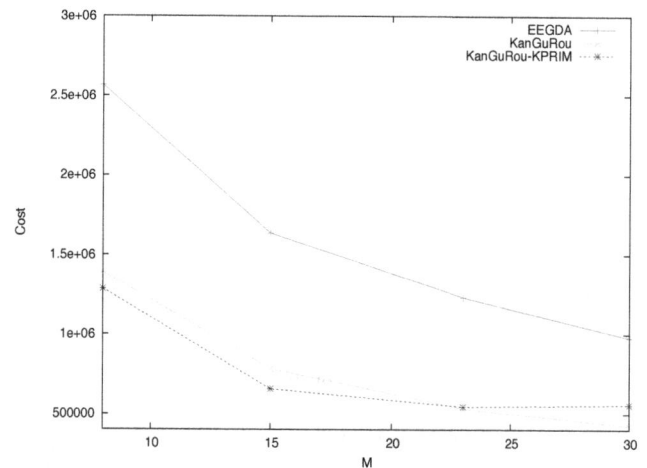

Figure 9. Algorithms performances for $k = 5$ and $N = 75$ nodes.

analysis. We also intend evaluate the properties of KanGuRou more deeply (robustness toward mobility, wireless instability, etc).

References

[1] B. Awerbuch, A. Brinkmann and C. Scheideler. Anycasting and multicasting in adversarial systems. Dept. of Computer Science, Johns Hopkins University, 2002

[2] P. Bose, P. Morin, I. Stojmenovic, and J. Urrutia. Routing with guaranteed delivery in ad hoc wireless networks. *Wireless Networks*, 7(8):609–616, 2001.

[3] S.C. Chen, C.R. Dow, S.K. Chen, J.H. Lin and S.F. Hwang. An efficient anycasting scheme in ad-hoc wireless networks. In *Consumer Com. and Networking Conf.*, 178:183, 2004.

[4] E. H. Elhafsi, N. Mitton, and D. Simplot-Ryl. Energy Efficient Geographic Path Discovery With Guaranteed Delivery in Ad hoc and Sensor Networks. In *IEEE PIMRC*, 2008.

[5] H. Frey, F. Ingelrest, and D. Simplot-Ryl. Localized mst based multicast routing with energy-efficient guaranteed delivery in sensor networks. In *WOWMOM*, 2008.

[6] H. Frey and I. Stojmenovic. On delivery guarantees of face and combined greedy-face routing in ad hoc and sensor networks. ACM MOBICOM, 2006

[7] W. Hu,N. Bulusu and S. Jha. A Communication Paradigm for Hybrid Sensor/Actuator Networks. In *Journal of Wireless Information Networks*, 12(1):47-59, 2005.

[8] T. Melodia, D. Pompili, V. C. Gungor and I.F. Akyildiz, A Distributed coordination Framework for Wireless Sensor and Actor Networks. In ACM Mobihoc, 2005,

[9] N. Mitton, D. Simplot-Ryl, and I. Stojmenovic. Guaranteed delivery for geographical anycasting in wireless multi-sink sensor and sensor-actor networks. In *IEEE INFOCOM*, 2009. Short paper.

[10] R.C. Prim. Shortest connection networks and some generalizations. *Bell System Technical Journal*, 36:1389–1401, 1957.

[11] V. Rodoplu and T. Meng. Minimizing energy mobile wireless networks. *IEEE JSAC*, 17:1333–1347, 1999.

[12] W. Wang, XY Li, and O. Frieder. k-anycast game in selfish networks. In *ICCCN*, 2004.

[13] X. Wang. Analysis and design of a k-anycast communication model in ipv6. *Comput. Commun.*, 31:2071–2077, June 2008.

[14] C.-J. Wu, R.-H. Hwang and J.-M. Ho. A scalable overlay framework for Internet anycasting service. In *Symp. on Applied Computing (SAC)*, 2007.

[15] B. Wu and J. Wu. k-anycast routing schemes for mobile ad hoc networks. In *IPDPS*, 2006.

[16] X. Xu, Y-L Gu, J. Du, and H.-y. Qian. A distributed k-anycast routing protocol based on mobile agents. In *WiCOM*, 2009.

4

Quorum system and random based asynchronous rendezvous protocol for cognitive radio ad hoc networks[*]

Sylwia Romaszko[1,*], Daniel Denkovski[2], Valentina Pavlovska[2], Liljana Gavrilovska[2]

[1]Institute for Networked Systems, RWTH Aachen University, Kackertstrasse 9, 52072 Aachen, Germany
[2]Faculty of Electrical Engineering and Information Technologies, Ss. Cyril and Methodius University in Skopje, Macedonia

Abstract

This paper proposes a rendezvous protocol for cognitive radio ad hoc networks, RAC²E-gQS, which utilizes (1) the asynchronous and randomness properties of the RAC²E protocol, and (2) channel mapping protocol, based on a grid Quorum System (gQS), and taking into account channel heterogeneity and asymmetric channel views. We show that the combination of the RAC²E protocol with the grid-quorum based channel mapping can yield a powerful RAC²E-gQS rendezvous protocol for asynchronous operation in a distributed environment assuring a rapid rendezvous between the cognitive radio nodes having available both symmetric and asymmetric channel views. We also propose an enhancement of the protocol, which uses a torus QS for a slot allocation, dealing with the worst case scenario, a large number of channels with opposite ranking lists.

Keywords: asymmetric channel view, asynchronous, cognitive radio ad hoc networks, quorum systems, rendezvous

1. Introduction

A common control channel (CCC) in multichannel Cognitive Radio Networks (CRNs) supports the transmission coordination exchange and cooperation between the active CR users. It is aimed to facilitate neighbor discovery, e.g., control signaling, exchange of local measurements, channel sensing etc. However, such CCC existence in CRNs may not be always feasible. When using the CCC notion, a channel needs to be found that is accessible by the majority of CR nodes and it is not interrupted over a long period of time. However, these tasks are not feasible in a CR environment without any imposed assumptions, since CR nodes can have a different view of the channels occupied by incumbents and/or other secondary users. Another issue that arises

when all nodes have chosen the same channel is the possibility of single channel bottleneck as well as the single point of failure.

Moreover, in Cognitive Radio Ad Hoc Networks (CRANs), the dynamic network topology, distributed multi-hop architecture, and time and location varying spectrum availability are the key factors [2]. Each Cognitive Radio (CR) user has a different spectrum availability according to the incumbent (Primary User, PU) activity, and it determines its actions based on its local observation. Therefore, rendezvous (RDV), the ability of two or more nodes to meet each other and establish a link, is a challenging task in CRANs.

This paper proposes a rendezvous protocol for CRANs that, firstly, utilizes the asynchronous and randomness properties of the RAC²E protocol [3]. Secondly, it investigates the suitability of different channel search orders that are based on:

(i) random selection utilizing weights and utility functions,

[*]Part of this work was presented at ICST Conference, ADHOCNETS 2012, Paris, France, October 2012 [1] published in Ad Hoc Networks, Lecture Notes of the Institute for Computer Sciences, Social Informatics and Telecommunications Engineering Volume 111, 2013, pp 135-148.
*Corresponding author. Email: sar@inets.rwth-aachen.de

(ii) grid quorum-based channel mapping (gQS-RDV) protocol [4, 5] taking into account channel heterogeneity (in terms of channel quality),

(iii) torus QS optimization of the gQS-RDV slot allocation method.

Finally, it gives an insight of the performance of the proposed protocol in terms of the time-to-rdv (TTR)[1] and the probability of RDV for the different channel search orders, as well as the inter-rendezvous time variance. Moreover, we evaluate the protocol in a symmetric channel view (homogeneous channel availability) and asymmetric channel view (heterogeneous channel availability, i.e., having different number of idle/unoccupied channels per CR) cases.

The paper is organized as following. Section 2 presents the related work. In Section 3 we describe a Quorum System (QS) concept and relevant properties. Section 4 presents the RAC^{2}E-gQS protocol, its phases and optimization of the channel mapping phase. In Section 5 we evaluate the proposed protocol with and without the proposed optimization. The last section concludes this study.

2. Related work

In [6] one can find a comprehensive guidance on the application of quorum systems in wireless communications, and rendezvous issues in decentralized CRNs. In general, rendezvous approaches can be divided into three branches, first, non-quorum based solutions representing blind or pseudo-random RDV techniques ([7–9] and more sophisticated [10, 11]). To the second branch belong protocols proposed for a multi-channel Medium Access Control (MAC) handling multi-rendezvous [12] (i.e., multiple transmissions pairs can accomplish handshaking simultaneously), missing receiver problem [13] or medium allocation in a hostile and jamming environment [14]. All these protocols are based on cyclic quorum systems. A cyclic QS is proposed in [15] and it is based on the cyclic block design and cyclic difference sets in combinatorial theory [16]. The last branch consists of either quorum-based, or difference set-based, or Latin Square-based protocols proposed for CRNs ([17–21]). However, an asynchronous channel view (ACHv) is not explicitly handled in quorum-based schemes[2] and the *channel heterogeneity* is not considered in any related work approaches to the best of our knowledge.

The asynchronous operation of the cognitive radio networks and its effect on the rendezvous phase is not well investigated subject. The synchronization establishment in cognitive radio networks, especially in the distributed case, is a time and power consuming task, and therefore, the assumption that the CR nodes operate in time synchronized manner is not always justified. There are only a few papers considering the asynchronism of the CR nodes during the rendezvous phase. Most of them use probabilistic models to generate the channel hopping sequences of the operating CR devices. In the modular clock algorithm (MC) [8] and its modified version MMC [8] each CR node picks a proper prime number P and randomly selects a rate r which is less than P. Based on the two parameters, the user generates its channel search sequence via pre-defined modulo operations. The channel rendezvous sequence (CRSEQ) algorithm [23] uses a method based on triangle numbers and modular operations to calculate the channel hopping order. The ring-walk (RW) algorithm [10] represents each channel as a vertex in a ring. The CR nodes sweep through the ring visiting the vertices (channels) with different velocities and the rendezvous is guaranteed since the nodes with lower velocities will sooner or later be caught by the ones with higher velocities. In [11] the authors propose a jump-stay rendezvous algorithm for blind rendezvous, using jump-pattern and stay-pattern channel hoping sequences in each round. The CR nodes continuously "jump" on available channels during the jump-pattern and "stay" on a specific channel during the stay-pattern. In [20] two systematic approaches (symmetric and asymmetric) for designing asynchronous channel hopping (ACH) protocols are proposed, which address the asynchronous rendezvous problem. An asymmetric ACH system that uses an array-based quorum system is introduced instead of utilizing Latin squares as done in the prior work [19]. This quorum-based approach generates a significantly greater number of CH sequences than the approach using Latin squares. In the recent study on the blind rendezvous for tactical networks [24], the performances of the MMC [8] and Random Channel Access (RA) [25] algorithms are compared on a testbed using USRP [26]. It is found out that added asynchronism can have a large beneficial effects reducing time to rendezvous.

Up to the best of the authors' knowledge there is only a couple of papers focusing on quorum based asynchronous rendezvous [6, 17, 20, 22]. However, these papers as well as the aforementioned work dealing with the asynchronism, do not handle the channel heterogeneity in the generation of channel hopping sequences and do not handle the details of asynchronous operation and rendezvous between the devices. On the contrary, the RAC^{2}E-gQS takes into consideration the *heterogeneity* in terms of the

---

[1]TTR is an amount of time, measured in slots, within which cognitive radios meet each other once they began hopping, or after the last rendezvous on a channel.

[2]Visiting unavailable channels is the most frequently used approaches while dealing with ACHv in the related work (e.g., [18, 22]).

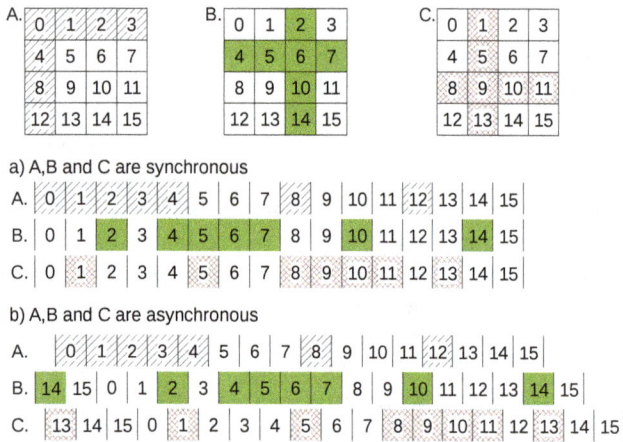

Figure 1. Example with grid-based quorums [4].

Table 1. 4x4 grid: Pair-On-Pair (PoP): quorum (0,0) in bold

0	5	11	15
4	1	7	13
10	6	2	9
14	12	8	3

Table 2. 4x4 grid: Diagonal (Diag): quorum (0,0) in bold

0	4	8	12
13	1	5	9
10	14	2	6
7	11	15	3

channel priority among the CR nodes. Furthermore, the proposed protocol covers the details of the operation of the nodes *prior the rendezvous* and *after the rendezvous*, i.e. the control channel operation.

3. Quorum systems

Quorum-based algorithms become popular, as the main asset of these algorithms is their resilience to node and network failures. The usual definition of a quorum system (QS) is given in [27]:

Definition 3.1. A *quorum system* Q under an universal set U, $U = \{0, 1, ..., n - 1\}$ with n being a cycle length (frequently used Z_n symbol referring to $U = Z_n$), is a collection of non-empty subsets of U, called quorums, satisfying the *intersection property* $\forall A, B \in Q : A \cap B \neq \emptyset$.

There are different types of QSs, within which a *grid*-based QS proposed by Maekawa [28], is widely utilized in power-saving (PS) protocols. In this system, sites (elements) are logically organized in a grid in the shape of a *square*. A quorum for a requesting site contains the union of a row and a column that the requesting site corresponds to. For PS nodes we can divide their beacon intervals into groups, where each group includes n consecutive intervals and is organized in a $\sqrt{n} \times \sqrt{n}$ array in a row-major manner. Quorum intervals are picked along an arbitrary row and column from this array, where the remaining intervals are non-quorum intervals. Figure 1 depicts an example for 16 beacon intervals and three nodes, A, B, C, selecting different quorum intervals. If the clocks of the nodes are synchronized (case a), the intervals of the nodes overlap twice, e.g., A and B meet in 2 and 4 slots. If their clocks are not synchronized (case b) the A's slots still overlap with B's and C's slots.

A QS, which satisfies the *Rotation Closure Property* (*RCP*), ensures that two asynchronous mobile nodes

selecting any two quorums have at least one intersection in their quorums. The grid QS satisfies the RCP:

Definition 3.2. For a quorum R in a quorum system Q under an universal set $U = \{0, ..., N - 1\}$ and $i \in 1, 2, ..., N - 1$, there is defined: $rotate(R, i) = (x + i) mod\ N | x \in R$. A quorum system Q has the *Rotation Closure Property* if and only if $\forall R', R \in Q, R' \cap rotate(R, i) \neq \emptyset$ for all $i \in 1, 2, ..., N - 1$.

In [4] *Grid-Pair-on-Pair* (PoP) way of forming a grid was proposed as shown and Equation 1 and in Table 1.

$$f(x, y) = \begin{cases} x & (x=y) \\ f(x-1, y-1) + 2 & (x=1..n-1, y=1..n-1 \\ & x!=y) \\ \lfloor \frac{n+1}{2} \rfloor * 2 & (x=1, y=0) \\ (\lfloor \frac{n}{2} \rfloor * 2) + 1 & (x=0, y=1) \\ f(1, 0) + ((x-1)n - \sum_{i=1}^{x-1} i) * 2 & (x=2..n-1, y=0) \\ f(0, 1) + ((y-1)n - \sum_{i=1}^{y-1} i) * 2 & (x=0, y=2..n-1) \end{cases}$$

(1)

However, this grid does not satisfy the RCP. Hence, a node using this grid will encounter problems with RDV when there is no cycle alignment.

For example, while a node selects a quorum $(0,0)$ as depicted in Table 1, a quorum $Q1 = \{0, 4, 5, 10, 11, 14, 15\}$ is chosen. $Q1$ does not satisfy the RCP, since for $i = 8$ there is no common element: $Q1 \cap rotate(Q1, i = 8) = \emptyset$, because of $\{0, 4, 5, 10, 11, 14, 15\} \cap \{8, 12, 13, 2, 3, 6, 7\} = \emptyset$. In all other cases there is at least one common element.

Therefore, *Grid-Diagonal* (Diag), shown in Table 2 and Equation 2, was designed in [4]. In this method the numbers are ordered according to the positive diagonal rule, i.e., elements are ordered according to

$$f(x, y) = ((y \times r) - ((r - 1) \times x))\% (r \times r) \quad (2)$$

where $x = 0, ..., n-1$, $y = 0, ..., n-1$, and $r = \sqrt{n}$. This grid *does* guarantee the RCP. For instance for $Q2 = \{0, 4, 7, 8, 10, 12, 13\}$ from Table 2, there is always at least one common element, i.e., $Q2 \cap rotate(Q2, i) \neq \emptyset$ for all $i \in 1, 2, ..., 15$.

4. RAC^{2}E-gQS protocol description

In this section we describe RAC^{2}E-gQS composed of two main parts, first, a rendezvous-MAC protocol for asynchronous cognitive radios, and second being an optimization of the first part by utilization of a sequence -based channel mapping while handling a rendezvous phase. The former allows to benefit from the asynchronous and randomness properties, originally proposed in [3] as part of the RAC^{2}E protocol. The latter uses a grid Quorum System (gQS) properties, originally proposed in [4, 5], in order to deal with a rendezvous guarantee in a single cycle.

An original version of the RAC^{2}E protocol is proposed for nodes operating in an asynchronous and cooperative manner, i.e., a CR user utilizes information from its spectrum map to determine the best channel to be its control channel. Hence, if two CR nodes want to establish a direct link for a communication, they must exchange their spectrum maps first in order to select the mutually best channel (i.e. the channel having the lowest level of interference for both nodes).

In this work, we use only a part of the RAC^{2}E protocol, namely, asynchronous and randomness properties, while creating *asynchronous cycles* (and slots), and handling a control channel operation (*after the rendezvous* phase, thus, message exchange between nodes). Moreover, CRs can work in a non-cooperative manner, while allocating a channel. However, the reader should note that spectrum maps can be still utilized to determine the priorities of the channels and as a consequence the order of the channel in the channel ranking list. From now on, while talking about the RAC^{2}E, we refer to the aforementioned part of the RAC^{2}E-gQS protocol, and not original one designed in [3].

The mapping of channels into time slots, in the *prior the rendezvous* phase of the RAC^{2}E-gQS protocol, is a paramount task. This can be done using several methods considering the channel *priorities* based on the channel ranking lists created in the sensing phase by each node independently. And therefore, as the main contribution in this work we investigate the suitability of different channel search orders that are based on:

(i) random selection of channels using utility functions for channels priority- *UP*,

(ii) gQS based channel mapping strategies [4, 5] taking into account channel heterogeneity (in terms of channel quality).

(iii) torus QS optimization of the gQS slot allocation.

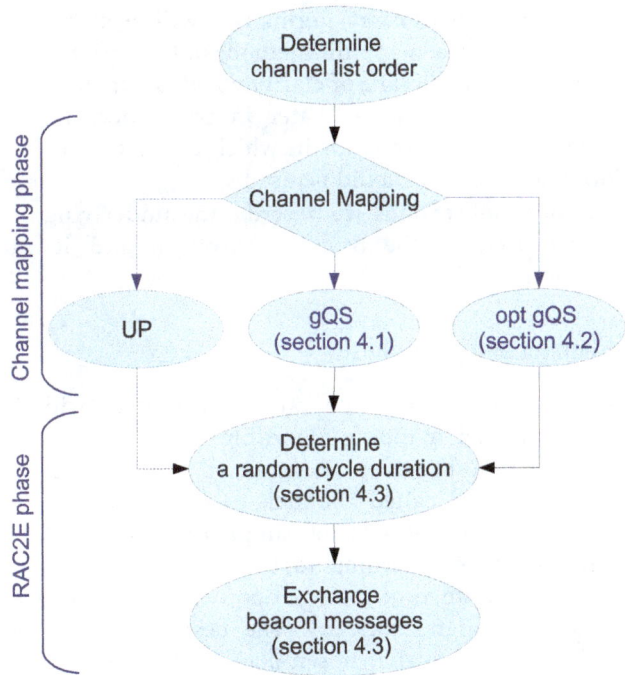

Figure 2. Asynchronous rendezvous-MAC protocol diagram with a choice of three different mapping methods, namely, UP, gQS, and optimized gQS.

The *UP* method randomly performs the channel mapping with probabilities guarantying that the channels, with respect to their priorities, will statistically be assigned to the same amount of slots per rendezvous cycle as the grid-quorum based methods. This channel mapping strategy is selected as a representative example since it probabilistically maps the channel to slots, oppositely on the regular grid mapping.

The combination of the RAC^{2}E protocol with the gQS mapping, RAC^{2}E-gQS, can yield a promising rendezvous protocol for asynchronous operation in a distributed environment assuring a rapid rendezvous between the CR nodes. However, in the worst case scenario, namely, a large number of of channels ($\sim$20 channels) and opposite channel raking lists, it works somewhat worse that the RAC^{2}E-UP combination, and therefore, we also proposed an optimization of RAC^{2}E-gQS in order to ensure its advantage in all case scenarios. This goal is accomplished by an enhancement of the channel mapping algorithm by utilizing a *torus* QS properties in a grid array. Moreover, note that the RAC^{2}E, proposed in [3], has already been implemented on a testbed platform (the results are presented in [3]), and the proposed protocol in this work, optimized RAC^{2}E-gQS, is under implementation and evaluation phase.

Figure 2 shows the steps of the RAC^{2}E-gQS protocol. The UP method is also visible in the diagram (with corresponding dashed arrows), however, since it is disadvantageous in all cases, normally it will probably not be a choice of a mapping method, but we plot it in the diagram for the sake of clarity which algorithms are analyzed in this work. Later in this work, we recommend the case scenarios in which either gQS or the optimized method should be used.

In the next subsections we describe the underlying algorithms used in the proposed protocol and its optimization.

4.1. Channel mapping phase

In this section we elaborate on a distributed grid QS based channel mapping algorithm. The outcome of the algorithm provides an input to the channel hopping sequences called channels-to-slots maps (or channel maps for the sake of simplicity). Each CR maps its channels according to the channel quality without any exchange of information, where the best channels get a higher priority. The best channel is mapped according to the chosen *quorum*. Hence, CRs that allocate a common best channel, while having the same number of available channels, will always meet as a result of the quorum intersection property (if satisfying the RCP they also always meet regardless cycle/slot misalignment). The period (cycle) N of a channel map depends on the number of channels r and it equals r^2 (selected from a $r \times r$ grid). Two channels-to-slots mapping methods are designed for three or larger number of channels (i.e. $r \geq 3$). In both methods, channels are mapped to grid indexes (Channel 1 (C1) is mapped to index 0, Channel 2 (C2) to index 1 etc.), each channel in a CR network has its own index known by nodes. A CR performs the channels-to-slots mapping based on the quality of the channels, e.g., node A has the channel-priority list C2/C4/C3/C1, and therefore C2 is the best channel, C4 is the second best channel etc.

In the first method, *Row-Column (RC)* mapping, in the first step (*Step 1* in Figure 3.(a)), a CR (with the channel-priority list C2/C4/C3/C1) selects its map in a row-column manner, where the row number (channel number) is always equal to the column number (channel number). The best A's channel is Channel 2, so it selects a quorum (*row=column=C2=index 1*). The set of elements, represented by (1,1) quorum, maps Channel 2 to {1, 4, 5, 9, 11, 13, 14} slots. Each time when a set of elements is chosen, a grid is cut to a sub-grid, together with the already mapped channel, i.e., each sub-grid maintains only the unallocated channels. A set of slots for consecutive channels (according to the quality) is mapped this way till we obtain a 2×2 sub-grid. Note that, each better quality channel has accordingly more assigned slots than a worse quality one. The last two channels are mapped to two slots in a diagonal manner. Analyzing the example (map 2/4/3/1) from Figure 3.(a), Channel 4 map (C4, C4)

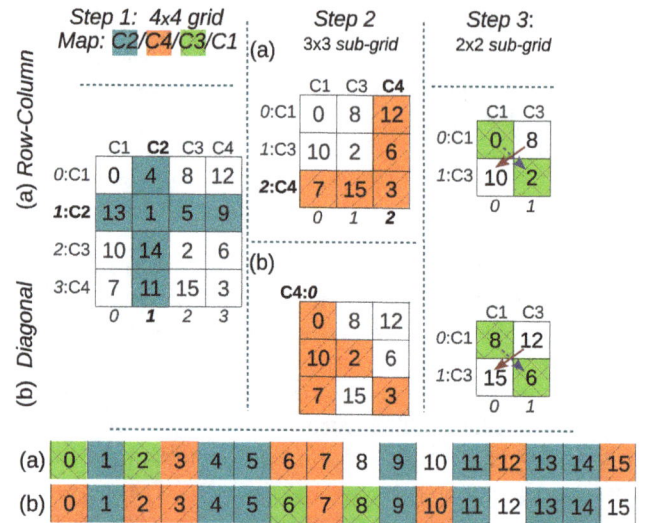

Figure 3. Three steps of the channels-to-slots mapping: (a) Row-Column, (b) Diagonal.

has set of slots: {3, 6, 7, 12, 15} (Step 2), Channel 3 is allocated in slot 0 and 2 and Channel 1 to slot 8 and 10 (Step 3).

The second method, *Diagonal mapping (CD)*, is similar to the Row-Column mapping till we obtain a 3×3 sub-grid. The next channel is mapped (and a sub-grid is cut accordingly) in a column-*diagonal* manner, selecting the first column and the main diagonal, e.g., Channel 4 is mapped to {0, 2, 3, 7, 10} slots (*Step 2* in Figure 3.(b)). The last two channels are allocated as in the first method., i.e, Channel 3 is mapped to slots 8 and 6, and Channel 1 is mapped to slots 12 and 15 (*Step 3* in Figure 3.(b)). Note that, a channel map with 3 available channels is an exception following the Row-Column mapping, since the first (the best) channel should always follow a quorum concept.

4.2. Channel mapping optimization

An enhancement of the gQS selection includes attributing the first channels less slots in advantage of the last channels. To do so without loosing the RCP property of the first channel (the reader should note that the property guarantees a RDV if two CRs have the same best channel) we use the *torus* quorum system selection. A torus-based QS [29] is similar to a grid-based QS [15], but normally adopting a rectangular array structure (instead of a $r \times r$ grid) called torus, i.e., wrap-around mesh, where the last row (column) is followed by the first row (column) in a wrap-around manner. The height r (number of rows, i.e. entire column) and width s (number of columns, i.e. entire row) are defined where $n = r \times s$ and $s \geq r \geq 1$. In order to understand the construction of a tQ we present below a standard definition.

Definition 4.1. A torus quorum in a $r \times s$ torus (grid) is composed of $r + \lfloor \frac{s}{2} \rfloor$ elements, formed by selecting any column c_j ($j = 1..s$) of r elements, plus one element out of each of the $\lfloor \frac{s}{2} \rfloor$ succeeding columns using end wrap-around. An entire column c_j portion is called the quorum's *head*, and the rest of the elements ($\lfloor \frac{s}{2} \rfloor$) its *tail*.

In our optimized approach, opt GD^{RC}, (for the sake of the RCP property we optimize the gQS mapping with the row-column selection and diagonal slot distribution) we use a *torus* concept allocating quorum slots for the best channels, but in a $r \times r$ grid, i.e., our torus-in-grid quorum is composed of $r + \lfloor \frac{r}{2} \rfloor$ elements, where a tail is selected in a *forward*-wrap manner (according the standard torus definition)[3]. The remaining elements of the column, which normally are also attributed to the best channels with the gQS mapping, are now equally distributed to the last worst channels. The reader should note that the remaining slots are picked in a backward-wrap manner and assigned to the worst channels so that a worse channel has not more attributed slots than a better channel. Figure 4 presents the opt GD^{RC} method, depicting a tQ forward selection and slot's reassignment for CRs with $C1/C2/C3/C4/C5$ and $C5/C4/C3/C2/C1$ channel priority orders, respectively. The best channel gets seven slots in total, where two remaining elements are assigned to the worst two channels, e.g., Channel 4 and 5 with $C1/C2/C3/C4/C5$ channel ranking list. In the *opt $GD^{RC}all$* algorithm, the full optimization, the second best and following channels must have lesser number of the assigned slots. Therefore, in this example one slot of the second best channel is reassigned to the second worst channel as can be seen in Figure 4. In a partial enhancement, *opt $GD^{RC}1$*, only slots of the best channel are reassigned. Note that while we use *opt GD^{RC}* without 1 or all we refer to the full optimization.

Let us analyze scenarios with a larger number of channels. In the scenario with a CR with 10 available channels, while reassigning the best channel slots, four slots are remaining, since a *torus* quorum has $10 + \lfloor \frac{10}{2} \rfloor = 15$ elements, thus four remaining slots, because the number of row plus column elements, in a 10×10 *grid* quorum, equals 19 slots (in the scenario with 20 free channels we get nine remaining slots). With 10 available channels Channel 9 and 10 (assuming $C1/.../C10$ channel ranking list) receive extra two slots from the best channel, and with 20 free channels the last two worst channels get four additional slots and Channel 18 receives one slot.

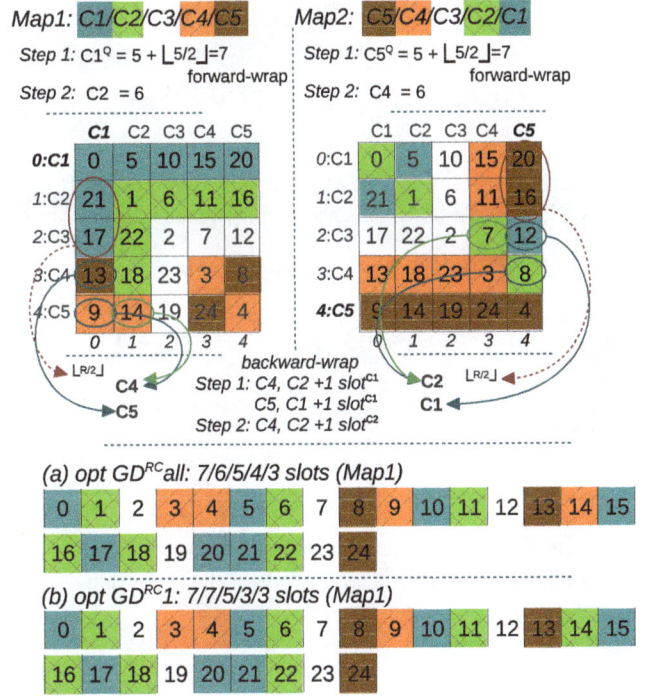

Figure 4. Torus (forward-wrap) way selection in a $r \times r$ grid: (a) opt $GD^{RC}all$ and (b) opt $GD^{RC}1$.

Summing up, the proposed optimization involves that better quality channels reassign their slots, using a torus instead of grid quorum selection, until there is no worse channel having more slots than a better one. In other words, with an increasing number of available channels the number of extra attributed slots to the last channels increases, bearing in mind a channel ranking.

4.3. RAC²E phase

The RAC²E phase encompasses a cycle (slot) size determination, and MAC process. The protocol relies on an asynchronous operation of the nodes, eliminating the need of synchronization establishment, which is a difficult task in the distributed environments. Moreover, it fosters even an additional randomization among the nodes to ensure a rapid rendezvous on a particular temporary unused channel from the primary system. The operation of the RAC²E phase is illustrated on Figure 5. The reader should note that the figure shows only asynchronism and randomization concept of the protocol cycles, and not exemplary maps of the proposed RAC²E-gQS protocol. Each CR aiming to establish a control channel independently selects a *random* rendezvous cycle duration of Tc_{i_j} (i^{th} cognitive radio, j^{th} cycle). Therefore, although CRs select the same channel map (the same channel ranking lists), they can still have different cycle durations thanks to the used randomization property. This time duration is selected uniformly in the range $[T_{min}, T_{max}]$, where

---

[3]One should note, that the RCP of a torus-in-grid quorum is guaranteed if and only if nodes select the same column and the same row. In the case of the proposed algorithm this is always the case, since a set of slots of a particular channel is selected from the predefined row and column.

Figure 6. Example of the random cycle duration with three CR nodes, with 5 unoccupied channels, using the optimized RAC^{2}E-gQS protocol: CR1 and CR2 have the same channel order, CR3 has an opposite channel order.

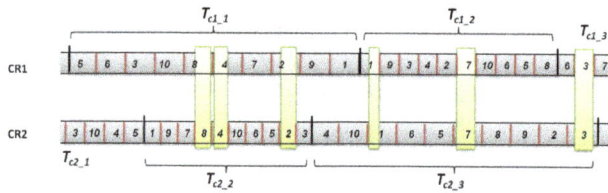

Figure 5. The random cycle duration and asynchronous operation provides overlapping between the both CRs in the free channels (ch_i, $i = 1, ..., 10$).

$T_{min} = T_c - \frac{\Delta T}{2}$, $T_{max} = T_c + \frac{\Delta T}{2}$ and T_c represents the mean rendezvous cycle duration while $\Delta T = kT_c$ is the randomization interval. The chosen Tc_{i_j} interval is further segmented into N time slots, with each slot (having a duration of $\tau_{i\_j} = \frac{Tc_{i\_j}}{r}$) assigned to a particular channel unoccupied by the primary users. As illustrated on Figure 5, the randomization ensures that overlapping at the same channels occurs randomly in wider or narrower time intervals.

Figure 6 depicts an example of the optimized RAC^{2}E-gQS nodes, where CR1 and CR2 have five unoccupied channels with the same channel order, and CR3 has also five free channels, but with an opposite channel order (as shown exemplary in Figure 4). It is easily visible, that there is quite a number of overlaps on different channels, also in the case of the opposite channel orders (e.g., CR2 and CR3).

In each slot interval τ_{i_j}, the CR sends a short beacon message at the beginning and end of the slot to signalize its presence in the channel. These particular times of beaconing are selected since they provide the highest probability of RDV between the CR nodes. In the meantime, between the both beacon messages, the RDV node aims to capture the beacons coming from the other CRs operating on the current channel. As Figure 7 illustrates, the randomization (i.e. asynchronous operation of the both nodes) guarantees that at least one of the beacon messages will be received by other nodes tuned to the same channel at the moment. This justifies the preference of a random Tc_{i_j} duration (Figure 5), which provides a more successful delivery of the beacon messages, in comparison to the synchronous case. A RDV occurs when two nodes are tuned to the same channel and they exchange at

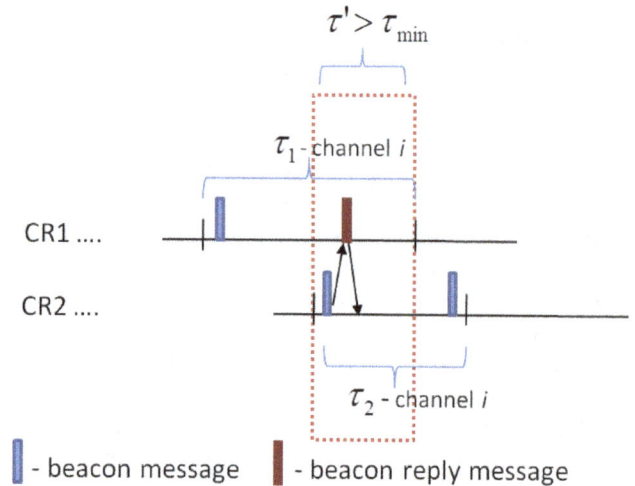

Figure 7. Rendezvous at channel i event.

least one beacon and one beacon reply message. The condition $\tau' > \tau_{min}$ must be fulfilled for the rendezvous to occur, where τ' is the overlapping duration and τ_{min} is the minimal required time for exchange and processing of both, the beacon and the beacon reply message (Figure 7). Generally, the τ_{min} duration is influenced by the used sample rates of the CR nodes and the length of the beacon and the beacon reply messages. Since there is not much information to transfer, the length of these messages can be in order of few bytes.

5. RAC^{2}E-gQS performance evaluation

This section evaluates the RAC^{2}E-gQS protocol with the different channel mapping strategies introduced before. The performance analyses exploit four grid quorum strategies for channel mapping [4, 5] (considering $N = r^2$ number of slots for r available channels): the Row-Column - RC and Column Diagonal - CD channel mapping, for the both grid forming methods: Pair-on-Pair-Grid (PoP) and Grid-Diagonal (GD), as well as the UP approach.

The simulation analyses envision a scenario with two CRs aiming for a RDV on a certain common channel. Two main cases are evaluated:

(i) the best case (same), when CRs have the same channel ranking lists, e.g., both have $C1/.../C5$ as a priority map for 5 free channels.

(ii) the worst case (opposite), when both CRs have completely different channel ranking list, e.g., CR1 has $C1/.../C5$ while CR2 has $C5/.../C1$ in the case of 5 unoccupied channels.

These two cases are taken as representative examples, since they provide the two extremes of rendezvous performances, i.e., they are the best and the worst case scenarios. Note, that later we also investigate the cases with random channel ranking lists, while comparing the proposed optimization. The detail of the corresponding cases are presented in the respective subsections.

One performance metric of interest in the analysis is the *average number of potential RDVs (channel matchings) per cycle* which is in inverse proportion to the time-to-rendezvous (TTR). Note that the TTR performance for slot synchronized CRs is expressed in slots. In an asynchronous environment, where we evaluate the combined protocol, the TTR is expressed in seconds, because of the nature of the RAC^{2}E protocol. The second evaluated performance metric is the *inter-rendezvous time variance*, representing the variance between two potential consecutive rendezvous, calculated with the Formula 3 and Formula 4:

$$\sigma_{irdv}^2 = \frac{1}{N-1} \sum_{i=2}^{N} ((t_i - t_{i-1}) - \mu_{irdv})^2 \qquad (3)$$

$$\mu_{irdv} = \frac{1}{N-1} \sum_{i=2}^{N} (t_i - t_{i-1}) \qquad (4)$$

where σ_{irdv}^2 is the inter-rendezvous time variance, μ_{irdv} is the mean inter-rendezvous time, N is the total number of rendezvous, while t_i is the time of rendezvous i. For the same number of average potential RDVs per cycle, a higher variance means that channel matchings occur in bursts, leaving longer gaps between bursts, while the lower variance represents the case when channel matchings are more regularly distributed in time. The *lower variance* case is better since it would assure that two CRs going online would not be stuck into the long no-RDV gaps before a successful RDV.

Monte Carlo simulations were made to test the performance of the RAC^{2}E-gQS protocol, for 5, 10 and 20 channels. A total of 10000 trials (RDV cycles) with random start times of the CR nodes were made for each case for statistical correctness. The simulations were performed for a mean rendezvous cycle duration $T_c = 1s$ and duration of $\tau_{min} = 1\mu s$. This τ_{min} duration roughly maps to a case when we have 10Msps sampling rate, 1 byte of beacon and beacon reply message lengths and 4-QAM modulation. Different randomization intervals were evaluated, for k ($k = \frac{T_c}{\Delta T}$) ranging from 1/4 up to 2 with step size of 1/4.

In the first subsection we evaluate the case with slot synchronized CRs using gQS only, followed by the subsection with an analysis of asynchronous CRs using the combined protocol, RAC^{2}E-gQS.

5.1. Slot synchronized CRs using gQS

In order to justify the need of randomization introduced by RAC^{2}E, the grid-QS mapping schemes were tested for a scenario of slot synchronized CRs aiming for RDV. Slot shifts are likely to occur since both CRs do not start the RDV phases simultaneously. Table 3 presents the performances of the grid-quorum schemes in terms of the minimum, the maximum and the average number of potential RDVs per cycle with respect to the slot shifts. As evident slot shifts can cause a high TTR even in the case when both CRs have the same channel ranking lists. The opposite ranking lists and several slot shifts between can result in no RDV between the CRs.

5.2. Asynchronous CRs using RAC^{2}E-gQS

Table 4 presents the average number of potential RDVs per cycle for the RAC^{2}E-gQS protocol, for the same and opposite channel ranking lists of the CRs. It is evident that the case of the same channel ranking lists of the both CRs, results in a higher average number of potential RDVs per cycle than the case with different channel ranking lists. RAC^{2}E improves the RDV performances of the grid-quorum channel mapping schemes, as evident comparing Table 3 and Table 4 results. The channel matching percentage, calculated as the average number of potential rendezvous per cycle divided by the number of slots, is about 52%, 26% and 13.25% for 5, 10 and 20 number of channels, respectively, for the same channel ranking lists case and two times lower for the case with opposite channel ranking lists.

All inspected grid channel mapping methods (PoP^{RC}, GD^{RC}, PoP^{DC}, GD^{DC}), for the particular channel ranking cases and the particular numbers of available channels, provide the same average number of potential RDVs per cycle. The UP mapping method provides the same performances as the grid channel mapping methods when the CRs channel ranking lists are the same. In the case of opposite channel rankings the UP performances differ from the grid-based methods: lower number of free channels results in worse performances compared to the gQS methods; higher number of available channels results in better performances than the gQS schemes.

Although most of the methods experience the same or similar average number of potential RDVs per cycle, they differ in the inter-rendezvous time variance, as demonstrated on Figure 8. It presents the dependence of the variance between consecutive RDVs of both users from the factor of randomization k ($k = \frac{T_c}{\Delta T}$), for the

Table 3. Minimum (Min), Maximum (Max) and average (Mean) number of potential RDVs per cycle for gQS schemes in slot synchronized CRNs; No.c/s stands for Number of channels / slots; List^{Rk} is the channel ranking lists

No.c/s	List^{Rk}	Metr.	PoP^{RC}	GD^{RC}	PoP^{DC}	GD^{DC}
5/25	same	Min	1	3	1	3
5/25	same	Mean	6.52	6.52	6.52	6.52
5/25	same	Max	25	25	25	25
5/25	opposite	Min	0	0	0	0
5/25	opposite	Mean	3.56	3.56	3.56	3.56
5/25	opposite	Max	7	7	7	7
10/100	same	Min	1	3	1	3
10/100	same	Mean	13.28	13.28	13.28	13.28
10/100	same	Max	100	100	100	100
10/100	opposite	Min	0	0	0	0
10/100	opposite	Mean	6.74	6.74	6.74	6.74
10/100	opposite	Max	20	28	30	28
20/400	same	Min	0	3	0	3
20/400	same	Mean	26.645	26.645	26.645	26.645
20/400	same	Max	400	400	400	400
20/400	opposite	Min	0	0	0	0
20/400	opposite	Mean	13.36	13.36	13.36	13.36
20/400	opposite	Max	158	108	160	108

Table 4. Average Number of potential RDVs per cycle for the RAC^{2}E-gQS; No.c/s stands for Number of channels / slots; List^{Rk} is the channel ranking lists

No.c/s	List^{Rk}	PoP^{RC}	GD^{RC}	PoP^{DC}	GD^{DC}	UP
5/25	same	13.042	13.042	13.037	13.043	13.041
5/25	opposite	7.1065	7.1207	7.0994	7.1045	3.6729
10/100	same	26.563	26.557	26.554	26.558	26.559
10/100	opposite	13.409	13.408	13.434	13.356	15.207
20/400	same	53.263	53.243	53.283	53.325	53.296
20/400	opposite	26.543	26.424	26.515	26.504	31.128

cases with the same and opposite channel ranking lists and for 5, 10 and 20 channels. UP achieves a lower variance between potential consecutive RDVs and outperforms the gQS strategies, only for the cases of higher number of free channels and different channel ranking lists, and lower number of free channels and the same ranking lists. The gQS strategies (RC-PoP, RC-GD, DC-PoP and DC-GD) perform better in all other cases. Among the grid quorum strategies, the DC-PoP and DC-DG achieve the lowest variance between RDVs, for the cases with large number of channels, different channel ranking lists and small number of channels, same ranking lists.

Regarding the randomization factor k, it is evident that there is an optimal setting providing the lowest variance between potential RDVs. The optimal k depends on the number of available channels, the difference between the channel ranking lists and the employed channel mapping method (Figure 8).

The fact that gQS strategies encounter somewhat worse performance for the cases of a *larger* number of

channels and *opposite* channel ranking lists (e.g., in the case with 20 channels the UP inter-rendezvous variance is ∼ 0.008 better than that of gQS) is as expected, since in gQS each better quality channel has accordingly more mapped slots than a worse quality one, but always in a *regular* distributed manner. Although UP channels are also assigned to the same amount of slots as in gQS, the *random* mapping increases an amount of RDVs as well as decreases an inter-rendezvous time variance with a larger number of available channels.

The opposite behavior, i.e., a *large* number of channels but the *same* channel ranking lists can justify this reasoning, since regular mapping is noticeable better approach (e.g., in the case with 20 channels the gQS inter-rendezvous variance can be even ∼ 0.05 better than that of UP), but it is favorable to have the same amount of assigned slots for the same channel while having a large number of free channels.

Moreover, while having a *smaller* number of channels in a set (e.g., 5 channels) and opposite channel ranking lists, the regular grid mapping is definitely better (it

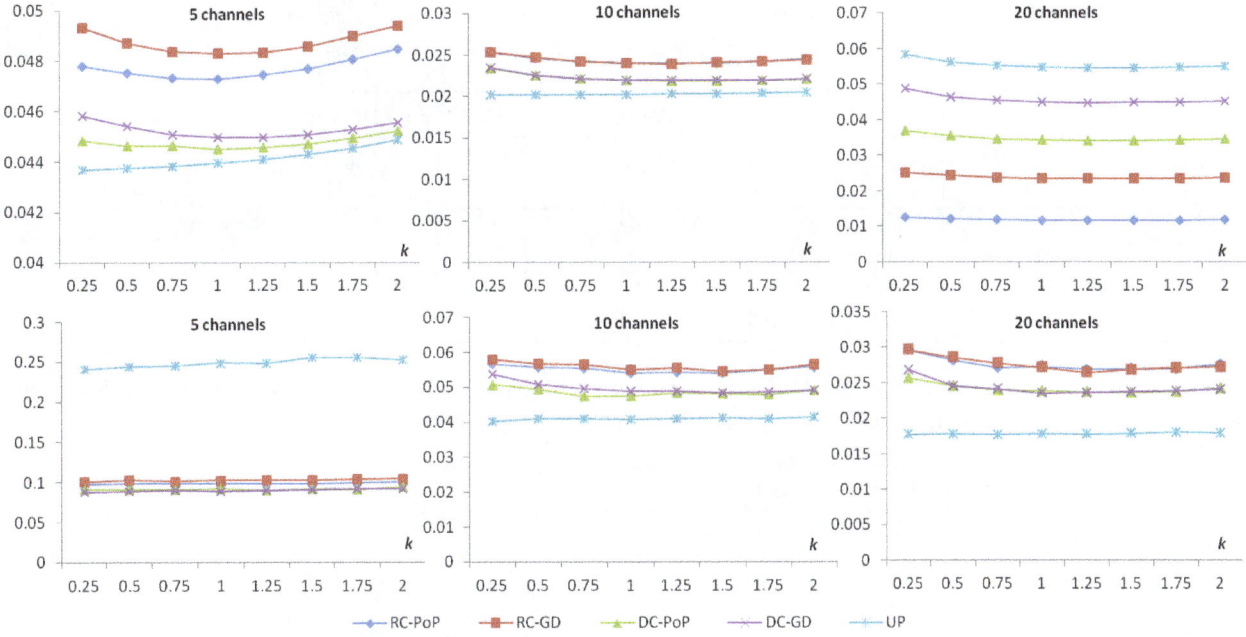

Figure 8. Inter–rendezvous time variance σ^2_{irdv} [sec^2] vs randomization coefficient k, first row: same channel ranking case, second row: different channel ranking case

decreases the inter-rendezvous variance around 0.15), as a difference of the amount of assigned slots of better quality-channels and worse quality-channels is not so drastic as in scenarios with a large number of free channels.

However, since there is indeed a disadvantage of the RAC^{2}E-gQS protocol in the worst case scenario, a large number of channels and opposite channel ranking lists, we apply the proposed optimization (Section 4.2) and compared to the gQS channel mapping. We also show that already a small enhancement (namely, reassigning slots from the first channel only) increases a performance in the scenario in question.

In the next subsection we present the comparison of the gQS channel mapping with its optimization in the slot synchronized case, with the same, opposite and random channel ranking lists, followed by a subsection presenting results of the optimized RAC^{2}E-gQS protocol in an asynchronous environment. The last subsection elaborates on the scenario with asymmetric CRs having heterogeneous channel availability.

5.3. Slot synchronized optimized CRs

Table 5 shows the results of the opt GD^{RC} method with the same and opposite channel ranking lists for 5, 10 and 20 channels in slot synchronized CRNs. As expected in the best case scenario, the RDV performance of the optimized (opt GD^{RC}) approach is degraded in comparison with the GD^{RC} (Table 3), since the number of attributed slots for the best channels

is diminished, thus the chance for RDVs on the most frequently visited channels is also diminished. On the other hand, in the worst case scenario the opt GD^{RC} has noticeable improved performance, since worse channels have now more assigned slots, thus more chance to have rendezvous. Since the best case scenario is rather unlikely in CRNs, the opt GD^{RC} can also improve the performance in asynchronous scenarios.

Before going into the details in an asynchronous environment, we show below some results in Table 6 with randomly chosen priorities for 5, 10, and 20 channels in total, ordered as follows:

1. random5: $C2/C4/C3/C1/C5$ and
   $C5/C1/C3/C4/C2$,

2. random10: $C1\ldots C10$ and
   $C9/C7/C4/C10/C8/C6/C2/C5/C3/C1$,

3. random20: $C20/\ldots/C1$ and
   $C12/C1/C7/C6/C13/C2/C5/C20/C4/C16/C15/$
   $C14/C3/C10/C19/C9/C11/C17/C8/C18$.

As Table 6 shows, the optimized GD^{RC} increases RDV's mean and decreases TTR (measured in slots) of slot synchronized CRs. However, the reader must note, that we can also find random scenarios where the previous version of GD^{RC} has a better performance, since the first best channels are (rather) similar, so the case close to the best case scenario.

Therefore, the further simulation results in an asynchronous environment of the complete RAC^{2}E-gQS protocol are presented in the next subsection.

Table 5. Slot synchronized CRs: Minimum (Min), Maximum (Max) and average (Mean) number of potential RDVs per cycle and average TTR (slots) for the opt gQS scheme; No.c stands for Number of channels; List^{Rk} is the channel ranking lists

Algorithm	No.c List^{Rk}	Min RDV	Max RDV	Mean RDV	avg TTR (slots)
opt GD^{RC}	5 same	1	25	5.4	4.66
opt GD^{RC}	5 opposite	0	8	4.6	5.26
opt GD^{RC}	10 same	1	100	11.06	9.05
opt GD^{RC}	10 opposite	0	31	8.96	10.61
opt GD^{RC}	20 same	1	400	21.81	18.34
opt GD^{RC}	20 opposite	0	119	18.25	21.54

Table 6. Slot synchronized CRs, random scenarios: Minimum (Min), Maximum (Max), average (Mean) number of potential RDVs per cycle, and average TTR (slots) for gQS and optimized gQS schemes; List^{Rk} is the channel ranking lists

Algorithm	List^{Rk}	Min RDV	Max RDV	Mean RDV	avg TTR (slots)
GD^{RC}	random5	0	8	3.56	6.82
opt GD^{RC}	random5	1	11	4.6	5.48
GD^{RC}	random10	0	28	7.56	13.11
opt GD^{RC}	random10	1	34	9.17	10.92
GD^{RC}	random20	1	164	17.39	23.01
opt GD^{RC}	random20	3	124	19.40	20.62

5.4. Asynchronous optimized CRs

Table 7 depicts the results for the RAC^{2}E-gQS protocol in the same random scenarios (random5, random10 and random20) as in the synchronized case presented in the previous subsection. In order to see better the difference between GD^{RC} and its optimization, we show also the results while the slots are reassigned from the first best channel only (opt GD^{RC}1, example in Figure 4). The reader should note that this small enhancement already improves the average number of potential RDVs per cycle and average TTR.

The results justify the need of having more attributed slots for the worst channels in comparison with the grid QS mapping (GD^{RC}). While allowing of reassignment of all channels (opt GD^{RC}all) the average RDVs significantly increases and the mean TTR decreases.

In the next section we present the results, obtained by using cognitive radios having heterogeneous channel availability in order to verify the protocol behavior in more realistic scenarios.

5.5. Asymmetric channel view

The simulation analyses envision scenarios with two CRs aiming for a RDV on a certain channel while having asymmetric channel views. Four cases are evaluated:

(i) the best case 1 (same1): when CRs have the same channel ranking lists, i.e., CR1 has $C1/C2/C3/C4/C5$ as a priority map for 5 free channels and CR2 has $C1/C2/C3/C4/C5/C6/C7$ as a priority map for 7 unoccupied channels.

(ii) the worst case 1 (opposite1): when both CRs have opposite channel ranking list, e.g. CR1 has $C1/C2/C3/C4/C5$ priority map of 5 free channels, while CR2 has $C7/C6/C5/C4/C3/C2/C1$ in the case of 7 unoccupied channels.

(iii) the best case 2 (same2): when CRs have the same channel ranking lists, i.e., CR1 has $C1/C2/.../C14/C15$ as a priority map for 15 free channels and CR2 has $C1/C2/.../C19/C20$ as a priority map for 20 free channels.

(iv) the worst case 2 (opposite2): when both CRs have opposite channel ranking list, e.g. CR1 has $C1/C2/.../C14/C15$ for 15 free channels, while CR2 has $C20/C19/.../C2/C1$ in the case of 20 unoccupied channels.

All cases are taken as representative examples, since they provide the two extremes of rendezvous performances, i.e. they are the best and the worst case scenarios with a lower and higher number of channels.

The UP method randomly performs the channel mapping with probabilities guarantying that the channels, with respect to their priorities, will statistically be assigned to the same amount of slots per rendezvous cycle as all grid (torus)-quorum based methods.

Table 8 shows the average number of potential RDVs per cycle for the RAC^{2}E-gQS protocol against the optimized RAC^{2}E-gQS protocol (optimization for the first channel only, opt GD^{RC}1, and all, opt GD^{RC}all), for the *same* channel ranking lists and *opposite* channel ranking lists. It is natural that the case of the same

Table 7. Asynchronous CRs: Average (Mean) number of potential RDVs per cycle and average TTR (seconds) for optimized RAC^2E-gQS with optimized slot allocation (a) for the best channel only (opt $GD^{RC}1$) (b) for all channels (opt GD^{RC}all) in comparison with GD^{RC} from Section 4; No.c/s stands for Number of channels / slots; $List^{Rk}$ is the channel ranking lists

Algorithm	$List^{Rk}$	No.c/s	Mean RDV	Mean TTR (s)
GD^{RC}	random5	5/25	7.1165	0.1405
opt $GD^{RC}1$	random5	5/25	8.7188	0.1147
opt GD^{RC}all	random5	5/25	9.1990	0.1087
GD^{RC}	random10	10/100	15.1179	0.0661
opt $GD^{RC}1$	random10	10/100	17.1173	0.0584
opt GD^{RC}all	random10	10/100	18.3362	0.0545
GD^{RC}	random20	20/400	34.7735	0.0288
opt $GD^{RC}1$	random20	20/400	35.1575	0.0284
opt GD^{RC}all	random20	20/400	38.8063	0.0258

Table 8. Asynchronous CRs with asymmetric channel views: Average (Mean) number of potential RDVs per cycle and average TTR (seconds) for optimized RAC^2E-gQS with optimized slot allocation (a) for the best channel only (opt $GD^{RC}1$) (b) for all channels (opt GD^{RC}all) in comparison with GD^{RC} from Section 4; No.c/s stands for Number of channels / slots; $List^{Rk}$ is the channel ranking lists

Algorithm	$List^{Rk}$	No.c/s	Mean RDV	Mean TTR (s)
GD^{RC}	same1	5/25, 7/49	15.9947	0.0626
opt $GD^{RC}1$	same1	5/25, 7/49	13.8252	0.0724
opt GD^{RC}all	same1	5/25, 7/49	12.5381	0.0799
GD^{RC}	opposite1	5/25, 7/49	5.3882	0.1861
opt $GD^{RC}1$	opposite1	5/25, 7/49	7.3473	0.1363
opt GD^{RC}all	opposite1	5/25, 7/49	8.6083	0.1163
GD^{RC}	same2	15/225, 20/400	46.8691	0.0214
opt $GD^{RC}1$	same2	15/225, 20/400	44.3698	0.0226
opt GD^{RC}all	same2	15/225, 20/400	36.9728	0.0271
GD^{RC}	opposite2	15/225, 20/400	15.6793	0.0639
opt $GD^{RC}1$	opposite2	15/225, 20/400	18.5895	0.0539
opt GD^{RC}all	opposite2	15/225, 20/400	25.3545	0.0395

Table 9. Asynchronous CRs with asymmetric channel views: Average (Mean) number of potential RDVs per cycle and average TTR (seconds) for optimized RAC^2E with UP slot allocation optimized (a) for the best channel only (opt UP1) (b) for all channels (opt UPall) in comparison with UP from Section 4; No.c/s stands for Number of channels / slots; $List^{Rk}$ is the channel ranking lists

Algorithm	$List^{Rk}$	No.c/s	Mean RDV	Mean TTR (s)
UP	same1	5/25, 7/49	15.9926	0.0626
opt UP1	same1	5/25, 7/49	13.7718	0.0727
opt UPall	same1	5/25, 7/49	12.4826	0.0802
UP	opposite1	5/25, 7/49	3.5126	0.2866
opt UP1	opposite1	5/25, 7/49	4.7335	0.2123
opt UPall	opposite1	5/25, 7/49	5.5498	0.1809
UP	same2	15/225, 20/400	46.6890	0.0214
opt UP1	same2	15/225, 20/400	44.1094	0.0227
opt UPall	same2	15/225, 20/400	36.9216	0.0271
UP	opposite2	15/225, 20/400	21.5907	0.0465
opt UP1	opposite2	15/225, 20/400	21.6848	0.0462
opt UPall	opposite2	15/225, 20/400	23.9843	0.0418

channel ranking lists of the both CRs, results in a higher average number of potential RDVs per cycle than the case with opposite channel ranking lists. The opt GD^{RC} method decreases slightly the mean RDV and increases mean TTR while having the same channel ranking lists. On the other hand, it improves significantly the performance in the worst case scenario.

The next table, Table 9 shows the RAC^2E performance but with the UP method. It is clear that the grid (torus) QS slot allocation approach is (slightly) better almost in all cases, except the case with opposite channel ranking list with a higher number of channels, while allocating slots with either GD^{RC} or $GD^{RC}1$. However, while applying the full optimization (opt $GD^{RC}all$) the RAC^2E-gQS outperforms all other approaches including the use of the UP method.

These results justify that indeed the slot allocation should be more balanced than in the original version of grid QS mapping (GD^{RC}), and that the optimized RAC^2E-gQS improves definitely the performance with different channel ranking lists outperforming RAC^2E utilizing the UP method. The best case scenario is unlikely to occur, because in practice due to various propagation effects such as slow/fast fading, scattering etc., the nodes have different view of the surrounding environment. Therefore, the use of the optimized algorithm is to be recommended in situations where a heterogeneous channel view is to be expected. The RAC^2E-gQS protocol without the optimization can be used in environments where a homogeneous channel view is to be expected, such as in a controlled lab environment or small scale deployments, where all channels have the same or similar priority.

6. Conclusion

This paper proposes a RDV protocol for CRANs that (i) utilizes the asynchronous and randomness properties of the RAC^2E protocol and (ii) the grid quorum channel mapping (gQS-RDV) protocol taking into account channel heterogeneity. We showed that the combination of the RAC^2E protocol with the gQS mapping can yield a powerful RAC^2E-gQS rendezvous protocol for asynchronous operation in a distributed environment assuring a rapid RDV between the CR nodes. We also propose an optimization of the protocol enhancing its performance noticeable in the case of a large number of channels and different channel ranking lists. The case with asymmetric channel views has also been investigated showing significant improvement of the performance. In our future work, the algorithms will be implemented on a testbed platform with USRP nodes in order to evaluate both approaches in real conditions.

Acknowledgements. We would like to thank Professor Petri Mähönen for fruitful discussions. This work was funded by FP7-ICT NoE ACROPOLIS project [30]. We also thank a partial financial support from Deutsche Forschungsgemeinschaft and RWTH Aachen University through UMIC-research centre.

References

[1] ROMASZKO, S., DENKOVSKI, D., PAVLOVSKA, V. and GAVRILOVSKA, L. (2012) Asynchronous rendezvous protocol for cognitive radio ad hoc networks. In *Proceedings of the 4th International Conference on Ad Hoc Networks (ADHOCNETS)*. doi:10.1007/978-3-642-36958-2_10.

[2] AKYILDIZ, I.F., LEE, W.Y. and CHOWDHURY, K.R. (2009) Crahns: Cognitive radio ad hoc networks. *Elsevier International Journal of Ad Hoc Networks* 7(5): 810–836. doi:10.1016/j.adhoc.2009.01.001.

[3] PAVLOVSKA, V., DENKOVSKI, D., ATANASOVSKI, V. and GAVRILOVSKA, L. (2010) RAC2E: Novel rendezvous protocol for asynchronous cognitive radios in cooperative environments. In *the 21st Annual IEEE International Symposium on Personal, Indoor and Mobile Radio Communications (PIMRC)*: 1848–1853. doi:10.1109/PIMRC.2010.5671630.

[4] ROMASZKO, S. and MÄHÖNEN, P. (2011) Grid-based channel mapping in cognitive radio ad hoc networks. In *22nd Annual IEEE International Symposium on Personal, Indoor and Mobile Radio Communications (PIMRC)*: 438 – 444. doi:10.1109/PIMRC.2011.6139999.

[5] ROMASZKO, S. and MÄHÖNEN, P. (2011) Quorum-based channel allocation with asymmetric channel view in cognitive radio networks. In *6th ACM Performance Monitoring, Measurement and Evaluation of Heterogeneous Wireless and Wired Networks Workshop (PM2HW2N)*: 67–74. doi:10.1145/2069087.2069097.

[6] ROMASZKO, S. and MÄHÖNEN, P. (2012) Quorum systems towards an asynchronous communication in cognitive radio networks. *Journal of Electrical and Computer Engineering, Article ID 753541* **2012**: 22. doi:10.1155/2012/753541.

[7] SILVIUS, M.D., GE, F., YOUNG, A., MACKENZIE, A.B. and BOSTIAN, C.W. (2008) Smart radio: spectrum access for first responders. In *Wireless Sensing and Processing III (SPIE)*, **6980**: 698008–698008–12. doi:doi:10.1117/12.777678.

[8] THEIS, N.C., THOMAS, R.W. and DASILVA, L.A. (2010) Rendezvous for cognitive radios. *IEEE Transactions on Mobile Computing* **10**: 216–227. doi:10.1109/TMC.2010.60.

[9] GANDHI, R., WANG, C.C. and HU, Y.C. (2012) Fast rendezvous for multiple clients for cognitive radios using coordinated channel hopping. In *IEEE International Conference on Sesing, Communication, and Networking (SECON)* (Seul, Korea): 434–442. doi:10.1109/SECON.2012.6275809.

[10] LIU, H., LIN, Z., CHU, X. and LEUNG, Y.W. (2010) Ring-walk based channel-hopping algorithms with guaranteed rendezvous for cognitive radio networks. In *International Workshop on Wireless Sensor, Actuator and Robot Networks (WiSARN2010-FALL), in conjunction with IEEE/ACM CPSCom* (China): 755–760. doi:10.1109/GreenCom-CPSCom.2010.30.

[11] LIN, Z., LIU, H., CHU, X. and LEUNG, Y.W. (2011) Jump-stay based channel-hopping algorithm with guaranteed rendezvous for cognitive radio networks. In *IEEE International Conference on Computer Communications (INFOCOM)* (China): 2444–2452. doi:10.1109/INFCOM.2011.5935066.

[12] CHAO, C.M., TSAI, H.C. and HUANG, K.J. (2009) A new channel hopping MAC protocol for mobile ad hoc networks. In *Wireless Communications and Signal Processing (WCSP)*. doi:10.1109/WCSP.2009.5371543.

[13] CHAO, C.M. and WANG, Y.Z. (2010) A multiple rendezvous multichannel MAC protocol for underwater sensor networks. In *IEEE Wireless Communications and Networking Conference (WCNC)* (Australia). doi:10.1109/WCNC.2010.5506099.

[14] LEE, E.K., OH, S.Y. and GERLA, M. (2010) Randomized channel hopping scheme for anti-jamming communication. In *IFIP Wireless Days (WD) conference*. doi:10.1109/WD.2010.5657713.

[15] LUK, W.S. and WONG, T.T. (1997) Two new quorum based algorithms for distributed mutual exclusion. In *17th International Conference on Distributed Computing Systems (ICDCS)*. doi:10.1109/ICDCS.1997.597862.

[16] HALL, J.M. (1986) *Combinatorial Theory* (John Wiley and Sons).

[17] BIAN, K., PARK, J.M. and CHEN, R. (2009) A quorum-based framework for establishing control channels in dynamic spectrum access networks. In *15th annual international conference on Mobile computing and networking (MobiCom)*. doi:10.1145/1614320.1614324.

[18] HOU, F., CAI, L.X., SHEN, X. and HUANG, J. (2011) Asynchronous multichannel MAC design with difference-set-based hopping sequences. *IEEE Transactions on Vehicular Technology* 60(4): 1728 – 1739. doi:10.1109/TVT.2011.2119384.

[19] BIAN, K., PARK, J.M. and CHEN, R. (2011) Control channel establishment in cognitive radio networks using channel hopping. *IEEE Journal on Selected Areas in Communications* 29: 689–703. doi:10.1109/JSAC.2011.110403.

[20] BIAN, K. and PARK, J.M. (2012) Maximizing rendezvous diversity in rendezvous protocols for decentralized cognitive radio networks. *IEEE Transactions on mobile computing* 12(7): 1294–1307. doi:10.1109/TMC.2012.103.

[21] ROMASZKO, S. (2013) A rendezvous protocol with the heterogeneous spectrum availability analysis for cognitive radio ad hoc networks. *Journal of Electrical and Computer Engineering, Article ID 715816* **2013**: 22. doi:10.1155/2013/715816.

[22] BIAN, K. and PARK, J.M. (2011) Asynchronous channel hopping for establishing rendezvous in cognitive radio networks. In *IEEE International Conference on Computer Communications (INFOCOM), Mini-Conference* (China). doi:10.1109/INFCOM.2011.5935056.

[23] SHIN, J., YANG, D. and KIM, C. (2010) A channel rendezvous scheme for cognitive radio networks. *IEEE Communications Letters* 14(10): 954–956. doi:10.1109/LCOMM.2010.091010.100904.

[24] ROBERTSON, A., TRAN, L., MOLNAR, J. and FU, E.H.F. (2012) Experimental comparison of blind rendezvous algorithms for tactical networks. In *IEEE CORAL2012 in conjunction with IEEE WoWMoM2012*. doi:10.1109/WoWMoM.2012.6263760.

[25] BALACHANDRAN, K. and KANG, J. (2006) Neighbor discovery with dynamic spectrum access in adhoc networks. In *IEEE 63rd Vehicular Technology Conference, VTC 2006-Spring*, 2: 512–517. doi:10.1109/VETECS.2006.1682877.

[26] Universal software radio peripheral (USRP). URL http://www.ettus.com. Accessed March 15, 2013.

[27] JIANG, J.R., TSENG, Y.C., HSU, C.S. and LAI, T.H. (2003) Quorum-based asynchronous power-saving protocols for IEEE 802.11 ad hoc networks. In *IEEE International Conference on Parallel Processing (ICPP)*: 257–264. doi:10.1109/ICPP.2003.1240588.

[28] MAEKAWA, M. (1985) A p n algorithm for mutual exclusion in decentralized systems. *ACM Transactions on Computer Systems (TOCS)* 3(2): 145–159. doi:10.1145/214438.214445.

[29] LANG, S. and MAO, L. (1998) A torus quorum protocol for distributed mutual exclusion. In *International Conference on Parallel and Distributed Systems (ICPADS)*.

[30] FP7-ICT NoE ACROPOLIS project. URL www.ict-acropolis.eu. Accessed November 04, 2013.

5

Scalable Stream Processing with Quality of Service for Smart City Crowdsensing Applications

Paolo Bellavista, Antonio Corradi, and Andrea Reale

DISI - University of Bologna, Italy,
{paolo.bellavista, antonio.corradi, andrea.reale}@unibo.it

Abstract

Crowdsensing is emerging as a powerful paradigm capable of leveraging the collective, though imprecise, monitoring capabilities of common people carrying smartphones or other personal devices, which can effectively become real-time mobile sensors, collecting information about the physical places they live in. This unprecedented amount of information, considered collectively, offers new valuable opportunities to understand more thoroughly the environment in which we live and, more importantly, gives the chance to use this deeper knowledge to act and improve, in a virtuous loop, the environment itself. However, managing this process is a hard technical challenge, spanning several socio-technical issues: here, we focus on the related quality, reliability, and scalability trade-offs by proposing an architecture for crowdsensing platforms that dynamically self-configure and self-adapt depending on application-specific quality requirements. In the context of this general architecture, the paper will specifically focus on the Quasit distributed stream processing middleware, and show how Quasit can be used to process and analyze crowdsensing-generated data flows with differentiated quality requirements in a highly scalable and reliable way.

Keywords: Stream Processing, Scalability, Quality of Service, Support Frameworks

1. Introduction

The vision of smart cities as urban areas where people, places, environment, and administrations become closer and get connected through novel ICT services and networks, is becoming reality at an increasingly faster pace. Thanks to disruptive technologies such as location-based services and ubiquitous connectivity via multiple interfaces, cities are promising candidates to become, in the near future, a central development and deployment platform for a novel and increasingly important set of services. Confirming this trend, in the last years there have been several initiatives led by governments and industries directed toward the study and development of smart urban areas. Examples of these initiatives are the many European Digital Agenda funded projects, such as European Digital Cities, InfoCities, IntelCity roadmap, Intelligent Cities, and EUROCITIES [15], or industry-led activities, such as the

IBM Smarter Cities project[1], or the Intel Collaborative Research Institute for Sustainable Connected Cities[2]. At the same time, the widespread diffusion of smartphones and tablets with heterogeneous connectivity and rich sensing capabilities creates novel opportunities for extending the reach of smart city services. Among them, one of the most interesting and still open directions is the exploitation of the new sensing paradigm that has been usually defined as *crowdsensing*.

Crowdsensing shifts the principles of the more traditional crowdsourcing processes — mostly diffused in static Internet scenarios (see, for example, Wikipedia[3]

---

[1] http://www.ibm.com/smarterplanet/us/en/smarter_cities/, last visited in August 2013.
[2] https://www.intel-university-collaboration.net/?page_id=1420, last visited in August 2013.
[3] http://www.wikipedia.org/, last visited in August 2013.

or Amazon Mechanical Turk[4]) — to cyber-physical urban spaces, by leveraging the unprecedented monitoring capabilities of mobile citizens and coupling them with distributed actuation tasks to be completed collaboratively, for instance by moving data or physical items (e.g., bikes of a smart city bike sharing service) to maximize the targeted objective functions (e.g., uniform coverage of traffic monitoring or uniform bike availability at bike pickup points). This way, relatively complex distributed goals can be achieved thanks to the weakly-organized, and massive-scale cooperation of the mobile *crowd*: in a typical crowdsensing application, people are asked to perform simple and often very fine-grained geo-based tasks that usually involve tracing some physical real-world measure through their mobile devices, such as registering noise or environmental pollution in some area, or taking photos of specific locations.

Connecting participatory sensing with crowd-based actions in what we call the *crowdsensing loop* is a fundamental, continuous step of real-time and large-scale data monitoring and analysis: by aggregating and processing the flowing streams of data coming from collaborative citizens, it is possible to obtain significant information about the current status of the city and its inhabitants, and to build models that can accurately forecast city dynamics, which, in turn, can help decision-makers to determine the appropriate actions and policies that improve the overall urban quality-of-life. The very peculiar characteristics of these new processing scenario pose several fundamental challenges to existing data processing platforms, and call for novel models/architectures that can satisfy strong requirements of scalability, quality, reliability, and cost-effectiveness.

This paper introduces a general and novel crowdsensing architecture, whose ultimate goal is to guide the realization of scalable and reliable crowdsensing platforms that exploits the concept of *quality* at two complementary levels, i.e., i) at *task generation/assignment* level and ii) at *data processing* level. After briefly introducing our ongoing work to implement this architecture in a working middleware-level solution, we concentrate the focus on the main subject of this paper by describing in detail Quasit, a novel data stream processing model and middleware implementation specifically designed for scenarios with strong scalability and quality requirements. Quasit is built to run effectively on large clusters of commodity hardware and to automatically handle various types of failures. Originally, Quasit allows to annotate its processing elements with QoS specifications, which are leveraged at runtime to adapt their behavior to both dynamic load conditions and user-defined quality requirements. We present preliminary results obtained with our under-development prototype (open-source and freely available for download[5]), showing that out system combines the easy definition of processing functionalities and QoS requirements, with automatic scalability to the available processing resources.

Let us note that the Quasit stream processing model and framework can also be used to application domains and scenarios other than smart city crowdsensing applications described here. We claim that any big-data stream-oriented application benefiting from QoS-aware task assignment and data processing could fruitfully exploit the QoS-aware scalability of Quasit, with significant performance improvements if compared with traditional, non-QoS-aware stream processing frameworks. However, in this paper, for the sake of description focus, we will describe only how Quasit may be usefully adopted in the wide domain of crowdsensing applications.

The remainder of the paper is organized as follows. In Section 1 we present our novel crowdsensing architecture, by introducing the challenges that it tries to solve and explaining the solution approach that it proposes. In the context of this architecture, we introduce Quasit, whose processing model is presented in Section 3. A description of the design of the Quasit prototype and some central implementation insights is given in Section 4, followed, in Section 5, by a set of preliminary evaluation results that show the feasibility of the approach and the effectiveness of our prototype implementation, also compared to a solid, state-of-the-art alternative as Apache S4 [26]. Section 6 overviews the work in the literature sharing common characteristics with Quasit, and clearly points out the original aspects and technical elements of our proposal. Conclusive remarks and directions of ongoing research work end the paper.

This paper is an extended version of [9]. It originally introduces Quasit within our broader vision of crowdsensing in Smart Cities (Section 2), and includes an extended discussion of the supported QoS policies (Section 3) and additional experimental results in the performance evaluation of our prototype (Section 5).

2. The Crowdsensing Loop

A crowdsensing platform supports the execution of several sensing applications, each typically having one high-level goal, which is decomposed in several small geo-located sensing *tasks* to be executed by the crowd by opportunistically exploiting people movements through the city. The types of applications and tasks

---

[4]https://www.mturk.com/, last visited in August 2013.

[5]http://lia.deis.unibo.it/research/quasit, last visited in August 2013.

managed by the platform can be highly heterogeneous, and can involve capturing, harvesting, and processing very different *physical features* of the real world, as well as performing actuation tasks that change the status of the cyber-physical space.

A critical aspect to consider, in order to develop a scalable crowdsensing platform, is the important trade-off between the obtained sensing/actuation accuracy and the crowdsensing cost. The execution of a task instance by a citizen is intrinsically *unreliable*: a person may simply refuse to execute a task, or be unable to complete it; more importantly, the quality of the crowd-sensed data can have a high variability due to the poor accuracy/precision of the sensors embedded into personal devices. Task *replication*, i.e., assigning copies of the same task to several people in order to have higher confidence on the collected data or to achieve monitoring reliability through multiple source participation, is the normal solution to this series of challenging issues. However, the efficient determination of how much to replicate a task, or to whom to assign different tasks in a real crowdsensing urban scenario is still an open problem. Moreover, as the result of the execution of thousands, possibly replicated, sensing tasks from different applications, large volumes of sensing data are continuously produced and need to be processed effectively and in almost real time. Notwithstanding the recent advances in data-center/cloud data elaboration technologies, managing this unprecedented volume can still present unacceptable costs and scalability limitations if the peculiar characteristics of crowd-sensed data are neglected by the processing framework.

Our crowdsensing architecture, shown in Figure 1, aims at tackling these fundamental quality/cost trade-offs by putting the task generation/assignment and the data processing phases in a closed loop, and by leveraging at the same time user-provided and autonomously-inferred quality requirements to optimize the platform runtime execution. By processing the data received from the crowd, the data processing component builds two important artifacts. On the one hand, it learns models that incorporate the history of the sensed features and that can be queried by the user to monitor or predict the status of the real world aspects of interest for her sensing applications. On the other hand, it builds and constantly updates *profiles* of users, regions, and sensing features. These profiles are the original core of our architecture, since they represent the elements that close the crowdsensing loop.

This way, cross knowledge of user histories, of the characteristics of different geographical regions, and of the importance of the contribution that different sensing features have on the output models, can be leveraged by the platform to self-regulate the allocation of the available human and computational resources,

also by taking into careful consideration application-level quality/reliability requirements. For example, task assignment strategies can be specifically tailored to the citizens' habits or can use personalized incentive types [29] in order to maximize the chance of obtaining the desired sensing accuracy while minimizing costs. Similarly, data-processing components should use the combination of the learned profiles to prioritize the analysis of data streams that can potentially give a more important contribution to the output models, while delaying or discarding less critical information. For instance, considering environmental pollution monitoring, the number of replicas for the monitoring task of a given area can be minimized by assigning them to people who habitually or recently frequented that area. Similarly, previous knowledge about the fact that collecting traffic information in a green area has a minor influence on the air pollution model than the humidity information, can be used to configure the data analysis phase to give priority (and hence more resources) to the processing of humidity crowd-sensed data, by possibly discarding less important monitored features.

2.1. A Scalable Platform for Quality–aware Crowdsensing

By following the ideas, model, and architecture presented above, we are developing a middleware-level crowdsensing platform, based on the convergence of the McSense[6] middleware for crowdsensing task management and the Quasit[5] [10] stream processing engine for quality-aware data analysis.

McSense, developed in collaboration with the New Jersey Institute of Technology, is a middleware for the management of crowdsensing flows; it consists of a central control component, responsible of quality-aware task generation and assignment, and a Software Development Kit (SDK) (only available for the Android platform at the time of writing) that can be used to develop mobile sensing applications that support users in accomplishing all the phases of their crowdsensing tasks. The control component offers ad-hoc tools to describe geo-based sensing tasks and to specify their specific quality requirements, such as the desired reliability (as the expected probability of completing a task) or task completion time. These quality parameters, in conjunction with the profiles knowledge base, are leveraged by a pluggable task assignment algorithm that allocate the appropriate amount of human resources by concretely assigning and dispatching task instances to citizens. Currently, two algorithms can be alternatively chosen: one based on the *recency* of people's last visit to given target locations,

---

[6]http://lia.deis.unibo.it/research/McSense/, last visited in August 2013.

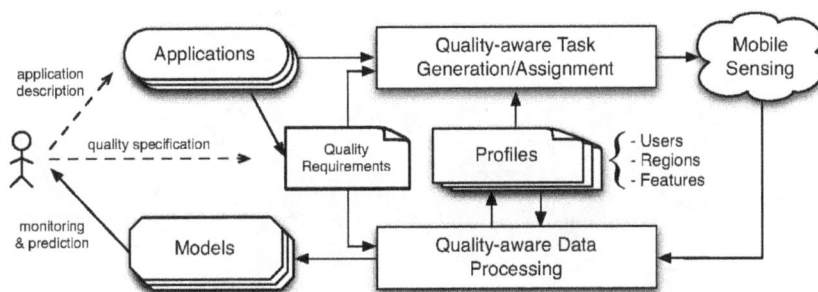

Figure 1. A general high-level architecture for quality-based crowdsensing platforms.

the other based on people habits of *attendance* to certain places. Further details on the McSense platform can be found in [11] or on the project Web site[6].

The other core component of the architecture, i.e., the data back-end processing platform, is being developed by integrating the Quasit framework into the crowdsensing flow. Quasit is a quality-aware stream processing framework for data-center environments designed to handle large volumes of streaming data by seamlessly scaling to the available computing/memory/network resources. It gives developers a simplified processing model that describes streaming problems as the composition of processing graphs (called SIGs, see the following), each made of an interconnected set of reusable *operators*, which define the transformations that, applied to the input data, produce the desired output. Originally, Quasit offers advanced configuration/customization features, by permitting to associate QoS specifications to all the elements of its application models (operators, communication channels, or entire graphs). The set of supported QoS specifications permits to configure the system through several QoS parameters, ranging from high-level indications (e.g., output priorities) to low-level ones (e.g., detailed set-up of network buffers). In addition, Quasit lets developers define and reuse their custom stream processing *operators*, by supporting their easy dynamic arrangement in graphs to be automatically deployed on the infrastructure of available computational resources. The design of Quasit operators supports a functional-like programming style that clearly separates operator behavior and state, thus making it easier for the runtime framework to support different and sophisticated strategies for QoS provisioning.

The original features of Quasit to auto-configure its behavior depending on application-dependent QoS requirements fit the needs of our crowdsensing vision. Crowdsensing quality requirements, either explicitly specified at the application level or autonomously inferred through users, regions, and features profiles, are automatically mapped to Quasit QoS specifications in

order to set up the most appropriate data processing quality. Let us note that, for instance, the possibility to consider only a subset of all available data sources under critical congestion situations is particularly useful in crowdsensing applications due to their intrinsic nature with redundant data streams, typically with similar values, concurrently originated by different smart city "observers". In these cases, proper QoS management of stream processing can allow to achieve the most suitable trade-off between latency and completeness, only to mention a simple example. Developers of crowdsensing applications define their own data aggregation/processing algorithms by arranging prebuilt or custom operators into Quasit SIGs, and associate specific QoS parameters directly to the graphs elements, this way expressing their application-specific requirements (for example, to indicate fine grained low-level resource needs of the various parts of their data analysis steps). In the following sections, we describe the Quasit processing model and the related prototype, and we thoroughly analyze the unique design characteristics that make the platform scalable, easy to use, and capable to effectively support the specification of quality requirements with arbitrary level of detail.

3. The Quasit Stream Processing Model

Quasit is used to process multiple input data streams concurrently, to perform arbitrary transformations on them, and to produce other data streams as output, which can be fed to other systems for storage or further processing. A Quasit data stream is modeled as a temporal sequence of data samples, whose content is a set of key-value attributes. Any stream is associated with one data type that defines the keys and types of the attributes of its samples.

The basic modeling unit in Quasit is the *Streaming Information Graph* (SIG), a weakly connected acyclic and directed graph that represents the information flow and the transformations that, applied to one or more input streams, produce an output data stream. The nodes of a SIG represent data transformation stages, while

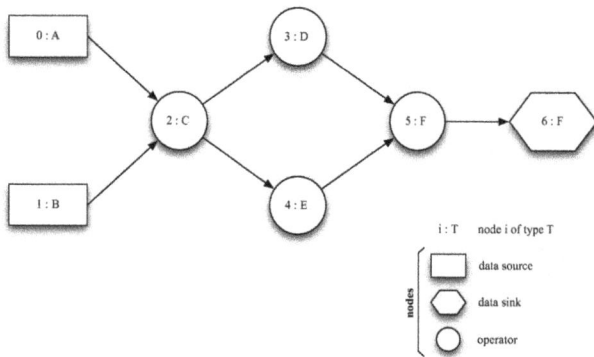

Figure 2. Simple SIG example, with two data source nodes, one sink node, and four operator nodes. source0 and source1 respectively produce a data stream of typeA and typeB; operator2 receives them as input and produces a typeC data stream, received by operators 3 and 4, producing respectively typeD and typeE data streams. Finally, the typeF data stream generated by operator5 goes into data sink6, of the same type.

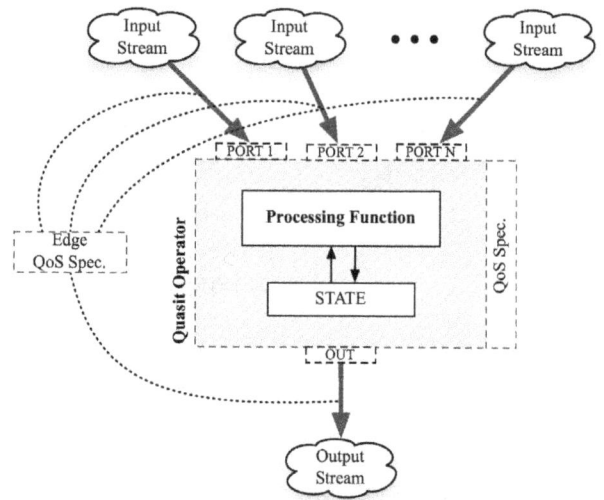

Figure 3. Structure of a Quasit *simple operator*.

its edges model communication dependencies. Figure 2 depicts a simple example of SIG.

Three different kinds of SIG nodes are possible: *data source, data sink,* or *operator*. A *data source* node identifies a data stream that is conceptually out of the SIG and its role is to abstract from the actual nature of the stream producer; it can represent either an external stream source or the output of another Quasit SIG. A *data sink* node, conversely, represents the destination of the data stream that is the output of the SIG; data sinks can be used either to redirect output streams to other systems for additional processing steps or storage, or to connect the output of a SIG with the input of another SIG. An *operator* node associates with one or more input data streams and *generates* exactly one output stream. SIG edges model communication *channels* between nodes.

Every element of a SIG (either node or edge) may be labeled with a QoS specification: QoS specifications allow users to enrich their processing graphs with additional information about non-functional quality requirements. Given the centrality of QoS specifications and their runtime support in Quasit, we will devote Section 3.2 to their discussion; but, before that, let us first present the basic building block of our SIG, i.e., the *operator* component, based on which developers can model their stream processing issues in terms of composition of simple transformation stages.

3.1. Operators

An *operator* performs arbitrary operations on the data samples it receives as input, and produces samples for its output stream. We designed Quasit operators having in mind three main goals. First, an operator should be *concurrency friendly*: whenever the application semantics allow it, the execution of different operators should be parallelized across all the available processing resources; this should require few or no effort at all for the developer defining the operator. Second, operators should be *easily manageable* in order to allow the Quasit framework to effectively control their execution at runtime, e.g., by moving them from a processing node to another, saving and restoring their processing state, or transparently recovering them from failures. Third, the operator abstraction should favor *maximum reusability* in order to let developers model their problems in terms of SIGs by writing as less new code as possible.

Quasit operators can be *simple* or *composite*, and both types can be either *stateful* or *stateless*, depending on whether they need a processing state to be kept or not. A *simple operator* logically consists of several sub-components, as shown schematically in Figure 3. It always has one or more *input ports* and exactly one output port: input ports model the input requirements of the operator, while the output port represents its output contract. The behavior of the operator depends on the combination of its *state* and *processing function*, or solely on the *processing function* in the case of stateless operator.

The processing function is a user-defined function that the Quasit framework invokes asynchronously as data samples are available at input ports. If the operator is stateless, the function takes one parameter, which is bound at runtime to the incoming data samples; if it is stateful, a further parameter is present and is bound to the current state of the operator. The output of the processing function is a tuple made of two optional components: if present, the first is the data sample to send to the output port; the second, always absent for stateless operators, represents the new state the operator will assume. In other words, by defining an operator's processing function, developers specify the set of transformations that, applied to the input, produce its output and state transitions.

Quasit adopts an asynchronous and event-based processing approach, according to which an operator produces output and/or changes its state only in response to incoming data; this permits a large number of operators to share processing resources very efficiently, by enabling high execution *concurrency* in multi-processor and multi-core environments. Furthermore, the sharp separation between the behavior of the operator, expressed through its (stateless) processing function, and its processing/communication state gives Quasit great *flexibility* in taking transparent management decisions at runtime, in order to effectively support the execution of operator components. For instance, Quasit can offer complex and differentiated state persistence/reliability policies, which would have been much more difficult to realize if state was kept mixed with processing logic.

To achieve *maximum reusability*, Quasit introduces a mechanism that permits to use already defined operators as building blocks for creating more complex and powerful ones, i.e., *composite operators*. Developers can define composite operators by arranging existing operators (either simple or composite) into a special type of SIG that completely defines the execution characteristics of the composite operator, called *Operator Definition SIG* (OD-SIG). Operator composability permits to easily encapsulate complex behavior into composite operators, and leverage them to model many problems, with evident reusability advantages.

3.2. QoS Support in Quasit

One of the most original aspects of Quasit is its ability to let developers augment their stream processing models with very rich and differentiated QoS specifications, to be used at runtime to guide the Quasit framework in the management of system behavior and resource allocation according to the desired quality requirements. Related to the design of Quasit QoS-related features, our main goal is to support a wide spectrum of QoS policies, ranging from simple and high-level quality indications (allowing developers to express their requirements quickly and with as few effort as possible) to richer and lower-level parameters, to be used for finer performance tuning when a deeper and more QoS-aware control over processing is needed.

In particular, any SIG element can be augmented with an optional *QoS Specification*, defining a set of non-functional configuration parameters or constraints. Depending on its target, a QoS specification can consist of several *QoS Policies*, each policy influencing

Table 1. List of Quasit *QoS Policies*.

Element	QoS Policy	Possible values
Data Sink	Output Priority	*Priority value*
Operator	Processing Cap	*Time threshold*
Operator	State Fault Tolerance	*Replication factor*
Operator	State Consistency	*Lazy, Snapshot, Strong*
Operator	Queueing Spec.	*Input queues size, Scheduling policies*
Operator	Input Ordering	*No order, Causal*
Channel	Delivery Semantics	*Best Effort, At most once, At least once, Exactly once, Probabilistic*
Channel	Deadline	*Time threshold*

a different quality aspect. Table 1 reports the list of Quasit QoS policies, by concisely showing their applicability scope and their possible values.

In order to provide readers with a high-level overview of the practical aspects that can be regulated through QoS augmentation of SIGs, in the following we will give a short description of Quasit policies, also by putting them into their practical applicability context by presenting examples of their possible use within a simple crowdsensing scenario. The considered scenario is that of a smart-city application that combines car-sharing services with urban pollution monitoring. A fleet of cars is equipped with air-pollution meters and are put on disposal to citizens, who can request, use, or share them through a mobile application. While moving through the city, cars report their position and real-time pollution data (through their 3G radio) to a data back-end application running on a Quasit deployment. Similarly, the back-end processes and properly matches citizen requests for car trips with car availability.

- *Output Priority.* By using this QoS policy, it is possible to differentiate the way Quasit assigns resources to parts of SIGs that contribute to produce different outputs. In our reference scenario, users may be provided with gold, silver, or bronze services according to their service membership level: these levels can be mapped on different priorities for the operators responsible for matching their requests with possibly available cars.

- *Processing Cap.* This QoS policy determines a hard constraint on the time available to an operator to process a data sample. When this constraint is violated, the computation is interrupted and a default action executed. For instance, the operator that classifies incoming car requests and maps them to the appropriate subgraphs, needs to complete this process fast, or else assign requests to a default class; in this case default assignment is considered better than too late but more precise assignment.

- *State Fault Tolerance and Consistency.* Both policies determine how the state of an operator is handled by the Quasit platform. A weaker state consistency strategy/replication factor can save resources when partial state loss can be tolerated. For instance, the loss of partial updates on some tiles of the urban pollution map is acceptable in many related applications, especially given the supposed high-update frequency.

- *Queueing Specifications.* This low-level policy controls the way Quasit manages the operators input queues. In our scenario, this policy could be used to set blocking behavior for the input queue of an operator that dispatches matched user-car requests and, at the same time, to define an ordering function that prioritizes only the samples related to requests from gold members.

- *Input Ordering.* It determines whether samples entering an operator marked with this policy are to be processed in their arrival order, or if their processing order should satisfy *happened-before* [23] relations established by preceding operators. For example, a request to modify the destination for a car-shared trip should always be handled after the related request for the trip.

- *Delivery Semantics.* This QoS policy, which may be attached to SIG edges connecting a pair of operators, configures the communication protocol used by the Quasit platform to enable data exchange between the two operators. For instance, for reasons that are similar to the ones discussed above for the State Fault Tolerance policy, pollution-map updates can be transferred using a best-effort communication protocol.

- *Deadline.* This policy controls how the samples in an operator output queue are handled. In particular, by setting a time-based deadline on a channel, the Quasit network management layer is instructed to adopt a network-scheduling policy that tries to ensure that every tuple is transferred from source to destination within a required time threshold after its generation. In our example scenario, a deadline could be set on the graph path that manages application critical operations, such as the management of payments for the car-sharing service via users' credit cards.

As far as we know, the rich variety of QoS modeling options available in Quasit is unique in the stream processing literature. Let us remark again that a proper tuning of the various QoS Specifications attached to SIG elements permits to flexibly adapt the Quasit runtime to different application scenarios, by deeply influencing its strategies for effectively allocating and scheduling the dynamically available processing resources; some details about how the Quasit framework effectively puts into execution the Quasit SIG elements and manages them at runtime are presented in the following part of the paper about Quasit framework design and implementation.

4. The Quasit Framework Prototype

In the following, we present the results of our research work of design, implementation, experimental validation, and quantitative evaluation of a first prototype of the Quasit framework, which implements the Quasit stream processing model previously described; let us remark once again that the source code of our framework is freely available for download, evaluation, and extension at our project Web site[5].

This section is structured in three parts: in the first (Section 4.1) we present the Quasit architecture; in Section 4.2 we overview how QoS is achieved and controlled at runtime, while in Section 4.3 we provide some implementation insights about the current Quasit prototype.

4.1. Distributed Architecture

Like other systems for data management and processing in data-centers [14, 17, 21, 26], the Quasit distributed architecture follows a simple *master-workers* model, where a logically centralized node (the *master*) implements management and coordination tasks, while a possibly large number of *worker* nodes perform data processing tasks. In particular, Quasit SIGs are deployed and executed by a set of computing nodes called *Quasit Runtime Nodes* (QRNs), which are monitored and managed by one *Quasit Domain Manager* (QDM), as shown in Figure 4. The set of QRN nodes and the QDM that manages them are collectively called *domain*. A domain runs one or more SIGs, providing advanced runtime services, such as tolerance to operator/QRN failures, and QoS-based management of SIG execution. New SIGs can be added to the domain dynamically at runtime. We assume that QRNs are connected through a high-speed local area network (LAN), as typically occurs in data-center scenarios.

In order to distribute the workload and leverage all the dynamically available resources, Quasit decomposes arbitrarily complex user SIGs in smaller units, which are then assigned to individual worker nodes. The granularity of work decomposition and distribution is determined by the defined *simple operators*.

Clients submit SIGs to the QDM, which is responsible of planning and monitoring their distributed execution. As soon as a new SIG is received, the QDM must decide an initial partitioning, in order to determine its distributed execution among the available QRNs. The QDM takes this decision by running an *operator*

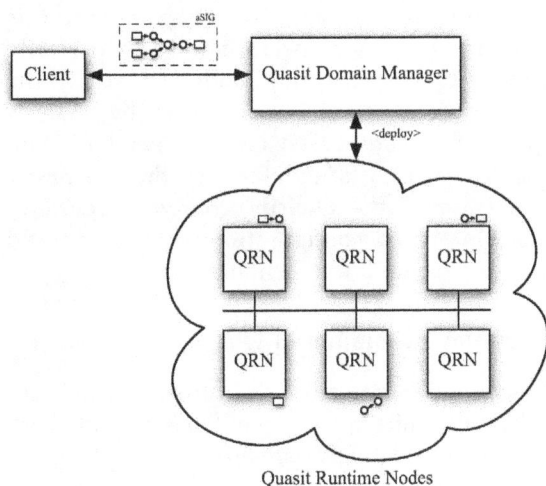

Figure 4. A Quasit domain includes one QDM (conceptually centralized entity with monitoring and management responsibilities) and several QRNs as middleware instances performing the actual stream processing.

placement algorithm that exploits information about the current status of the QRNs in the domain (e.g., the list of operators already running and their resource availability) to *optimize the execution cost* of the SIG according to the enforced QoS-aware cost function. The development of a proper cost function and placement algorithm is one of our main research challenges: in the current prototype we are exploring a greedy algorithm, called *affinity placement*, which sequentially assigns every operator to the QRN that minimizes its local execution cost, and two additional more trivial algorithms, primarily used as comparison references, i.e., *uniform* and *random* placement, which respectively distribute the operators uniformly (according to a topological ordering of graph vertices) and randomly on the QRNs. Although conceptually centralized (and currently implemented in a centralized way), let us point out that the QDM does not represent a bottleneck for the Quasit architecture, because it is not directly involved either in data processing or in any data transfer. Moreover, we plan to implement resilience to QDM failures through traditional replication techniques applied to the only QDM entity [18].

A QRN implements a QoS-aware execution container for Quasit operators and is responsible for offering them scheduling and communication support. Reflecting the operator model, the QRN execution model is *asynchronous and event-based*. Communication between operators is managed by the set of distributed QRNs according to a PUB/SUB interaction model: every output port of operators (or data sinks) running on a QRN associates with a named endpoint; QRNs *subscribe*

to all the endpoints associated with the input ports of operators (and data sinks) that they are running, and store the samples from these subscriptions in event queues associated with the input ports. A pool of executor threads is used to pick samples from the queues, dispatch them to their destination operators, and execute the associated processing function.

4.2. QoS Management

QoS policies defined at model-level on Quasit SIGs are enforced at runtime thanks to a two level QoS-management architecture, realized through the interaction of one *domain QoS manager*, running within the QDM, and several *node QoS managers*, one for each QRN. The domain QoS manager performs global admission control and QoS-based system configuration, while node QoS managers leverage the computational resources of the QRNs on which they execute to implement and enforce the requested QoS policies on locally running operators and I/O ports.

In order to provide deeper and more detailed insights about this QoS management scheme, let us briefly examine its role in the process of deployment and execution of a SIG. At *deployment time*, the domain QoS manager, after having checked whether the QoS policies applied to the SIG are self-consistent, performs a translation phase, during which user-level QoS policies are transformed to implementation specific configuration parameters, which are sent to QRNs inside operator deployment commands. For example, QoS policies on channels, such as the *delivery semantics* policy, are translated into configuration parameters for the PUB/SUB protocol and for the network queues used by the ports corresponding to the channel endpoints. Node QoS managers use these data to provide an initial configuration for the instances of operator and ports they are responsible of. At *execution time*, QoS monitoring tasks are cooperatively performed by domain and node QoS managers: node managers continuously collect data about the behavior of their locally running components, and try to autonomously adjust their configuration to avoid possible QoS violations; for example, they can reallocate their local resources by giving a greater share to operators with higher priority. This way, most QoS management decisions are taken and enforced locally by node QoS managers, thus relieving the central domain QoS manager from this load and improving the system scalability. Actions by the domain QoS manager are only necessary in a limited set of situations, when global knowledge is needed or when the adaption actions of single local managers are no longer sufficient to avoid QoS violations. For example, it is up to the global QoS manager to decide whether to move an operator from a QRN to another in case of system overload. This design aims at avoiding the domain QoS manager to represent a bottleneck for system scalability, by demanding as many decision as possible to the local managers. We are planning to perform through measurements to quantitatively evaluate the effectiveness of this solution (see Section 7). However, it is possible that, for very large scale scenarios (e.g, with thousands of nodes), the QoS management duties on the single QDM could become overwhelming: to deal with such cases a possible solution could be a further hierarchical partition of management responsibilities, where the set of all QRNs is divided into separate QoS-management clusters, each supervised by a different cluster head.

4.3. Implementation Insights

Our QDM and QRN components are realized using the Scala[6] programming language. Scala has been preferred to other possible alternatives for three main reasons. First, the language runtime comes with a rich library that offers an excellent support for writing concurrent and multi-threaded applications. Second, its elegant and concise syntax allows us to simplify the design of the user API through which developers model their stream processing problems. Third, Scala code, once compiled, is executed on the solid and widely supported Java Runtime Environment.

Quasit PUB/SUB interactions are instead realized on top of the OMG Data Distribution Service (DDS) [27] middleware, which is used as the basis for both reliable group membership management and inter-QRN SIG channels. The choice of using a DDS-based communication middleware grants several benefits. First, DDS message dissemination uses an IP-multicast-based protocol that well fits the typical one-to-many communication patterns of Quasit operators and perfectly adapts to network characteristics of data-centers where nodes are commonly arranged in a hierarchy of Ethernet segments, connected by layer2 switches. Second, the DDS standard defines a rich set of QoS parameters, that can be used to configure and personalize many low-level details of the communication middleware: using DDS to implement our PUB/SUB communication layer has provided us with a solid ground on which we build our ad-hoc QoS enforcement mechanisms, especially those relative to channels. Whenever possible, in fact, we exploit mappings between high-level Quasit QoS policies and possible configurations of the various DDS QoS parameters, and set up the QRN networking layers according to them.

Finally, the scheduling of actors and the management of their queues is currently implemented using the Scala Actors framework [20]: every operator is represented by an actor instance, which perfectly suits our event-based processing model. Currently, the scheduling of these actors is taken care by a *work-stealing* pool of threads based on the Java Fork/Join framework [24]. This scheduler, in the currently available version of the Quasit prototype, does not permit any QoS-based configuration: we plan to add this feature as a future implementation step.

5. Experimental Evaluation of Quasit Performance

In this section we present some first preliminary results collected while testing our Quasit framework prototype in a relatively small-scale deployment environment. Although the deployment does not reflect the characteristic of our target scenarios completely, its simplicity permits to easily measure and evaluate basic system characteristics, such as the effectiveness of the platform communication and threading mechanisms. We believe that the reported results demonstrate the feasibility and the effectiveness of our approach, and represent an important starting point for a future, large scale evaluation campaign on real-world use cases.

The selected and simple test scenario consists of an external source producing a periodic stream of image frames. For instance, this stream could correspond to the sequence of key frames of a video produced by a security camera. These image samples are transformed through a series of manipulation steps, and then streamed again to an external destination. The samples generated from the source correspond to the repetition of a 192x128 24bpp PNG image, which is a scaled version of one of the photos from a Kodak public test set[7]. The size of each sample is approximately 43 KB.

We have modeled the image manipulation process as a pipeline of Quasit operators, whose processing function is implemented as stateless OpenCV[8]-based transformations. The combination of these operators forms a 30 steps pipeline-shaped SIG (as shown in Figure 5) deployed and ran on top of the Quasit framework prototype. All the stages of this pipeline have approximately the same computational complexity. Let us note that this simple scenario is anyway highly representative because i) pipeline-shaped patterns are very common in more complex SIGs and ii) the number of involved operators (30) is relatively high and close to the real size of many SIGs of practical application interest.

The testbed Quasit domain consists of one machine running the QDM component, plus from one up to

---

[6] http://www.scala-lang.org/, last accessed in August 2013.

[7] kodim23.png, publicly available at http://r0k.us/graphics/kodak/, last accessed in August 2013.
[8] OpenCV, http://opencv.willowgarage.com/wiki/, last accessed in August 2013.

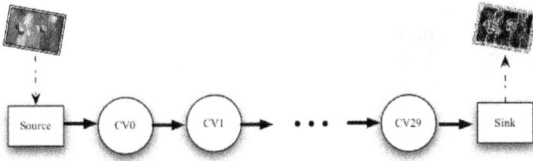

Figure 5. The simple and pipeline-shaped SIG used in this experimental evaluation.

Table 2. Hardware and software configuration of QRN nodes.

Host: Intel Pentium Dual-Core E2160 @ 1.80GHz
Main Memory: 2 GB
Network Interface: Gigabit Ethernet
OS: Ubuntu 11.04 (Linux kernel 3.0.0)
DDS: OpenSplice DDS 5.4.1 Community Edition
Scala: 2.9.1-final
JVM: OpenJDK 64-bit Server VM (IcedTea7-2.0 build 147)
JVM Flags: -Xms128M -Xmx512M -Xss4M

four different physical nodes having the role of QRNs. The QRNs are interconnected through one Ethernet segment, while the QDM, although in the same IP subnet, is separated from the QRNs by two switches. The machine hosting the QDM is also used as the external source and sink of the image frames. The hardware and software configuration of the machines is shown in Table 2.

In each experiment run, we feed the deployed SIG with 500 image samples, not counting "warm-up" and "cool-down" sets of samples processed when the SIG pipeline is not full. For each configuration, we have collected the results of 15 to 50 runs of the same experiment (depending on the variability of results).

The experimental results reported in the following aim at discussing three main aspects that we have measured on our testbed:

- The management overhead with respect to an ideal parallel processing scenario.

- The ability to scale horizontally, by dynamically adding QRNs to one Quasit domain.

- A preliminary performance comparison with Apache S4, a state-of-the-art stream processing engine by the Apache Software Foundation.

5.1. Comparison with Ideal Parallel Processing

In order to quantitatively evaluate the overhead imposed by the Quasit middleware (if compared with the maximum possible improvement of stream processing performance thanks to parallel execution), we have also designed a very simple simulator that models our scenario but omits all the overhead associated with middleware-level management of

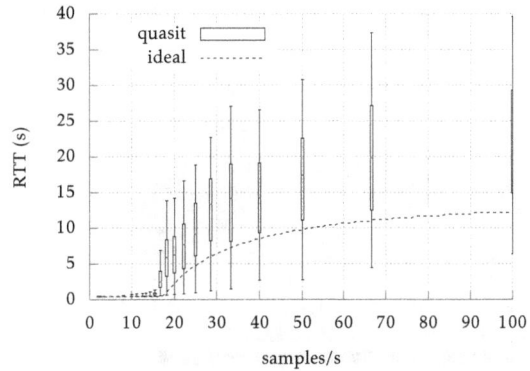

Figure 6. Distribution of sample processing time with 4 QRNs and uniform operator placement. The dashed line represents the performance upper bound in ideal conditions.

operators (including operator scheduling) and inter-QRN network communication. The simulator models a group of parallel workers arranged in a pipeline; their number reflects the number of available CPUs across all the QRNs. OpenCV transformations of the original SIG are distributed evenly among workers, and each of them executes sequentially, for each incoming sample, the transformations it is responsible for, before forwarding it to the next worker. In the simulations, we measure the average time needed to perform a complete processing of an image sample by varying the rate at which new samples are produced, and we compare the results with the performance data obtained on a real deployment environment with 4 QRNs in a Quasit domain (operators deployed according to the uniform placement strategy). In the real deployment environment, image processing time is measured as the sample *round trip time* (RTT), i.e., the time interval between the generation of a new frame and the reception of the processed version of that frame (recall that the external source/sink of the input/output streams coincide in our simple pipeline-shaped test SIG). Figure 6 shows the distribution of the measured RTTs while increasing generation rates in the real deployment and the average processing time in the "ideal" simulated scenario.

Clearly, in both cases, the processing time increases abruptly as soon as our Quasit framework is no longer able to keep up with image production rate and the input queue of the first operator (worker) starts filling up. For low sample rates, Quasit performance is very close to the ideal one, thus demonstrating the very limited platform overhead in unloaded conditions; the difference tends to grow as the input rate increases; we experienced that this is mainly due to the overhead introduced by operator scheduling, which is completely neglected in the simplified simulated scenario.

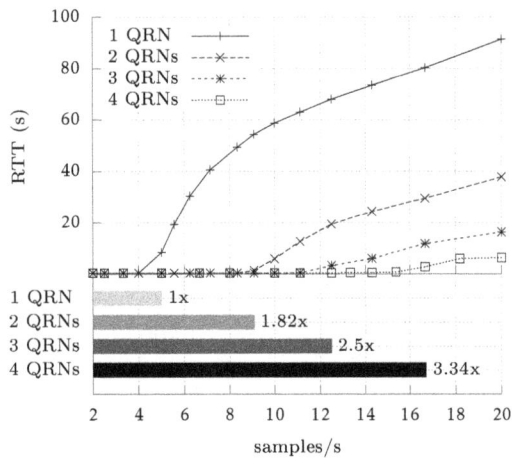

Figure 7. Comparison of average processing times using 1, 2, 3, or 4 QRNs and uniform placement.

Table 3. Critical input rates and speed-up with different numbers of QRNs

# of QRNs	critical input rate	speed-up
1	5 samples/s	1
2	9.10 samples/s	1.82
3	12.5 samples/s	2.5
4	16.7 samples/s	3.34

5.2. Quasit Horizontal Scalability

About our second evaluation goal of verifying the ability of Quasit to scale as additional QRNs are added to a domain, we have deployed the same test pipeline-shaped SIG on four different execution environments, with respectively one, two, three, or four QRNs. In all cases we have deployed the graph using the uniform placement strategy.

Figure 7 shows the results. The trend of the curves is the same in all the examined domains: as long as the production rate does not exceed the maximum processing rate in unloaded conditions, the average sample RTT is constant and low (around 450 milliseconds); as soon as Quasit is no longer able to keep up with the sample arrival rate, the average processing time starts to grow. However, the results show that by adding processing resources to one Quasit domain, it is seamlessly possible to increase the Quasit ability to serve more aggressive input rates, with reasonably limited overhead. Table 3 shows how the *critical sample-rate* (i.e. the data rate at which the system starts to be overloaded and accumulate samples at the operator queues) varies by adding additional QRNs. Clearly, the speed-up values do not grow with a perfect linear trend with the number of QRNs, because of the overhead due to management and network communication, but still the performance degradation is very limited. However, the system ability to scale horizontally also

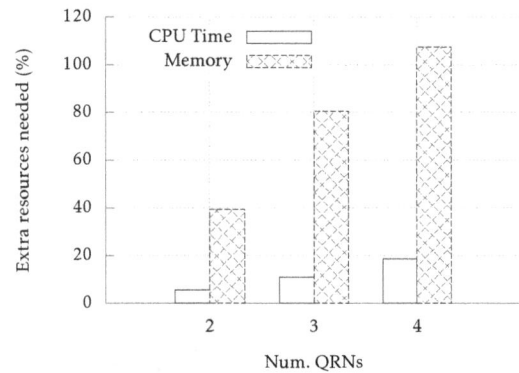

Figure 8. Overhead (in terms of extra resources needed) when adding additional QRNs

depends strongly on the characteristics of the SIGs being executed: for this reason, Quasit fosters a SIG design made of many fine grained components sharing no state, giving the framework many parallelization opportunities to be exploited according to the required QoS level and resource availability.

In order to estimate the cost of the management overhead when a Quasit deployment is scaled horizontally, we focused on the scenario with the lowest input rate (i.e. 2 samples/s). Note that, in this scenario, a single QRN has enough resources to keep up with the data input rate: by measuring the amount of extra resources needed to process the same data stream in deployments with 2, 3 and 4 QRNs we can effectively estimate the management overhead cause by the additional running processing nodes. Figure 8 shows the percentage of extra CPU Time and memory needed to process the same stream of 500 image samples, when the Quasit deployment is over-provisioned with additional QRNs. The amount of CPU Time required, which in the 1 QRN case amounts in average to 201.81 seconds, increases up to 239.40 seconds in the case of 4 QRNs (less than 20% more). The increase in the amount of memory consumed, instead, is remarkably more significant: if 105.86 MB are consumed on average on a 1 QRN deployment, the 4 QRNs one consumes on average, in the same scenario, about 223 MB, i.e. more than double the resources. This is not really surprising, since the extra nodes running the additional QRNs need to instantiate their own Java Virtual Machines, which in turns load all the classes and instantiate all the objects needed for the management of the QRN itself.

5.3. Preliminary Performance Comparison with Apache S4

Finally, we report here a set of results that compare the performance of our system with Apache S4. The Apache

S4 project[8], initially developed and maintained by Yahoo! [26], is probably the research effort closest to our Quasit proposal. As Quasit, S4 lets users freely define stream analysis graphs and PEs; in addition, inspired by MapReduce, S4 permits to partition streams according to user defined *keys*. The platform instantiates PEs based on the graph layout and on the keys dynamically found in the data, guaranteeing that, within a stream, samples with the same key are always processed by a unique PE instance. According to the project Web site, S4 has been used in several production systems at Yahoo! before being released to the public under open-source license in October 2010; by the end of 2011 it was accepted under the Apache Incubator project umbrella. In the experiments presented here, we used the 0.6 release, code-named *piper*, which we pulled from the project's git repository.

Through the S4 API we modeled the same pipelined OpenCV image processing scenario we implemented in Quasit, and executed it on our testbed. Unfortunately, we were not able to control the placement algorithm used by S4 to deploy operators on different nodes, and we had to adopt the default algorithm, which assigns PE instances to nodes according to a hash function applied to stream *keys*; in our pipeline scenario, where the key of each edge connecting consecutive processing steps is constant, this means that S4 will create one PE instance for each processing step. Note that, being the placement of these instances based on the result of a hashing function, it will be, in general, totally unaware of the graph communication characteristics. For this reason, to avoid an unfair comparison, for the following set of results we configured Quasit to use a *random* placement algorithm, which is equivalently unaware of any graph characteristic. As in the previous group of experiments, in each run, we feed the pipeline application deployed on S4 with 500 samples and measure the sample processing time. Again, before starting the measurements, we perform an initial warm-up by generating a preliminary low-rate input sequence. All the reported results are average values over 10 runs for each configuration.

In Figure 9 we show the results for the cluster configuration with four QRNs/S4 nodes, and in Figure 10 we summarize the variation in the average sample RTT for all the tested deployment configurations. It can be observed that Quasit outperforms S4 for what concerns the average sample processing time, thus showing that our prototype exhibits a very limited overhead. Moreover, the difference between the two system is largely more marked in the deployment with just one processing node. It is likely that this is the consequence of

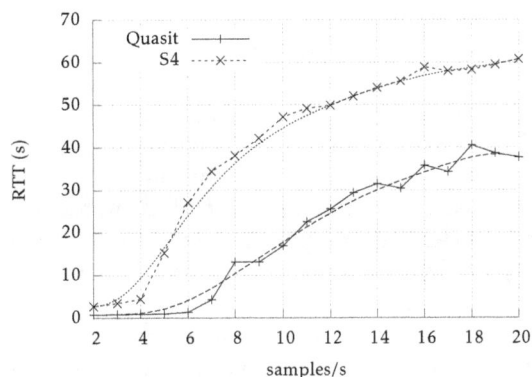

Figure 9. Quasit vs. S4: average sample processing time for increasing data rates with 4 processing nodes and random placement (Quasit) or hash-based placement (S4).

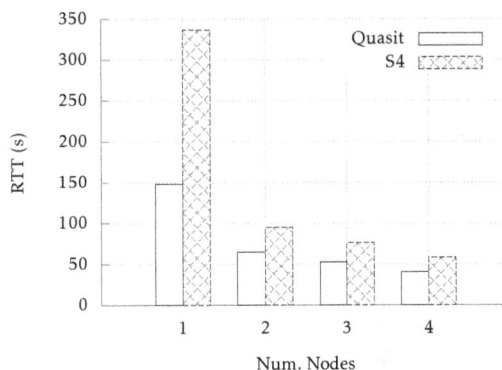

Figure 10. Quasit vs. S4: average sample processing time with 1, 2, 3, or 4 processing nodes and random placement (Quasit) or hash-based placement (S4). The input sample rate is 20 samples/s.

the different threading architecture of the two systems: while Quasit leverages a pool of threads whose size is proportional to the available CPU cores (two on the machines in our testbed) and independent from the number of locally deployed operators, S4 creates a new thread for each data stream in the application graph (in our scenario this corresponds to one thread per local PE instance, plus one on the node where the sink is deployed). This causes higher contention for the available processing resources and greater thread scheduling overhead, in the common cases where the number of "active" components on a single host (operators for Quasit, streams for S4) is significantly bigger than the number of available processing units. Our DDS-based networking solution should also give us some advantage, in terms of serialization space efficiency (DDS serialization format is based on the OMG CDR standard, while S4 uses a custom solution based

---

[8]Apache S4 project Web site, http://incubator.apache.org/s4/. Last visited in August 2013.

on the Kyro[9] serialization framework), but also, and most importantly, in scenarios presenting several *one-to-many* communication patterns: in these situations, the implementation of operator channels over IP multicast could provide significantly reduce network overhead compared to the TCP based solution used by S4 (internally based on the JBoss Netty framework[10]). We are planning extended comparative analysis focusing on the usage of the network in different scenarios to validate the above claims.

As a final remark, it is important to consider that, while in this scenario we used very simple placement strategies (*uniform* and *random*) due to the simplicity of the pipelined processing scenario, in a more general scenario Quasit could effectively exploit additional application-level knowledge, provided in the form of QoS specification attached to part of user graphs, for example by using smarter placement strategies, or by dynamically modifying its thread scheduling mechanisms (e.g., enlarging the thread pool size if operators perform many I/O operations).

6. Related Work

In recent years, there has been an increasing trend shifting the processing load of complex application and services inside data-centers [8] thanks to the ease at which cheap storage and computational resources can be reached on the cloud [5], but also thanks to the flourishing of highly parallel, scalable, and fault-tolerant hardware and software architectures and data processing paradigms. The most popular model for processing large datasets inside data-centers is certainly MapReduce [14], which has received a lot of attention thanks to its ease of use and the diffusion of open source implementations, such as Apache Hadoop[11]. In MapReduce, developers have to model their processing problems in terms of *map* and *reduce* functions. Leveraging this constraint, the MapReduce runtime takes care of efficiently running the defined functions against input data while providing fault-tolerance and horizontal scalability. This programming model makes the simplifying assumption that input consists of static datasets stored in a distributed file system such as GFS [17], and, thus, is not appropriate for dynamic streaming processing scenarios where input data cannot be statically known.

Given the industrial success of MapReduce, several authors have tried to enhance it with more dynamic and advanced stream processing capabilities. For example,

[4, 19, 25] leverage a *map-reduce-merge* strategy (originally proposed by [28]) to run MapReduce jobs on datasets that are dynamically created as the result of *windowing* operations on data streams; partial output from these jobs is then joined through the additional *merge* step. DEDUCE [22] permits to define MapReduce operators through an extension of the SPADE language [16], and to use these operators within an IBM InfoSphere Streams processing graph; DEDUCE jobs can run on either static datasets or, as in the previously cited approaches, sliding windows over streaming data. In [13], instead, the authors propose HOP, a modified version of Hadoop that, by supporting intra- and inter-job pipelined communication between map and reduce tasks, permits to run continuous MapReduce jobs. All these examples show the interest in extending MapReduce to solve stream processing problems that can be modeled as a sequence of batch jobs working on slices of input streams. However, we claim that, by using a model that is inherently designed to work with static input, these solutions cannot offer the flexibility of a native stream-oriented programming model and are often inadequate to effectively deal with the dynamic characteristics of streaming data, such as highly variable sample rate.

Some existing solutions, similarly to Quasit, use directed graphs to model stream processing problems and to distribute processing responsibilities on available nodes. The Stanford Stream Data Manager [7], and the Aurora [12] projects are two early examples of data stream management systems; coming from the databases community, they introduced the concept of *continuous queries* over data streams by defining specific query languages and algebras [1]. The two systems have a centralized architectures that limit their ability to scale to large data-streams. The Borealis Stream Processing Engine [2, 3] extends its predecessor Aurora, and it leverages the resources of a set of distributed nodes to handle user-defined *query diagrams*, in which a limited set of pre-defined relational-like operators are arranged. Very interestingly, Borealis allows developers to define QoS specifications for the output of their query diagrams: it is possible to estimate the output quality as a function of *response times*, *event drops*, or specific (and user-defined) *event values*. Quasit adopts these solution guidelines by improving and extending them: Quasit users can additionally define their own operators by directly programming them, and acquire a more direct control of quality-related parameters of every part of the processing graph.

Dryad [21] by Microsoft Research also models computations as directed acyclic graphs. In Dryad graphs, vertices are mapped to native programs that are executed — each in its own process — by the Dryad framework: mainly because of the overhead associated to spawning and managing full processes, the grain

---

[9]Kyro project Web site, http://code.google.com/p/kryo/. Last visited in August 2013.

[10]Netty project Web Site https://netty.io/. Last visited in August 2013.

[11]http://hadoop.apache.org, last accessed in August 2013.

of Dryad computational components is coarser than Quasit operators, which, instead, are very lightweight objects confined in the Java Runtime Environment. In addition, while Quasit specifically targets continuous stream processing, Dryad, like MapReduce, seems more oriented to the execution of batch-like jobs where input datasets are fixed and known a priori.

Also SPC [6], the core of IBM Infosphere Streams [16], and S4 [26], a project initially developed by Yahoo! and now maintained by the Apache Foundation[8], share some similarities with Quasit in terms of goals and solution guidelines. Both let developers model their continuous stream processing problems as graphs of *Processing Elements* (PEs), which, similarly to Quasit simple operators, may be user-defined. The main difference between Quasit and these two projects is that our proposal is primarily focused on the support of a rich set of QoS-related parameters to customize stream processing behavior, while SPC and S4 do not allow rich QoS specifications.

7. Conclusive Remarks and Future Work

In this paper we have introduced Quasit, both a programming model and a framework prototype for scalable, reliable, and quality-aware stream processing. The design of Quasit was guided by the need of having a robust and cost-effective data processing layer, capable of well fitting large-scale deployment scenarios where the awareness of differentiated quality requirements could be exploited to take proper decisions about the most suitable dynamic trade-off among latency, completeness, precision, and accuracy. These characteristics make Quasit especially suited to crowdsensing application scenarios, with i) large variability and unpredictability of input load (with possible frequent peaks, also including very redundant information) and ii) variable quality of the data to analyze and of the contribution that these data can bring to different application goals.

Our first prototype of the Quasit runtime, although still partial, represents a concrete proof-of-concept of a possible implementation of the proposed model (available for extension and refinement to the community of researchers/practitioners in the field), is showing the feasibility of the approach, and is encouraging our further development efforts. In particular, we are concentrating our future work along two main directions. On the one hand, we will extend our prototype toward the implementation of a richer set of QoS policies for SIG operators and channels, and we will experiment alternative operator placement and management strategies. On the other hand, we are performing a more significant set of experiments to verify the ability of our Quasit model and prototype to sustain challenging large-scale deployment environments, with a special focus on dynamic differentiation of stream processing services depending on QoS requirements specified at the SIG level. In this context, we plan to extensively evaluate the effectiveness of our distributed and hierarchical QoS management architecture, especially when the scale of applications and data grow.

Acknowledgments. Special thanks go to Professor Cristian Borcea, NJIT. The fruitful and enjoyable discussions on McSense we had together have strongly inspired this paper and the application of Quasit to smart city crowdsensing applications.

References

[1] D. J. Abadi, D. Carney, U. Çetintemel, M. Cherniack, C. Convey, S. Lee, M. Stonebraker, N. Tatbul, and S. Zdonik: Aurora: a new model and architecture for data stream management. The VLDB Journal The International Journal on Very Large Data Bases, vol. 12, no. 2, pp. 120–139 (2003)

[2] D. J. Abadi, Y. Ahmad, M. Balazinska, U. Cetintemel, M. Cherniack, J.-H. Hwang, W. Lindner, A. S. Maskey, A. Rasin, E. Ryvkina, N. Tatbul, Y. Xing, and S. Zdonik: The Design of the Borealis Stream Processing Engine. In: 2nd Biennial Conference on Innovative Data Systems Research (CIDR), pp. 277–289, VLDB Endowment (2005)

[3] Y. Ahmad, N. Tatbul, W. Xing, Y. Xing, S. Zdonik, B. Berg, U. Cetintemel, M. Humphrey, J.-H. Hwang, A. Jhingran, A. Maskey, O. Papaemmanouil, and A. Rasin: Distributed operation in the Borealis stream processing engine. In: 2005 ACM SIGMOD international conference on Management of data (SIGMOD '05), pp. 882–884, ACM New York, NY, USA (2005)

[4] D. Alves, P. Bizarro, and P. Marques: Flood: elastic streaming Map-Reduce. In: 4th ACM International Conference on Distributed Event-Based Systems (DEBS '10), pp. 113–114, ACM New York, NY, USA (2010)

[5] M. Armbrust, A. Fox, R. Griffith, A.D. Joseph, R. Katz, A. Kowinski, G. Lee, D. Patterson, A. Rabkin, I. Stoica, and others: A view of cloud computing. Commun. ACM, vol. 53, no. 4, pp. 50–58 (2010)

[6] L. Amini, H. Andrade, R. Bhagwan, F. Eskesen, R. King, Y. Park, and C. Venkatramani: SPC: A distributed, scalable platform for data mining. In: R. Grossman and S. Connelly (eds.) 4th international workshop on Data Mining Standards, Services and Platforms (DM-SS), pp. 27–37, ACM New York, NY, USA (2006)

[7] A. Arasu, B. Babcock, S. Babu, J. Cieslewicz, K. Ito, R. Motwani, U. Srivastava, and J. Widom: STREAM : The Stanford Data Stream Management System, Technical report, Stanford InfoLab (2004)

[8] L. Barroso, J. Dean, and U. Holzle: Web search for a planet: The Google cluster architecture. IEEE Micro, vol. 23, no. 2, pp. 22–28 (2003)

[9] P. Bellavista, A. Corradi, and A. Reale: The QUASIT Model and Framework for Scalable Data Stream Processing with Quality of Service. In: 5th International Conference on Mobile Wireless Middleware, Operating

Systems, and Applications (MOBILWARE '12), pp. 92–107, Springer Berlin Heidelberg, Berlin, Germany (2012)

[10] P. Bellavista, A. Corradi, and A. Reale: Design and Implementation of a Scalable and QoS-aware Stream Processing Framework: the Quasit Prototype. In: 5th IEEE International Conference on Cyber, Physical and Social Computing (CPSCOM '12), pp. 458–467, IEEE, Besançon, France (2013)

[11] G. Cardone, L. Foschini, C. Borcea, P. Bellavista, A. Corradi, M. Talasila, and R. Curtmola: Fostering ParticipAction in Smart Cities: a Geo-Social CrowdSensing Platform. IEEE Commun. Mag, vol. 51, no. 6 (2013)

[12] D. Carney, U. Çetintemel, M. Cherniack, C. Convey, S. Lee, G. Seidman, M. Stonebraker, N. Tatbul, and S. Zdonik: Monitoring streams: a new class of data management applications. In: 28th international conference on Very Large Data Bases (VLDB '02), pp. 215–226, VLDB Endowment (2002)

[13] T. Condie, N. Conway, P. Alvaro, J. M. Hellerstein, K. Elmeleegy, and R. Sears: MapReduce Online. In: 7th USENIX conference on Networked systems design and implementation (NSDI '10), USENIX Association, Berkeley, CA, USA (2010)

[14] J. Dean and S. Ghemawat: MapReduce : Simplified Data Processing on Large Clusters. Commun. ACM, vol. 51, no. 1, pp. 107–113 (2008)

[15] EUROCITIES committee. Smart Cities Workshop (2009)

[16] B. Gedik, H. Andrade, K.-l. Wu, P. S. Yu, and M. Doo: SPADE: the System S declarative stream processing engine. In: 2008 ACM SIGMOD international conference on Management of data (SIGMOD '08), pp. 1123–1134, ACM New York, NY, USA (2008)

[17] S. Ghemawat, H. Gobioff, and S.-T. Leung: The Google File System. ACM SIGOPS Operating Systems Rev., vol. 37, no. 5, pp. 29–43 (2003)

[18] R. Guerraoui and A. Schiper: Software-based replication for fault tolerance. Computer, vol. 30, no. 4, pp. 68–74, (1997)

[19] J. Horey: A programming framework for integrating web-based spatiotemporal sensor data with MapReduce capabilities. In: ACM SIGSPATIAL International Workshop on GeoStreaming, pp. 51–58, ACM New York, USA (2010)

[20] P. Haller and M. Odersky: Scala Actors: Unifying thread-based and event-based programming. Theoretical Computer Science, vol. 410, no. 2–3, pp. 202–220 (2009)

[21] M. Isard, M. Budiu, Y. Yu, A. Birrell, and D. Fetterly: Dryad: distributed data-parallel programs from sequential building blocks. In: 2nd ACM SIGOPS/EuroSys European Conference on Computer Systems, vol. 41, no. 3, p. 59–72, ACM New York, NY, USA (2007)

[22] V. Kumar, H. Andrade, B. Gedik, and K.-L. Wu: DEDUCE: at the intersection of Map-Reduce and stream processing. In: I. Manolescu, S. Spaccapietra, J. Teubner, M. Kitsuregawa, A. Leger, F. Naumann, A. Ailamaki, and F. Ozcan (eds.) 13th International Conference on Extending Database Technology (EDBT '10), pp. 657–662, ACM New York, NY, USA (2010)

[23] L. Lamport: Time, clocks, and the ordering of events in distributed systems. Commun. ACM, vol. 21, no. 7, pp. 558–565 (1978)

[24] D. Lea: A Java fork/join framework. In: ACM 2000 conference on Java Grande (JAVA '00), pp. 36–43, ACM New York, NY, USA (2000)

[25] D. Logothetis and K. Yocum: Ad-hoc data processing in the cloud. In: Proceedings of the VLDB Endowment, vol. 1, no. 2, pp. 1472–1475, VLDB Endowment (2008)

[26] L. Neumeyer, B. Robbins, A. Nair, and A. Kesari: S4: Distributed Stream Computing Platform. In: 2010 IEEE International Conference on Data Mining Workshops (ICDMW '10), pp. 170–177, IEEE Los Alamitos, USA (2010)

[27] Object Management Group: Data Distribution Service for Real-time Systems, version 1.2. Technical report, Object Management Group (2007)

[28] H.-c. Yang, A. Dasdan, R. Hsiao, and D. Parker: Map-reduce-merge: simplified relational data processing on large clusters. In: 2007 ACM SIGMOD international conference on Management of data, pp. 1029–1040, ACM New York, NY, USA (2007)

[29] D. Yang, G. Xue, X. Fang, and J. Tang: Crowdsourcing to smartphones: incentive mechanism design for mobile phone sensing. In: 2012 International Conference on Mobile Computing and Networking (Mobicom '12), pp. 173–184, ACM New York, NY, USA(2012)

6

Reliability-based server selection for heterogeneous VANETs

Seyedali Hosseininezhad*, Victor C. M. Leung

Department of Electrical and Computer Engineering, The University of British Columbia, Vancouver, BC, Canada

Abstract

Heterogeneous wireless networks are capable of providing customers with better services while service providers can offer more applications to more customers with lower costs. To provide services, some applications rely on existing servers in the network. In a vehicular ad-hoc network (VANET) some mobile nodes may function as servers. Due to high mobility of nodes and short lifetime of links, server-to-client and server-to-server communications become challenging. In this paper we propose to enhance the performance of server selection by taking link reliability into consideration in the server selection mechanism, thereby avoiding extra client-to-server hand-offs and reducing the need of server-to-server synchronization. As a case study we focus on location management service in a heterogeneous VANET. We provide a routing algorithm for transactions between location servers and mobile nodes. We assume that location servers are vehicles equipped with at least one long-range and one short-range radio interfaces, whereas regular nodes (clients) are only equipped with a short-range radio interface. The primary goal of our design is to minimize hand-offs between location servers while limiting the delays of location updates. Taking advantage of vehicle mobility patterns, we propose a mobility-aware server selection scheme and show that it can reduce the number of hand-offs and yet avoid large delays during location updates. We present simulation results to show that proposed scheme significantly lowers the costs of signaling and rate of server hand-offs by increasing the connection lifetimes between clients and servers.

Keywords: *ad hoc* routing, heterogeneous networks, service discovery, VANET

1. Introduction

Vehicular ad-hoc networks (VANETs) are emerging as one of the most important practical applications of mobile ad-hoc networks (MANETs). As the demand for pervasive computing is increasing, providing services to nodes in the network is also becoming critical. For instance, location management is becoming one of the most important modules in vehicular networking. Multimedia streams, news broadcasting, entertainment and other applications which require Internet connectivity, peer-to-peer applications, local advertisements, vehicle pooling and local cab services are some examples of the broad range of feasible applications when vehicles are equipped with positioning and communication capabilities [1, 2].

1.1. Service discovery in MANETs

In MANETs with various devices in the network, a service is a facility or capability that is provided by some of the nodes that can be used by other mobile nodes. These services are usually implemented as an application inside the nodes. Service discovery has been solved in wired networks and single hop wireless networks. DEAPSpace [3] is a service sharing mechanism produced by IBM research for short-range, single hop wireless networks. However, in multihop mobile networks, the physical proximity and mobility of nodes are issues that impact service discovery. Therefore existing approaches may not be suitable for multihop MANETs.

Service discovery protocols consist of four components:

- *Service Description* provides an abstraction of the services being provided. These information help clients

*Corresponding author. Email: seyedali@ece.ubc.ca

to choose between provided services and connect to the server which is providing required services. Usually a service description is composed of server identification, server characteristics and service characteristics.

- *Service Registration and Advertisement* includes the procedure of storing service descriptions in nodes and advertisement of the services in the network. Depending on the approach, advertisement and registration could be simple flooding of information or based on a hierarchical directory storage approach.

- *Service Discovery* is the method employed by requesters to find services. Service discovery can also include the decision making mechanism to choose between available services. Service discovery and service advertisements are complementary to each other in a system. Therefore, more complexity in one could be traded off for more simplicity in the other. When the services are discovered, the service selection is done based on some criteria.

- *Routing* includes the process of relaying service advertisements, service discovery, service requests and replies. Routing could be done based on different criteria as well. Since we are focusing on VANETs with very high node mobility, the performance of routing is very sensitive to decision making mechanisms used in service discovery. A combination of good advertisement and service discovery methods could decrease the number of needed packet relays and lower the rate of route changes in the network.

There are several service discovery methods designed for MANETs [4–6]. Based on literature, service discovery mechanisms are divided into directory-based vs. directory-less approaches. In networks with high mobility, less reliance on topology information would lead to a better performance because the topology is changing very fast and updating would be very costly. Therefore instead of directory-based methods like [5], directory-less methods have been proposed [4].

In a survey by Mian *et al.* [7] service discovery protocols are categorized based on their performance under various network conditions. Based on this comparison, there is a trade-off between the extent of covered area and supported mobility. There is no one-size-fits-all method that is capable of handling high mobility while the size of network is relatively large.

Service discovery inspired by field theory. Lenders *et al.* [4] defined an approach for efficient and robust service discovery. This concept is similar to anycast routing, which is supported in IPv6 [8]. In anycast routing, an address is associated with more than one interface belonging to distinct nodes that are similar in nature. As it is preferable for clients to get service from the nearest among several potential servers, use of anycasting would allow the desired server to be reached easily.

From electromagnetic field theory, the point potential of a spot is related to its distance to the maxima potential charge. In wireless networks, the most commonly used definition for distance is based on hop count; nonetheless, geographical distance is also applicable. In [4], hop count is considered as the distance between nodes:

$$\varphi(n) = \sum_{i=1}^{N} \frac{Q_i}{\text{dist}(n, n_i)}, \qquad (1)$$

where Q_i is the potential assigned to server i and $\varphi(n)$ is the total received potential by node n from all servers. The amount of potential assigned to each server could be a factor of their capacity or quality of service (QoS) metrics.

Reliability consideration in route selection. Whereas in highly mobile networks, hop-based distance metrics may be unreliable and unstable, and route maintenance may cause extra signaling overhead and delay, route reliability can be a more suitable metric for route selection. Longevity of a route is introduced in [9] as a new metric for route assessment, measured by the association between two nodes. It is assumed that a link is reliable if the association between the two end nodes is higher than a certain threshold. A route-lifetime assessment based routing (RABR) algorithm is proposed in [10], in which route selection is based on hybrid criteria of route lifetime and path length. The route lifetime is measured by the link affinity, which is calculated based on the received signal strength. Since in practice the signal strength varies over time, RABR can make wrong decisions especially in an urban environment with rapid signal fading. It is desirable to utilize the concept of link lifetime as a decision factor in routing, but a new measure of link lifetime tailored for the variable conditions of VANETs is needed.

1.2. Location services and server selection problem

With increasing demand for pervasive computing, location management is becoming an essential function in VANETs. Location management is a service for determination of each vehicle's location and making this information available to other nodes. A location management system is generally composed of location servers and clients. Location servers are responsible to hold the location information of clients and to retrieve the information to requesters. Nowadays localization technologies have become widespread and it is a justifiable assumption that all vehicles in a VANET can determine their positions using onboard global position system (GPS) receivers. Therefore we assume that each vehicle can determine its geographic coordinate and report it to the corresponding location server(s). Since connection to stationary resources may not be guaranteed in VANETs, vehicles should form a location management overlay on the infra-structure-less network. When mobile clients need to register, update or query about location information of other nodes, they should choose a server based on proper decision factors and send their request to it. Choosing a server for location service can be considered a specific case of server selection in service discovery. We shall next provide a short review of current location management approaches and their limitations. We shall consider location management as a specific scenario of service discovery, and propose a method to select location servers in a heterogeneous VANET.

Location management methods in MANETs. Several protocols have been designed to handle mobility of nodes [11–18]. Location servers are responsible for handling geographic location information of nodes in the vicinity and provide them to others when needed. Different categories of location management have been classified: flooding-based location management is the most straight-forward method for passing location information. Due to high redundancy overhead, researchers have strived to decrease unnecessary packet relays. The hypothesis of methods like distance routing effect algorithm for mobility [12] is that relative locations of closer nodes are changing faster compared to nodes far away from each other. Therefore location updates are sent more frequently to location servers close by than those that are farther away. Quorum-based location management is another category that is based on assuring a rendezvous between queries and updates. A localized quorum-based location service is proposed in [13], which disperses the location of every node horizontally and vertically. As the authors have stated, this method is proper for networks without a significant relative motion. Vehicular ad-hoc networks with high relative speeds are not suitable environments for this class of location management systems.

The GLS method [14] for distributed location service management divides the area into different degrees of grids in a way that in every grid around the node there is a fixed number of servers that collect location information about that node. As grids grow larger, the probability of a server being chosen for other nodes decreases. This method is not very flexible for highly variant environments like VANETs. Hierarchical methods [17, 18] for server allocation are highly scalable because the rates of location updates are reduced for servers in higher levels. However, in VANETs, the number of servers may not be very high and forming a hierarchical structure may not be feasible.

In high mobility networks such as VANETs, keeping track of location information would in general be a huge overhead on the network if location information is saved in all location servers. Therefore having the record of every node in network while nodes are rapidly changing their locations can be performed better if we are able to save

these information in a specific set of nodes. Moreover, to reduce the detrimental effect of mobility, we focus on hop-by-hop packet relaying rather that finding a deterministic route from clients to servers. We consider location management as a service to vehicles, which is offered by some specific nodes in the area. Therefore we need a service discovery mechanism to find the service providers.

Many different wireless access technologies can be employed in VANETs. Short-range technologies include wireless local area networks (WLANs) and its variant called Dedicated Short-range Radio Communications targeting specifically vehicular communications. Long-range wireless technologies include cellular networks and wireless metropolitan area networks such as WiMAX. Vehicular ad-hoc networks employing short-range radio access face problems in area coverage and fast hand-offs between nodes. Because of high mobility speed, rate of hand-offs in the network becomes a limiting factor in location registration and updates. In the presence of these difficulties, heterogeneity can come to the rescue for services like location management. Using long-range wireless access as a higher layer of communications, we can interconnect location servers together as a logical mesh network. We can assume that a connected graph of location servers can exchange signaling messages through this logical mesh network. We shall base our work on utilizing available long-range wireless connections to facilitate location management in VANETs.

In Section 2 we propose an architecture for a heterogeneous VANET to provide location service. After clarifying the problems and challenges in this server selection scenario, we present the proposed method for reliability-based server selection in Section 3. In Section 4 we give the methodology for performance evaluations, present the results and discuss their significance. Section 5 concludes the paper.

2. Location management over heterogeneous VANET—the architecture

Figure 1 depicts a heterogeneous VANET architecture with partial Internet connectivity. In this system, nodes

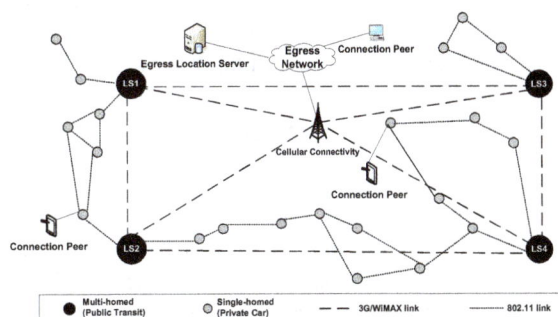

Figure 1. Heterogeneous VANET architecture for location management.

equipped for heterogeneous wireless access are connected to each other and edge gateways using their long-range wireless access capability.

The requirements and assumptions in aforementioned architecture are:

- All vehicles are considered to have a mechanism to extract their own geographic location, e.g. using an onboard GPS receiver.

- All nodes are equipped with at least one short-range radio (e.g. 802.11a/b/g/p).

- Some special nodes with heterogeneous wireless access capability are equipped with all valid short-range communication interfaces and one long-range communication interface (e.g. WiMAX). These nodes function as location servers that are interconnected to each other in a logical mesh network to exchange their location records, and to stationary gateways for Internet access.

- In consideration of valuable licensed spectrum used for long-range wireless access, the use of long-range radio should be minimized. Therefore it is our goal to reduce the numbers of queries between servers and server hand-offs for vehicles.

- Location queries and updates should not be propagated more than a certain number of hops.

- Server advertisements should not be rebroadcasted more than a certain limit.

In one scenario of this architecture, a public transportation system provides a wireless Internet relay service inside an urban area. Public transit vehicles are equipped with multiple radios and they are tasked to provide connectivity and related services to other vehicles. They utilize their long-range radios to relay local data network traffic to stationary gateways and to provide a location management service to vehicles in their vicinity by exchanging location information with other location servers.

For location servers to advertise location service and receive updates and queries from vehicles, we propose a service discovery mechanism to find routes to location servers in the area with the best matching mobility pattern. We shall evaluate the effect of this service discovery method with different scenarios of urban and highway mobility.

3. Mobility-aware service selection and packet relay

Vehicular mobility patterns (urban or highway) generally follow roadways with random change of directions at intersections. We assume that every vehicle responsibly sends its location information to a location server. This location information can be used by other vehicles or service providers to present location-based services. When

a vehicle and its location server move away from each other and the distance grows more than a certain hop distance, path delay and high link breakage probability make their interactions ineffective. Therefore the client has to hand off from the old server to another server that is better in terms of delay, robustness and lifetime. Every hand-off between two location servers is comprised of several 'server-to-server' and 'server-to-client' signaling interactions. However, server-to-server interactions are more expensive because they use licensed spectrum to communicate.

Based on expectations and assumptions in the architecture of Figure 1, if we want to use the field theory approach explained before, clients should send their location management packets toward other relays or servers in the vicinity, which have the highest potential. Signaling for a location update comprises of a primary phase of registration between client and server. After the registration, the client is able to update the server periodically or when triggered by specific events. If a client is unable to send updates to the designated server, a new registration with an available server is required. Based on our proposed architecture and location management procedure, the field theoretic method reviewed in Section 1.1.1 has the following deficiencies that should be addressed:

- The measure for distance between nodes is unrealistic, since mobility pattern is not considered in server selection. In our case the relative speed between a client and its server defines the connectivity lifetime and we prefer to choose a server that has a higher connectivity lifetime as long as the path delay is less than a certain limit.

- The server selection is stateless. Service discovery would lead to a set of choices for each relay to forward the packet. However there is no guarantee that a packet will be relayed to the same server to which the former packet is sent. It is desirable for a client to send location updates to a server that it has already registered in. It means that if a client selects a server with highest potential as its location server, all the relay nodes should be notified to relay the packet from that client toward the same server. Consequently a server hand-off does not happen unless the delay threshold is exceeded or disconnection occurs.

By modifying the service discovery method proposed in [4], we propose a location management method that minimizes hand-offs, which is applicable to geographical and topological location management.

3.1. Reliability versus distance

Hop distance is a simple and effective criterion for route selection, but in cases with high mobility this measure is very unstable. To avoid this problem we propose to use link stability and usability (also known as reliability) as the route selection criteria instead. Denote the set of links in the chosen path between s and d as $P(s, d)$. We want to account for reliability of each link $l \in P(s, d)$ and choose a path with highest aggregated reliability. Reliability of a link is directly related to the estimated link lifetime. However calculation of reliability includes error and an unmeasured factor of future alternative connections. For instance, a weak link could be replaced in the future by a new relay node that is not present at the moment. Due to this factor, it is not rational to underrate a path by considering the reliability of the weakest link in path as decision factor. On the other hand, we cannot rely on arithmetic average because strong links in the path would cause overestimation of path reliability.

Since we want the factor to tend toward the most realistic reliability value of the path (to mitigate the impact of links with excellent reliability in calculation of total path reliability), we desire to have the smallest average value as the measuring factor. Instead of using arithmetic mean, we use harmonic mean to calculate the reliability factor. Harmonic means tend toward the smallest values compared to geometric and arithmetic means. Hence, with r_l being reliability of link l and the number of hops $|P(s, d)| = n$ we have:

$$\frac{1}{\frac{1}{n} \sum_{l \in P(s,d))} \frac{1}{r_l}} \leq \sqrt[n]{\prod_{l \in P(s,d)} r_l} \leq \frac{1}{n} \sum_{l \in P(s,d)} r_l .$$

To apply the value of reliability in routing decision, we define the distance factor in (1) as:

$$D(s, d) = \frac{1}{n} \sum_{l \in P(s,d)} \frac{1}{\text{reliability}(l)} . \qquad (2)$$

In every calculation period, each node will predict the locations of current neighbors and based on estimated path-loss exponent of environment and foreseen mobility patterns, calculate the link lifetime. Path prediction and lifetime estimation are two major ongoing research topics and they are explained more in Section 3.2. Intuitively, a longer link lifetime leads to a more reliable path between nodes and location servers. Notwithstanding, due to high error rates in prediction mechanisms [19], we cannot rely solely on measures of one link. Therefore we will define *reliability factor* for all (*node*, *server*) pairs to show how reliable the *node* could be to relay packets toward the *server*. Based on the assumption that a higher node density can make the route more reliable, we define the *reliability factor* as the probability of a packet being successfully relayed by a *node* to another which is closer to *server*.

3.2. Reliability measurement

We define the link reliability between two neighbors as the estimated expected remaining time of connectivity between the node pair. To calculate the link reliability we assume that each node will listen to data packets and beacons sent by its neighbors. Using sampled signal characteristics and location information, the receiver predicts how link condition will change in the future. Therefore, to calculate reliability factors, nodes need to have knowledge about future variations in link connectivity. In [20] a method for estimating link residual time and link stability has been proposed. In this method, after denoising and classification of the radio signal strength indications (RSSI) from neighbors, future lifetime is estimated. In [19], Euclidean distance information is utilized to estimate future trajectory. It seems that by using relative mobility between nodes and digital map information together, future estimations can become more precise [21]. A method to calculate the probability of turns in road intersections is proposed in [22], based on the theory that turning options that lead to more destinations in shorter times are more popular than those that lead to local areas or take more time to reach destinations. In [23], mobility behavior of nodes is used to classify their transportation mode. Moreover, using particle filter method they estimate parameters in a Bayesian network for path selection decisions. By learning these parameters, they try to estimate future velocities and turning selections.

Figure 2 shows an example of vehicular mobilities in two time steps. The variations in links caused by their relative speeds are visible. In this example, nodes **5** and **6** will get into the range of **S** while **1**, **3** and **4** move out of range. Therefore, the reliability of a connection with **5** from **S** is less than that with **k** or **2**.

To calculate the link reliability in a path toward a server, every node will calculate the cumulative probability of connectivity to next hop for each server. The next hop is defined as any node in communication range that has a loop-free path to the server. In practice every node can put a short hash value of its unique address in forwarded advertisement to avoid considering routes which

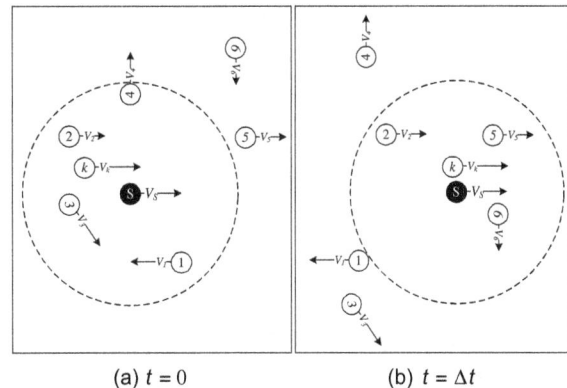

(a) $t = 0$ (b) $t = \Delta t$

Figure 2. Variation in neighbors regarding their relative speed.

are originated from itself. Hashing can reduce the length of a string to a few bits and yet avoid duplicate indexes with a high probability.

We assume that a link connectivity prediction method can provide a process consisting of connectivity probabilities during a prediction period. Suppose for a specific environment and mobility pattern, a prediction method is able to predict future motions and channel states for n time units. Moreover, we can extract the average percentage of errors in lifetime prediction (which can be empirically found) as $E(\hat{e})$. This error extraction could be enforced to the system as an *a priori* known measure or it could be readjusted based on observations from the past (by comparing prediction and real condition after it happened). Therefore $P_{l(i)}(t)$ is the probability of link $l(i)$ being connected from t_0 (now) until $t_0 + t$ and is equal to:

$$P_{l(i)}(t) = (1 - E(\hat{e}))P(\hat{L}_i, t), \quad (3)$$

where $P(\hat{L}_i, t)$ represents the link condition (alive/dead) which is calculated using the desired mobility prediction method. As mentioned before, each mobility prediction and link classification method has a distinct estimation capability in different environments. Therefore values of $E(\hat{e})$ and $P(\hat{L}_i, t)$ are highly dependent on the method of prediction being used. We will evaluate our method in Section 4 based on a simple linear prediction, but as long as a prediction method yields a prediction of link lifetime and an estimate of the error, it can be integrated into our approach.

Having estimates of the link lifetimes for all links in the path, the probability of having an undisrupted path from node k toward the server S for the next t time units (complement of the probability of no link being capable of relaying packets from k to S) is:

$$C_S^k(t) = 1 - \prod_{l(i) \in H^k(S)} (1 - P_{l(i)}(t)), \quad (4)$$

where $H^k(S)$ is the set of links between node k and its neighbors that have a loop-free path to server S. We use the cumulative distribution of $C^k(t)$ (for $t = 0 \ldots t_{max}$ with t_{max} equal to the maximum duration of predictability) as a factor which shows how reliable the node k is to pass the packets toward server S:

$$\text{reliability}(k) = \text{cdf}_{t_0}^{t_{max}}(C_S^k) = \sum_{t=0}^{t_{max}} C_S^k(t_0 + t). \quad (5)$$

We need to extract the reliability of a node for all servers being discovered. To avoid extra calculations, we define a maximum hop threshold for acceptable potentials received from neighbors. Intuitively it is obvious that information regarding far-away servers are not of any interest because of extra relay overhead and delay.

Finally we define the distance between client c and server S as:

$$D(c, S) = \sum_{k \in P(c,S)} \frac{1}{\text{cdf}_{t_0}^{t_{max}}(C_S^k)}. \quad (6)$$

Notice that as the hop number increases, the distance is affected and the chance of being chosen as best path decreases. This definition of distance would result in such a way that nomadic mobility patterns lead to higher potentials and connection between vehicles with opposite directions and/or sparse connectivity conditions causes less potential dispersion.

3.3. Potential assignment for path construction and server selection

Using (2) as the distance measure for (1), every node can receive a potential from servers based on the relative mobility and link condition of all nodes in the path from that server to the node. As described in Algorithm 1 every node advertises all valid server information received from neighbors to adjacent nodes. After receiving these advertisements, the node sets the current potential received from a server to the highest received value. These values are valid up to a certain time after the last advertisement. Whenever the node wants to send or relay a packet toward a server, it will choose the server with the highest potential. This policy leads to selection of a server that has the best mobility correlation with the transmitter.

3.4. Location update

To choose a new server for location updates, each node will select the accessible server with the highest potential. After choosing the best server, location updates are sent toward it using the neighbor who has the largest potential from that particular server. Using this approach instead of [4] we can make sure that location update packets will not face misroute to other servers which are not moving in favorable directions. Decision to hand over to another

Algorithm 1. Potential assignment.

1	Input *servers_advertisements*[ ]
2	for each *servers_info* in *servers_advertisements*
3	$L \leftarrow$ *servers_info.source*
4	predict_link_condition(L)
5	*servers*[L] $\leftarrow$ *servers_info*
6	for each S in *servers_info*
7	*rel_factor* $\leftarrow$ reliability (S)
8	$D(S.id) = \frac{S.\text{pot.original}}{\frac{S.\text{pot.original}}{S.\text{pot.received}} + \frac{1}{rel\,factor}}$
9	pot[$S.id$] $\leftarrow$ max $\left(\text{pot}[S.id], \frac{S.\text{pot.original}}{D(S.id)}\right)$
10	*next hop*[$S.id$] $\leftarrow$ *arg* max$_{l \in \text{neighbors}} l.servers[S.id].pot$
11	end for
12	end for

server is performed by a client when the hop distance to the current server has exceeded a certain threshold. Since it is assumed that location update messages are not in a high priority class and their packet size is reasonably small, packet relay in short-range wireless would still be more favorable compared to expensive long-range network. Anyhow, decision for when to hand off is still open to users and they can choose between prompt updates and lower cost. We will discuss about this trade-off in the next section. Packet relay to a chosen server is also done very easily by comparing received potential of the server from current alive links and therefore source routing is not necessary.

3.5. Location query

To find the location of another user, a requester would send a location query to the best available server at the moment (the one with the highest available potential). The packet relay mechanism would be similar to that for location updates. After receiving the query, the server looks up in its local database to see if it has up-to-date location entry. Otherwise it will send a query to its neighbors using long-range wireless. Since we assume that long-range wireless links form a connected graph topology, queries will have answer from one of the servers and this answer will reach the original server.

4. Performance evaluations

To evaluate performance of our proposed framework, we have modeled the system using the ns-2 network simulator [24]. We have added a new service discovery agent over the currently implemented network stack and added our logic as an application agent. Using application agent, we can use any routing algorithm for packet routing. We have tried our protocol on several test scenarios. These scenarios are based on realistic vehicular traffic generated by the SUMO vehicular traffic generator [25]. This microscopic vehicle traffic generator is able to create mobility patterns based on defined traffic flows, using street maps that are imported into SUMO to generate different test cases. The traces generated by SUMO yield a mobility log for vehicle movements based on road and traffic regulations. We have imported several maps with different key features for evaluations. Transmission range is set to 100 m with 95% confidence interval. This means that a transmission between two cars 100 m away would be successful with 95% probability. For short-range communication we use IEEE 802.11 with the minimum data rate (1 Mbps). We use log distance shadowing with parameters set to values provided in [27].

The first imported map is a 10 km long highway with two lanes in each direction. Two kinds of vehicles have been considered to commute on the road: private vehicles with short-range radios and public transit vehicles

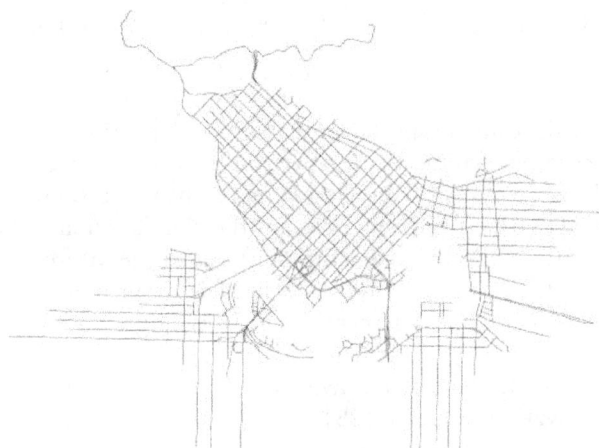

Figure 3. Extracted map of Vancouver downtown from OpenStreetMap [26].

equipped with long-range and short-range radios. The two categories of vehicles have different characteristics in speed limit, acceleration and deceleration. The second scenario is a realistic urban area extracted from actual street maps. These maps are extracted from free maps available in OpenStreetMap [26]. Figure 3 depicts an extracted map of the downtown area in Vancouver, BC. Obviously the most significant property of a downtown area is its high number of turning options and signaled intersections. After adding traffic lights to the map, we use SUMO to generate mobility traces for 10000 s. The procedure of map extraction and simulation is shown in Figure 4. After generation of mobility traces, they are fed into ns-2 as mobility scenario and simulation is performed by ns-2. Since we need prediction in our method and it is not performed in ns-2, we do the simulation twice; the first run is done to extract exact location of every vehicle during the simulation. Then we use these data in the next run as a precise prediction of future mobility patterns in network. To make the prediction more realistic, we add noise to location information. Since prediction precision is strongly dependent on prediction mechanism, we use one of the simplest predictions: in every second, each node predicts $x(t + 1) = v + x(t)$, where $x(t)$ is the location of node in time t ($0 \leq t \leq t_{max}$). The term t_{max} is the maximum time that prediction can be reasonably valid. We will set t_{max} to a value which:

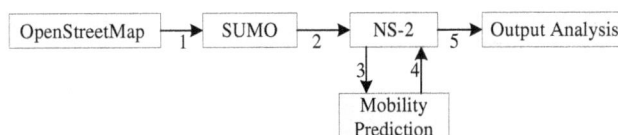

Figure 4. Simulation procedure.

$$E(Pr(C(t))Pr\left(\hat{C}(t)\right) + Pr(\overline{C(t)})Pr(\overline{\hat{C}(t)})) > \text{threshold},$$
$$1 < t < t_{max}. \qquad (7)$$

This is the sum of expected probability for having a true guess, whether positive or negative, on having a connection. This probability should be more than a certain threshold. To find this value we run the simulation and calculate the predicted and actual locations in the future. We consider a link as active if its RSSI is more than a threshold. Since measurement of RSSI is impacted by environmental clutters, it is impractical to deterministically define the link connectivity threshold. So we use the propagation model in [28]:

$$P_r(d) = P_t - PL(d) = P_t - (\overline{PL}(d) + X_\delta), \qquad (8)$$

where $\overline{PL}(d)$ is the log-distance path loss from the transmitter to receiver and X_δ is a zero-mean Gaussian distributed random shadowing effect with standard deviation δ. Values of path-loss exponent and δ are usually extracted from empirical data. We have borrowed these values from the experiment done by Otto *et al.* in [27]. Finally, the probability of RSSI being more than γ (dBm) in distance $d(m)$ is:

$$Pr[P_r(d) > \gamma(\text{dBm})] = Q(\frac{\gamma - \overline{P_r(d)}}{\sigma}). \qquad (9)$$

Figure 5 shows the estimated error in aforementioned prediction method. Results show that in the highway scenario prediction is close to reality and the connection condition after 40 s is predicted correctly with a 70% probability. However, in the suburban scenarios and downtown areas, nondeterministic stops and turning probabilities cause prediction errors to grow. For downtown scenario we find that predictions are 50% successful only for 20 s ahead. In suburban areas with less stops and turns compared to downtown, it is up to 35 s. We apply

Table 1. Variables definition in (10).

$d(i,S)$	Hop distance between i and server S_i
f_u	Frequency of location update messages
f_q	Frequency of location query messages
LU	Size of location update message
LQ	Size of location query message
LR	Size of location reply message
K	Usage cost/Kb for long-range network
N	Number of location servers
SYN	Size of synchronization message

these errors in calculating path reliability factor for each scenario.

To avoid excessive delay caused by late hand-offs we have to set a threshold for maximum allowable hop distance between nodes and servers. The trade-off is between location update cost (which is related to the amount of relayed data and type of media used for it) and end-to-end delay.

$$X = \begin{cases} d(i,S)\left(f_u.LU + f_q(LQ + LR)\right) & d(i,S) < \text{thr} \\ K*(N-1)*SYN+ \\ d(i, S_{\text{new}})\left(f_u.LU + f_q(LQ + LR)\right) & \text{otherwise.} \end{cases}$$
$$(10)$$

To calculate the proper threshold, we set our objective to minimize Cost * Delay for location update packets. Using (10) as the cost function and by knowing $d(i, S)$, the distance between a node and its second best server with eligible hop distance, every node can calculate the threshold as follows (for definition of variables in (10) see Table 1):

$$\text{thr} = \sqrt{d^2(i, S') + \frac{K*(N-1)*SYN}{f_u.LU + f_q(LQ + LR)}.d(i, S')}. \quad (11)$$

Here we assumed that delay is only based on hop distance and did not consider the delay caused by collisions.

After finding the estimation errors, we run the algorithm based on these estimation properties. We try to establish connections between servers nodes and regular nodes using the short-range wireless network.

We compare client–server path length and traffic cost with three other methods: [4] (shortest path is the metric for route selection), [10] (affinity based on signal-to-noise ratio) and [13] (quorum-based method with column-based advertisement and row-based query). Figure 6 shows the average lifetime of connections between mobile nodes and location servers in downtown mobility pattern. Hereafter we shall refer to our method as Life Time based method (LT). SP-1 represents shortest path anycasting based on [4]. In SP-2 we use the same method as SP-1 but whenever a server is selected for a node as a location

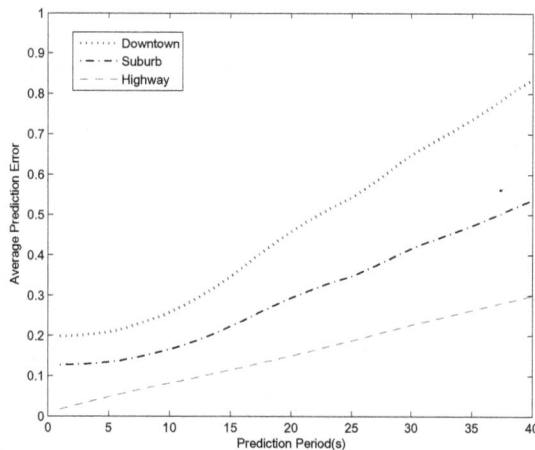

Figure 5. Linear prediction average error for three scenarios.

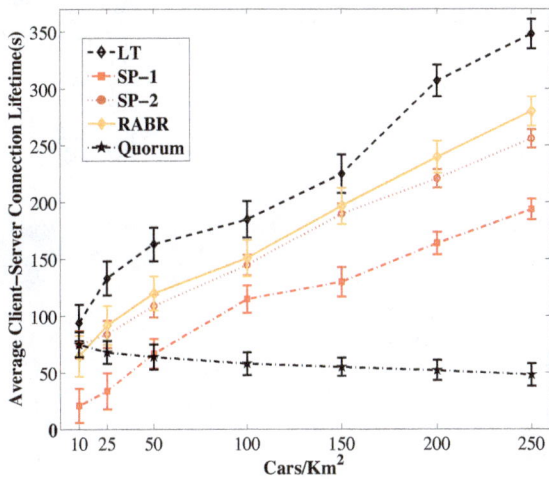

Figure 6. Average client–server connection lifetime (downtown).

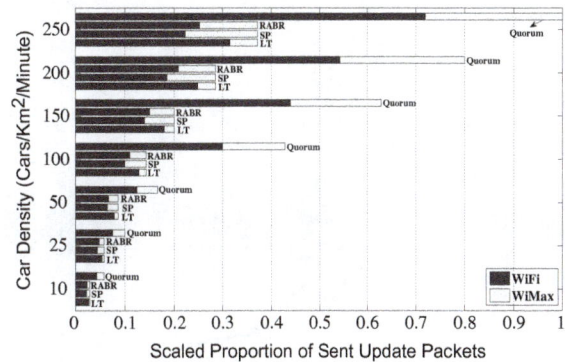

Figure 8. Normalized overhead of update packets being sent over WLAN and WiMax ($K = 100$).

server, it will remain chosen as long as the distance is less than the maximum hop distance. Results show that using lifetime as the distance metric results in significant connection lifetime improvements specially for higher node densities.

In affinity-based method, SNR is considered as the measuring factor for decision making. Therefore for downtown areas with highly volatile SNR conditions RABR cannot perform much better than SP-2. Since in quorum-based method, chosen servers are changed rapidly after any change in topology, connection lifetimes are not comparable to other methods.

Figure 7 shows the same measures as Figure 6 but for highway scenarios. Results show a 57% overall improvement in connection lifetime compared to SP-2. Especially in lower vehicle densities, our proposed method achieves more improvements compared to the shortest path method because of the steady mobility of vehicles which

leads to higher lifetimes if paths are selected from vehicles moving along the same direction. RABR performs better in highway scenarios duo to less perturbations in SNR. However, the quorum-based method exhibits the same behavior as in the downtown scenario. Figure 8 compares the overhead caused by location update packets. The quorum-based method uses several location servers, hence location updates become costly. Moreover, as mobility and interactions between nodes increase, the overhead of quorum-based method increases drastically. We have assumed that the quorum-based method uses WLAN and WiMax based on availability with no preferences. To compare the proportion of WLAN usage, we assumed that parameter K in (10) is equal to 100 (every transmission on WiMax is 100 time more expensive than WLAN). We can see that usage of WiMax in low traffic densities is significantly low and as mobility patterns grow more dynamic, the difference between LT and other three methods becomes noticeable.

Figure 9 depicts the normalized total cost of location management for different values of K. Since the cost of RABR is very close to that of SP and the cost of the

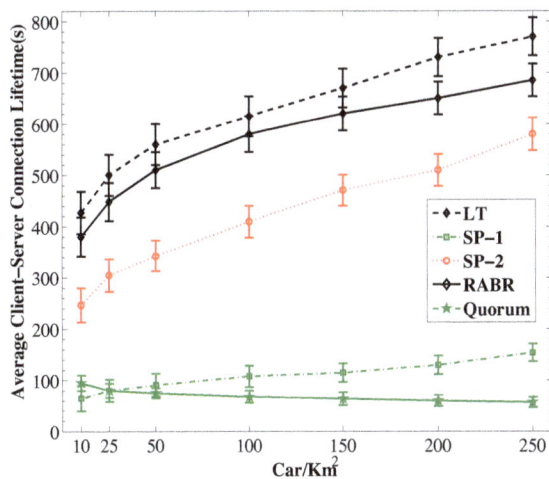

Figure 7. Average client–server connection lifetime (highway).

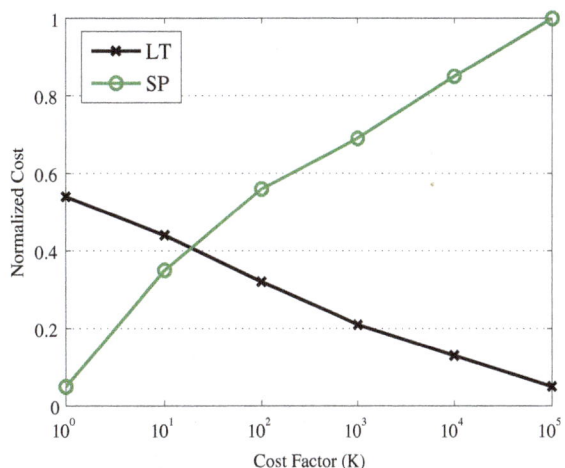

Figure 9. Signaling cost based on cost factor (K) in (10).

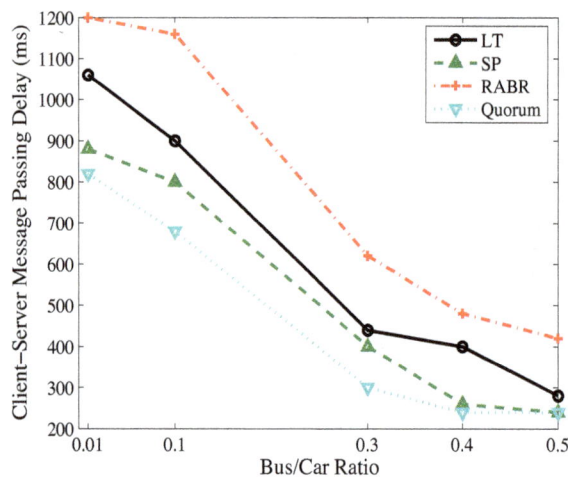

Figure 10. Average client–server message passing delay in downtown (only query/response messages are considered for quorum-based method).

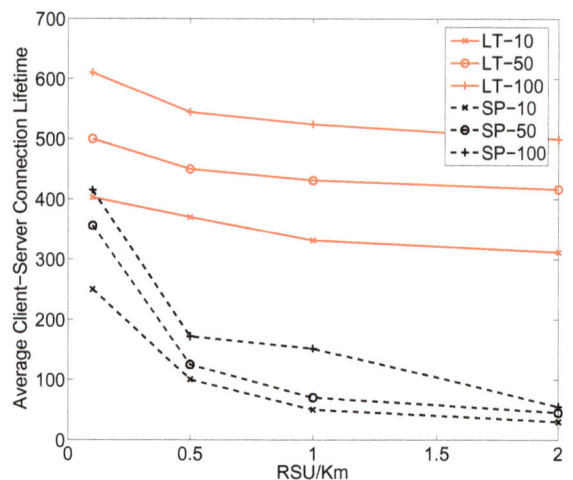

Figure 11. Average client–server connection lifetime with hybrid network in highway.

quorum-based method is significantly higher than those of other methods, we have only compared the cost of SP vs. LT. As one can observe, for $K \geq 100$ our method outperforms the shortest path method. K can be interpreted as a priority or preference parameter and could be tuned based on the trade-off between delay and cost efficiency. If providers prefer faster and more precise location updates, they can decrease K. In contrast, for better efficiencies (e.g. for less location-sensitive applications) higher values will lead to better spectrum conservation.

Figure 10 shows the average delay experienced by signaling packets (Location Update, Location Query and Responses) for the downtown scenario. It is important to note that the quorum-based method uses more than one location management server and always every immediate neighbor in the same row of the node is acting as a location server. As a result signaling delays for location updates and responses are very low. However, when it comes to location query, signaling delay is relatively higher than updates. Our method is always performing better than RABR in terms of delays but compared to SP, it suffers a 20% increase in delay. As the number of location servers (in this case described as buses) increases, the overall signaling delay for all methods decreases. We can see that if half of the vehicles in our system could act as location servers, delay would have become as low as SP and quorum-based methods.

4.1. Hybrid network scenario

Now we are going to consider a scenario in which infrastructure is available in some positions and road-side units (RSUs) are placed in some spots to provide services of interest. We evaluate the highway scenario with this configuration. To see the effect of including the wired network infrastructure, we place RSUs with specific distances between them. RSUs are considered to provide location service to mobile nodes. The selection of location servers will be identical to fully *ad hoc* mode. However location prediction for RSUs is accurate and estimation error is set to zero. We compare the effect of different RSU densities on our lifetime-based method and shortest path method.

In Figure 11 a comparison of average connection lifetime is shown. We have compared the average connection lifetime for three different vehicle densities. Clearly, when the density increases the connection lifetime increases as well. This means in denser traffic conditions the capability to relay the packets between servers and clients helps to increase scalability. Similar to *ad hoc* only network, the connection lifetime of the shortest path method is less than that of the LT method. As the number of RSUs increases, the connection lifetime starts to decrease. In shortest path method, this change is more visible because server selection is based on physical distance. This means when vehicles get close to RSUs, they prefer to change their connection to closer RSU. In contrast, connection lifetime is not dramatically changed in LT method.

Figure 12 compares our LT method with shortest path method in terms of network usage. When the number of utilized RSUs increases, not only utilization of wired network increases, but also usage of WiMax. The reason is that every synchronization caused by selecting a RSU as location server leads to extra synchronizations with WiMax servers. This comparison shows that even if we use RSUs to collaborate with other mobile servers, still there is a chance that selection of RSU as server would result in higher overheads. However, this is the price to pay in order to gain a better reliability. RSUs are essential when the network density is low and connectivity is weak.

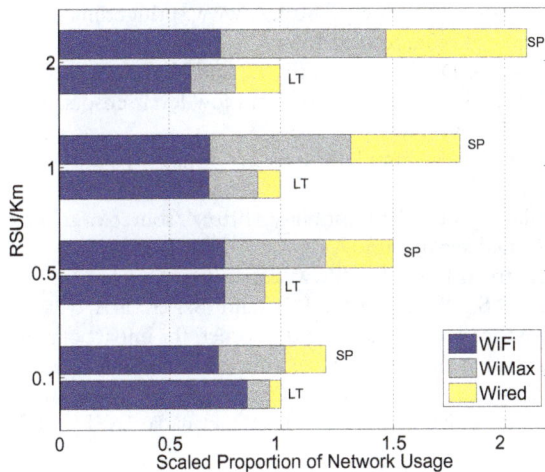

Figure 12. Scaled proportion of network usage.

5. Conclusion

Location management is a critical function in VANETs. In this paper we have assumed that some of the mobile nodes in a VANET are equipped with heterogeneous wireless connectivity. These vehicles are able to act as location servers for other vehicles and cooperate using their long-range radios. We have proposed a new server selection and packet relay mechanism that minimizes the rate of server hand-offs by relaying location update packets toward the server that has the lowest possibility of disconnection. This is done by proposing a new definition of distance. The proposed method has been evaluated by extensive computer simulations. The results show significant improvements in client–server connection lifetimes. Higher connection lifetimes lower the costs of server hand-offs, which require the use of long-range communications to update the record for the client at all the servers. We have provided a tuning factor which can be used for decision making based on tolerable delay and cost. The comparison has been made against three methods: associativity-based routing, shortest path selection and quorum-based location management. Results show that in scenarios with high mobility our method achieves the lowest costs and acceptable delays compared to other three methods. We have also compared our method to the shortest path method in a hybrid network scenario.

Acknowledgements. This work was supported by a grant from AUTO21 under the Networks of Centres of Excellence (NCE) Program of Canada.

References

[1] BALDESSARI, R., BÖDEKKER, B., BRAKEMEIER, A., DEEGENER, M., FESTAG, A., FRANZ, W., HILLER, A. *et al.* (2007) *C2C-CC Manifesto.* Technical Report, Car 2 Car Communication Consortium, http://www.car-to-car.org/fileadmin/downloads/C2C-CC_manifesto_v1.1.pdf.

[2] WILLKE, T., TIENTRAKOOL, P. and MAXEMCHUK, N. (2009) A survey of inter-vehicle communication protocols and their applications. *IEEE Commun. Surv. Tut.* **11**(2): 3–20.

[3] NIDD, M. (2002) Service discovery in DEAPspace. *IEEE Pers. Commun.* **8**(4): 39–45.

[4] LENDERS, V., MAY, M. and PLATTNER, B. (2005) Service discovery in mobile ad hoc networks: a field theoretic approach. *Pervasive Mob. Comput.* **1**(3): 343–370.

[5] KOZAT, U. and TASSIULAS, L. (2003) Network layer support for service discovery in mobile ad hoc networks. In *INFOCOM 2003. Twenty-Second Annual Joint Conference of the IEEE Computer and Communications. IEEE Societies* (IEEE), **3**: 1965–1975.

[6] TYAN, J. and MAHMOUD, Q. (2004) A network layer based architecture for service discovery in mobile ad hoc networks. In *Electrical and Computer Engineering, 2004. Canadian Conference* (IEEE), **3**:1379–1384.

[7] MIAN, A., BALDONI, R. and BERALDI, R. (2009) A survey of service discovery protocols in multihop mobile ad hoc networks. *IEEE Pervasive Comput.* **8**: 66–74.

[8] HINDEN, R. and DEERING, S. (1998) Ip version 6 addressing architecture, rfc 2373, http://ietfreport.isoc.org/rfc/rfc2373.txt.

[9] TOH, C. (1997) Associativity-based routing for ad hoc mobile networks. *Wireless Pers. Commun.* **4**(2): 103–139.

[10] AGARWAL, S., AHUJA, A., SINGH, J. and SHOREY, R. (2000) Route-lifetime assessment based routing (RABR) protocol for mobile ad-hoc networks. In *IEEE International Conference on Communications* (Citeseer) **3**: 1697–1701.

[11] FRIEDMAN, R. and KLIOT, G. (2006) *Location Services in Wireless Ad Hoc and Hybrid Networks: A Survey.* Technical Report, Department of Computer Science, Technion.

[12] BASAGNI, S., CHLAMTAC, I., SYROTIUK, V. and WOODWARD, B. (1998) A distance routing effect algorithm for mobility (DREAM). In *Proceedings of the 4th Annual ACM/IEEE International Conference on Mobile Computing and Networking* (ACM), 76–84.

[13] LIU, D., STOJMENOVIC, I. and JIA, X. (2006) A scalable quorum based location service in ad hoc and sensor networks. In *2006 IEEE International Conference on Mobile Ad hoc and Sensor Systems (MASS)* (IEEE), 489–492.

[14] LI, J., JANNOTTI, J., DE COUTO, D.S., KARGER, D.R. and MORRIS, R. (2000) Scalable location service for geographic ad hoc routing. In *Proceedings of the Annual International Conference on Mobile Computing and Networking (MOBICOM)* (ACM), 120–130.

[15] SALEET, H., LANGAR, R., BASIR, O. and BOUTABA, R. (2008) Proposal and analysis of region-based location service management protocol for VANETs. In *IEEE Global Telecommunications Conference (GLOBECOM)* (IEEE), 491–496.

[16] DIKAIAKOS, M.D., FLORIDES, A., NADEEM, T. and IFTODE, L. (2007) Location-aware services over vehicular ad-hoc networks using car-to-car communication. *IEEE J. Sel. Areas Commun.* **25**(8): 1590–1602.

[17] KIEß, W., FÜßLER, H., WIDMER, J. and MAUVE, M. (2004) Hierarchical location service for mobile adhoc networks. *ACM SIGMOBILE Mob. Comput. Commun. Rev. (MC2R)* **8**(4): 47–58.

[18] AHMED, S., KARMAKAR, G. and KAMRUZZAMAN, J. (2009) Hierarchical adaptive location service protocol for mobile ad hoc network. In *Proceedings of the 2009 IEEE Conference on Wireless Communications & Networking Conference* (Institute of Electrical and Electronics Engineers Inc.), 2932–2937.

[19] HAAS, Z.J. and HUA, E.Y. (2008) Residual link lifetime prediction with limited information input in mobile ad hoc networks. In *Proceedings of IEEE INFOCOM* (IEEE), 26–30.

[20] SOFRA, N. and LEUNG, K.K. (2009) Link classification and residual time estimation through adaptive modeling for vanets. In *IEEE VTC* (IEEE), 1–5.

[21] MENOUARAND, H., LENARDI, M. and FILALI, F. (2007) Movement prediction-based routing (MOPR) concept for position-based routing in vehicular networks. In *IEEE Vehicular Technology Conference* (IEEE), 2101–2105.

[22] KRUMM, J. (2009) Where will they turn: predicting turn proportions at intersections. *Personal and Ubiquitous Computing*, Springer, http://www.springerlink.com/content/n41v0m2q08g84554/.

[23] PATTERSON, D., LIAO, L., FOX, D. and KAUTZ, H. (2003) Inferring high-level behavior from low-level sensors. *Lect. Notes Comput. Sci.* **2864:** 73–89.

[24] The network simulator ns-2, http://www.isi.edu/nsnam/ns/.

[25] Simulation of urban mobility, http://sourceforge.net/apps/mediawiki/sumo/.

[26] OpenStreetMap, http://www.openstreetmap.org/.

[27] OTTO, J.S., BUSTAMANTE, F.E. and BERRY, R.A. (2009) Down the block and around the corner: the impact of radio propagation on inter-vehicle wireless communication. In *Proceedings—International Conference on Distributed Computing Systems* (Montreal, QC, Canada: IEEE), 605–614.

[28] RAPPAPORT, T. (2001) *Wireless Communications: Principles and Practice* (Upper Saddle River, NJ, USA: Prentice Hall PTR).

7

A Sensing Error Aware MAC Protocol for Cognitive Radio Networks*

Donglin Hu   and   Shiwen Mao*

Department of Electrical and Computer Engineering, Auburn University, Auburn, AL 36849-5201, USA

Abstract

Cognitive radios (CR) are intelligent radio devices that can sense the radio environment and adapt to changes in the radio environment. Spectrum sensing and spectrum access are the two key CR functions. In this paper, we present a spectrum sensing error aware MAC protocol for a CR network collocated with multiple primary networks. We explicitly consider both types of sensing errors in the CR MAC design, since such errors are inevitable for practical spectrum sensors and more importantly, such errors could have significant impact on the performance of the CR MAC protocol. Two spectrum sensing polices are presented, with which secondary users collaboratively sense the licensed channels. The sensing policies are then incorporated into p-Persistent CSMA to coordinate opportunistic spectrum access for CR network users. We present an analysis of the interference and throughput performance of the proposed CR MAC, and find the analysis highly accurate in our simulation studies. The proposed sensing error aware CR MAC protocol outperforms two existing approaches with considerable margins in our simulations, which justify the importance of considering spectrum sensing errors in CR MAC design.

Keywords: Cognitive radio, cross-layer design and optimization, dynamic spectrum access, medium access control, software defined radio, spectrum sensing.

1. Introduction

A *cognitive radio* (CR) is a frequency-agile wireless communication device with a monitoring interface and intelligent decision-making that enables dynamic spectrum access [2]. A CR can sense the radio environment and adapt to changes in the radio environment. The CR concept represents a significant paradigm change in spectrum regulation and utilization, i.e., from exclusive use of spectrum by licensed users (or, *primary users*) to dynamic spectrum access for unlicensed users (or, *secondary users*). The high potential of CRs has attracted considerable efforts from the wireless community recently, for developing more efficient spectrum management policies and techniques [2, 3].

Although the basic concept of CR is intuitive, it is challenging to design efficient cognitive network protocols to fully capitalize CR's potential. In order to exploit transmission opportunities in licensed bands, the tension between primary user protection and secondary user spectrum access should be judiciously balanced. Spectrum sensing and spectrum access are the two key CR functions. Important design factors include (i) how to identify transmission opportunities, (ii) how secondary users determine, among the licensed channels, which channel(s) and when to access for data transmission, and (iii) how to avoid harmful interference to primary users under the omnipresent of spectrum (or, channel) sensing errors. These are the problems that should be addressed in the medium access control (MAC) protocol design for CR networks. Although very good understandings on the availability process of licensed channels have been gained recently [4, 5], there is still a critical need to develop analytical models that take channel sensing errors into account for guiding the design of CR MAC protocols.

*Part of this work was presented at IEEE GLOBECOM 2009, Honolulu, HI, USA, November/December 2009 [1].

*Corresponding author. URL: http://www.eng.auburn.edu/~szm0001/, Email: smao@ieee.org.

In this paper, we present a channel sensing error aware MAC protocol for a CR network collocated with multiple primary networks. We assume primary users access the licensed channels following a synchronous time slot structure [2, 6]. The channel states are independent to each other and each evolves over time following a discrete-time Markov process [2, 4]. Secondary users use their software-defined radio (SDR)-based transceivers to tune to any of the licensed channels, to sense and estimate channel status and to access the channels when they are found (or, believed) to be available. We explicitly consider channel sensing errors in the design of the CR MAC protocol. It has been shown in prior work that generally there are two types of channel sensing errors: (i) *false alarm*, when an idle channel is identified as busy, thus a spectrum opportunity will be wasted, (ii) *miss detection*, when a busy channel is identified as idle, thus leading to collision with primary users, since CR users will attempt to use such "idle" channels. We consider both types of spectrum sensing errors in our CR MAC design, which have been shown to be unavoidable for practical spectrum sensors [2].

In particular, we develop two channel sensing polices, with which secondary users collaboratively sense the licensed channels and predict channel states. With the *memoryless sensing* policy, each secondary user chooses one of the M licensed channels to sense with equal probability. During the sensing phase, secondary users also exchange sensing results through a separate control channel. This sensing policy is further improved with a mechanism to spread out secondary users to sense different channels, therefore reducing the chance that a channel is not sensed by any of the users. When spreading out secondary users to the channels, the mechanism also considers the autocorrelation of channel processes to obtain more accurate sensing results. This is termed *improved sensing* policy.

These two sensing polices are then incorporated into the *p*-Persistent Carrier Sense Multiple Access (CSMA) mechanism to make sensing error aware CR MAC protocols. We analyze the proposed CR MAC protocols with respect to the interference and throughput performance and derive closed-form expressions. Primary user protection is achieved via tunning the channel access probability p of *p*-Persistent CSMA according to the interference analysis. The CR MACs also aims to maximize the CR network throughput while satisfying the primary user protection constraints. Through simulations, we find that the analysis is highly accurate as compared to simulation results. In addition, the proposed sensing error aware CR MAC protocols outperform two existing schemes with considerable gain margins, which justify the importance of considering channel sensing errors in CR MAC design.

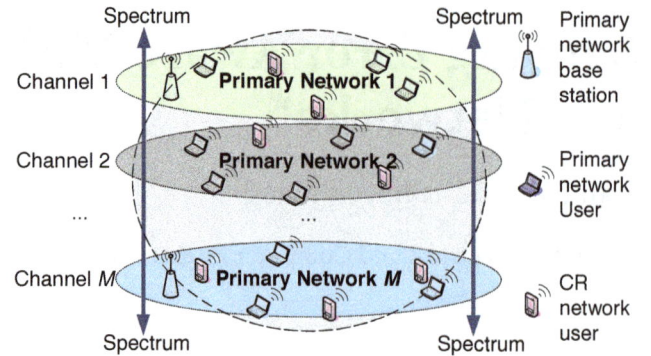

Figure 1. The CR secondary network is collocated with M primary networks.

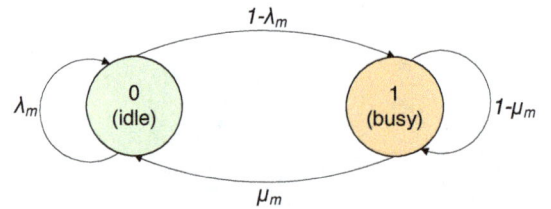

Figure 2. The discrete-time two-state Markov model for the state of channel m, S_m, for $m = 1, 2, \ldots, M$.

The remainder of this paper is organized as follows. We describe the network model and assumptions in Section 2. We then present the proposed CR MAC protocols in Section 3 and analyze their throughput and interference performance in Section 4. Our simulation studies of the proposed CR MAC protocols are presented in Section 5. Section 6 discusses related work and Section 7 concludes this paper.

2. Network Model and Assumptions

The network model considered in this paper is illustrated in Fig. 1. Consider M primary networks, each allocated with a licensed channel. We assume the primary users access the channels following a synchronous slot structure as in prior work [2, 6, 7]. The channel states are independent to each other and each of the M channels evolves over time following a discrete-time two-state Markov process, as shown in Fig. 2. Such channel model has been validated by recent measurement studies [2, 4, 6]. We define the *network state vector* in slot t as $\vec{S}(t) = [S_1(t), S_2(t), \ldots, S_M(t)]$, where $S_m(t)$ denotes the state of channel m, for $m = 1, 2, \cdots, M$. When channel m is idle, we have $S_m(t) = 0$; when channel m is busy, we have $S_m(t) = 1$.

Let λ_m and μ_m be the transition probability of remaining in state 0 and the transition probability from state 1 to 0 for channel m, respectively. Let $\eta_m = \Pr(S_m = 1)$ denote the *utilization* of channel m with respect to primary user transmissions. Let $\zeta_m = \Pr(S_m = 0)$ be the probability that channel m is idle (i.e., not being used by primary users). We then have

$$\eta_m = \lim_{T \to \infty} \frac{1}{T} \sum_{t=1}^{T} S_m(t) = \frac{1 - \lambda_m}{1 - \lambda_m + \mu_m} \qquad (1)$$

$$\zeta_m = 1 - \Pr(S_m = 1) = \frac{\mu_m}{1 - \lambda_m + \mu_m}. \qquad (2)$$

We assume a secondary network collocated with the M primary networks, within which N secondary users take advantage of the spectrum white spaces in M licensed channels for data transmissions. For protection of primary users, the probability of collision caused by secondary user transmissions to primary users should be upper bounded by a prescribed threshold γ_m, for $m = 1, 2, \cdots, M$.

As in prior work [4, 6, 8], we assume that each secondary user is equipped with two transceivers: a *control transceiver* that operates over a dedicated control channel, which we assume is always available (e.g., a channel in the industrial, scientific and medical (ISM) band), and a *data transceiver* that is used for data communications through the M licensed channels. The data transceiver consists of an SDR that can be tuned to any of the M licensed channels to transmit and receive data. Secondary users also use their transceivers for spectrum sensing and exchanging sensing results.

3. Sensing Error Aware CR MAC Protocol

For the CR network described in Section 2, we develop sensing aware MAC protocols for opportunistic spectrum access. The time slot structure of the proposed MAC protocols is shown in Fig. 3, which consists of a *sensing phase* and a *transmission phase*. The sensing phase is further divided into $\bar{K}$ mini-slots, within which each secondary user senses one of the licensed channels. CR users access the channels for data transmission during the transmission phase. Let T_s, T_{ms}, and T_{data} denote the duration of a time slot, a mini-slot, and the transmission phase, respectively (see Fig. 3), we have

$$T_s = \bar{K} \times T_{ms} + T_{data}. \qquad (3)$$

We first discuss the two key components of the proposed protocols, i.e., channel sensing and channel access, and then analyze their performance with respect to primary user protection and the expected throughput. Table 1 summarizes the notation used in this paper.

3.1. Sensing Phase

The first key element of the proposed MAC protocols is spectrum, or channel sensing. Although precise and timely channel state information is highly desirable for opportunistic spectrum access and primary user protection, contiguous full-spectrum sensing is both

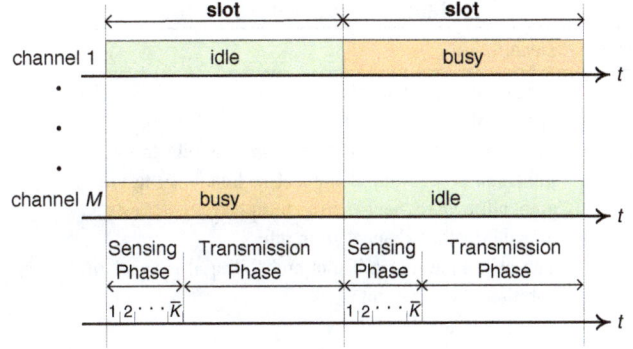

Figure 3. The time slot structure of the proposed sensing error aware CR MAC protocol.

energy inefficient and hardware demanding. Since we assume a secondary user is equipped with one transceiver for spectrum sensing, i.e., the data transceiver with SDR capability, only one of the licensed channels can be sensed by the secondary user at a time.

During the sensing phase (see Fig. 3), a secondary user picks a licensed channel and keeps on sensing it for one or multiple mini-slots. As discussed, two kinds of detection errors may occur: false alarm and miss detection. We assume all secondary users have the same probability of detection errors when sensing channel m, $m = 1, 2, \cdots, M$. Let ϵ_m and δ_m denote the probabilities of false alarm and miss detection on channel m, respectively. The spectrum sensing performance can be represented by the Receiver Operation Characteristic (ROC) curve, where $(1 - \delta_m)$ is plotted as a function of ϵ_m [2]. For a specific channel m in a certain time slot t, the sensing error probabilities can be written as:

$$\Pr(W_{m,i} = 1 \mid S_m = 0) = \epsilon_m, \text{for all } i = 1, 2, \cdots \quad (4)$$
$$\Pr(W_{m,i} = 0 \mid S_m = 1) = \delta_m, \text{for all } i = 1, 2, \cdots, (5)$$

where $W_{m,i}$ is the ith sensing result of channel m and S_m is state of channel m.

We assume that the sensing results from different users are independent and the sensing results in different mini-slots are also independent to each other. Suppose a secondary user continues to sense channel m for k mini-slots and obtains k sensing results. The conditional probability that channel m is available after the kth sensing mini-slot, denoted by $a_{m,k}$, can be

Table 1. Notation

Symbol	Definition
M	number of data channels
N	number of secondary users
λ_m	transition probability of channel m from idle to idle
μ_m	transition probability of channel m from busy to idle
η_m	probability that channel m is busy
ζ_m	probability that channel m is idle
γ_m	maximum allowable collision probability of channel m
T_s	duration of a time slot
T_{ms}	duration of a mini-slot
T_{data}	duration of the transmission phase
$a_{m,k}$	probability that channel m is idle conditioned on k sensing results
Θ_0, Θ_1	thresholds for channel decision $\Theta_0 < \Theta_1$
$\Psi_{m,k}^0$	set of 0 observations that make $a_{m,k}$ below Θ_0
$\Psi_{m,k}^1$	set of 0 observations that make $a_{m,k}$ above Θ_1
$\Psi_{m,k}^2$	set of 0 observations that make $a_{m,k}$ between Θ_0 and Θ_1
$S_m(t)$	state of channel m at time t
$W_{m,i}$	the ith sensing result on channel m
$\theta_{m,i}$	the ith observed sensing result on channel m (0 or 1)
ϵ_m	probability of false alarm on channel m
δ_m	probability of miss detection on channel m
K_m	stopping time in the sensing phase for channel m
p	transmission probability of a secondary user
P_m^{idle}	probability that no secondary user transmits on channel m
p_m^{succ}	probability that one secondary user wins channel m
p_m^{coll}	probability of collision on channel m
p^{idle}	probability that no secondary user transmits on control channel
p^{succ}	probability that one secondary user wins all the idle channels
p^{coll}	probability of collision on control channel
P_m^{intf}	probability of interference from secondary users on channel m
$\Lambda_m^1(u)$	throughput of channel m sensed by u users in Case 1
$\Lambda_m^2(u)$	throughput of channel m sensed by u users in Case 2
Ω_1	CR network throughput in Case 1
Ω_2	CR network throughput in Case 2
R_m	data rate of licensed channel m

derived as

$$
\begin{aligned}
a_{m,k} &= \Pr(S_m = 0 \mid W_{m,1} = \theta_{m,1}, \cdots, W_{m,k} = \theta_{m,k}) \\
&= \frac{\Pr(W_{m,i} = \theta_{m,i}, i = 1, \cdots, k | S_m = 0)\Pr(S_m = 0)}{\sum_{j=0}^1 \Pr(W_{m,i} = \theta_{m,i}, i = 1, \cdots, k | S_m = j)\Pr(S_m = j)} \\
&= \frac{\Pr(S_m = 0)\prod_{i=1}^k \Pr(W_{m,i} = \theta_{m,i}|S_m = 0)}{\sum_{j=0}^1 \Pr(S_m = j)\prod_{i=1}^k \Pr(W_{m,i} = \theta_{m,i}|S_m = j)} \\
&= \left[1 + \frac{\Pr(S_m = 1)}{\Pr(S_m = 0)}\prod_{i=1}^k \frac{\Pr(W_{m,i} = \theta_{m,i}|S_m = 1)}{\Pr(W_{m,i} = \theta_{m,i}|S_m = 0)}\right]^{-1} \\
&= \left[1 + \alpha_m^{d_m}\beta_m^{k-d_m}\frac{\Pr(S_m = 1)}{\Pr(S_m = 0)}\right]^{-1} \\
&\left(1 + \alpha^{d_m}\beta^{k-d_m}\frac{\eta_m}{\zeta_m}\right)^{-1}
\end{aligned}
\tag{6}
$$

where d_m is the number of observations whose sensing result is 0 on channel m, and α_m and β_m are defined as follows.

$$
\alpha_m = \frac{\Pr(W_{m,i} = 0|S_m = 1)}{\Pr(W_{m,i} = 0|S_m = 0)} = \frac{\delta_m}{1 - \epsilon_m}, \text{ for } \theta_{m,i} = 0 \tag{7}
$$

$$
\beta_m = \frac{\Pr(W_{m,i} = 1|S_m = 1)}{\Pr(W_{m,i} = 1|S_m = 0)} = \frac{1 - \delta_m}{\epsilon_m}, \text{ for } \theta_{m,i} = 1. \tag{8}
$$

For the secondary user, it is also possible that it obtains some of the k sensing results by local measurements, and receives the remaining sensing results from the control channel in the case that some other secondary users are sensing the same channel m. By abuse of notation, we also use $a_{m,k}$ to denote the conditional channel availability probability in this case, due to independence of the sensing results. We plot $a_{m,k}$ as a function of k for the channel idle and busy cases in Fig. 4, using the same parameters as one of the simulations (see Section 5). We have the following proposition for $a_{m,k}$.

Proposition 1. When channel m is idle, $a_{m,k}$ is a monotone increasing function of k; when channel m is busy, $a_{m,k}$ is a monotone decreasing function of k.

Proof. From the defintion of $a_{m,k}$ in (6), it follows that (9) holds true, where $\bar{W}_{m,i} = 1 - W_{m,i}$ and $\chi_m = \log((\frac{1}{\theta_1} - 1)\frac{\zeta_m}{\eta_m})$.

Since $\epsilon_m < 0.5$ and $\delta_m < 0.5$ for practical sensors, both $\log\left(\frac{1-\delta_m}{\epsilon_m}\right)$ and $\log\left(\frac{1-\epsilon_m}{\delta_m}\right)$ are positive. If $S_m(t) = 0$, we have that

$$
\Pr(W_{i,k+1} = 1) < \Pr(W_{i,k+1} = 0) = \Pr(\bar{W}_{i,k+1} = 1).
$$

It follows that $\Pr(a_{m,k} \geq \theta_1) < \Pr(a_{m,k+1} \geq \theta_1)$. That is, $a_{m,k}$ is a monotone increasing function of k.

Similarly, we can show that

$$
\Pr(a_{m,k} \leq \theta_0) < \Pr(a_{m,k+1} \leq \theta_0)
$$

when $S_m(t) = 1$. That is, $a_{m,k}$ is a monotone decreasing function of k when the channel is busy. $\square$

During the sensing phase, each secondary user chooses one channel to sense with equal probability at the beginning of the time slot. Secondary users also report their sensing results over the control channel, and share the corresponding channel sensing results during the mini-slots. Two threshold probabilities $\Theta_0 < \Theta_1$ are used for decision making.

- If the availability of channel m, i.e., $a_{m,k}$, is below Θ_0, the channel is believed to be busy and the secondary users will wait till the next time slot to start sensing again.

$$\Pr(a_{m,k} \geq \theta_1) = \Pr\left(\left[1 + \frac{\eta_m}{\zeta_m}\left(\frac{\delta_m}{1-\epsilon_m}\right)^{\sum_{i=1}^{k} \bar{W}_{m,i}}\left(\frac{1-\delta_m}{\epsilon_m}\right)^{\sum_{i=1}^{k} W_{m,i}}\right]^{-1} \geq \theta_1\right)$$

$$= \Pr\left(\left(\frac{\delta_m}{1-\epsilon_m}\right)^{\sum_{i=1}^{k} \bar{W}_{m,i}}\left(\frac{1-\delta_m}{\epsilon_m}\right)^{\sum_{i=1}^{k} W_{m,i}} \leq \left(\frac{1}{\theta_1}-1\right)\frac{\zeta_m}{\eta_m}\right) = \Pr\left(\sum_{i=1}^{k}\left[W_{m,i}\log\left(\frac{1-\delta_m}{\epsilon_m}\right) - \bar{W}_{m,i}\log\left(\frac{1-\epsilon_m}{\delta_m}\right)\right] \leq \chi_m\right).(9)$$

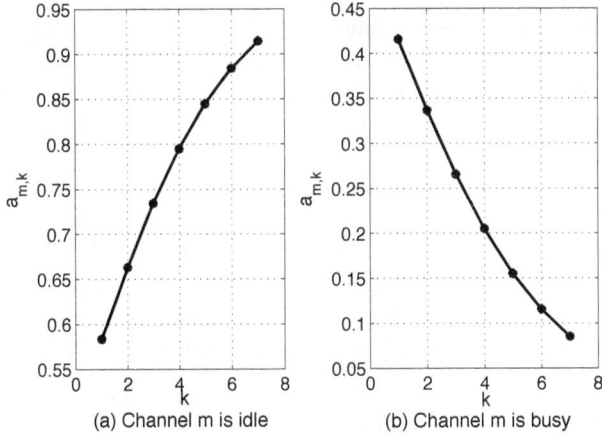

Figure 4. Illustration of $a_{m,k}$ as a monotone function of k, when $\epsilon_m = 0.3$, $\delta_m = 0.3$, and $\bar{K} = 7$.

- If the availability of channel m is between Θ_0 and Θ_1, secondary users will keep on sensing the same channel to obtain more sensing results for more accurate estimation of the channel state, until the maximum number of mini-slots, $\bar{K}$, is reached.

- If the availability of channel m exceeds Θ_1, the channel is believed to be idle and the secondary users stop sensing and prepare to access the channel (see Section 3.2).

The *stop time* K_m when secondary users stop sensing channel m, is a random variable that takes value between 1 and $\bar{K}$, the maximum number of mini-slots that can be used for sensing (see Fig. 3). If we have $\Theta_0 < a_{m,k} < \Theta_1$ by the end of the sensing phase, then channel m state is not identified due to lack of time (or sensing results) and the channel will not be accessed.

When there are k sensing results available (e.g., one user senses channel m for k mini-slots, or it senses channel m for less than k mini-slots and receives some channel m sensing results from other secondary users), we define three sets of estimates for the state of channel

m, as:

$$\Psi_{m,k}^0 = \{d_m \mid a_{m,k} \leq \Theta_0, \forall\ 0 \leq d_m \leq k\} \quad (10)$$

$$\Psi_{m,k}^1 = \{d_m \mid a_{m,k} \geq \Theta_1, \forall\ 0 \leq d_m \leq k\} \quad (11)$$

$$\Psi_{m,k}^2 = \{d_m \mid \Theta_0 < a_{m,k} < \Theta_1, \forall\ 0 \leq d_m \leq k\}$$

$$= \overline{(\Psi_{m,k}^0 \cup \Psi_{m,k}^1)}, \quad (12)$$

where d_m is the number of observations whose sensing result is 0 on channel m. We then present two channel sensing policies based on this classification in the following.

Memoryless Sensing Policy. In this section, we first present a memoryless sensing policy, with which secondary users cooperatively sense the licensed channels. We call the policy "memoryless" since it does not consider the channel sensing and access results in the previous time slot for simplicity. With this memoryless policy, each secondary user chooses one of the M licensed channels to sense with equal probability, i.e., $1/M$. Furthermore, channel selections of the N secondary users are independent and identically distributed (i.i.d.).

Let U_m be the random variable representing the number of secondary users that select channel m to sense. The probability that u_m secondary users choose channel m to sense is

$$\Pr(U_m = u_m) = \binom{N}{u_m}\left(\frac{1}{M}\right)^{u_m}\left(\frac{M-1}{M}\right)^{N-u_m} \quad (13)$$

The joint distribution that there are u_1 secondary users sensing channel 1, u_2 secondary users sensing channel 2, $\cdots$, and u_M secondary users sensing channel M, is

$$\Pr(u_1, u_2, \cdots, u_M)$$
$$= \begin{cases} \prod_{m=1}^{M} \Pr(U_m = u_m), & \text{if } \sum_{m=1}^{M} u_m = N \\ 0, & \text{otherwise.} \end{cases} \quad (14)$$

We next derive the conditional probability that secondary users compete for the channel after the sensing phase stops at the end of mini-slot $K_m < \bar{K}$. The stop time $K_m < \bar{K}$ has two implications. First, it means that secondary users stop sensing channel m after mini-slot K_m. Second, it indicates that the estimated availability of channel m, $a_{m,k}$, has already exceeded the threshold Θ_1. Thus these secondary users think channel m is idle and are ready to access the channel for data transmission. Note that a secondary user also stops sensing a channel m when $a_{m,k} < \Theta_0$ (when it is sure that the channel is busy). We are not interested in this case, since the secondary user will back off until the next time slot. Thus K_m is defined with regard to the event $a_{m,k} > \Theta_1$.

There are U_m users sensing channel m and $U_m K_m$ observations are available after mini-slot K_m, which is also a random variable. We first derive the conditional probability for event $K_m = 1$, as

$$\Pr(K_m = 1 \mid U_m = u,\, S_m = 0) = \Pr(a_{m,u} \geq \Theta_1)$$

$$= \sum_{d_m^1 \in \Psi_{m,u}^1} \binom{u}{d_m^1}\left[(\epsilon_m)^{u-d_m^1}(1-\epsilon_m)^{d_m^1}\right], \qquad (15)$$

where d_m^1 is the number of observations whose sensing result is 0 in the first mini-slot.

Following similar reasoning as in (15), we can obtain the conditional probability for the event that the stop time $K_m = 2$ as

$$\Pr(K_m = 2 \mid U_m = u,\, S_m = 0)$$

$$= \Pr\left[(\Theta_0 < a_{m,u} < \Theta_1) \cap (a_{m,2u} \geq \Theta_1)\right]$$

$$= \sum_{D_m^2 \in \Psi_{m,2u}^1} \sum_{d_m^1 \in \Psi_{m,u}^2} \binom{u}{d_m^1}\binom{u}{d_m^2}\left[(\epsilon_m)^{2u-D_m^2}(1-\epsilon_m)^{D_m^2}\right], \quad (16)$$

where $\Psi_{m,k}^2$ is defined in (12) and $D_m^2 = d_m^1 + d_m^2$. In the general case, we can derive the conditional probability for the event that the stop time is $K_m = k$ as:

$$\Pr(K_m = k \mid U_m = u,\, S_m = 0)$$

$$= \Pr\left[(\Theta_0 < a_{m,u} < \Theta_1) \cap (\Theta_0 < a_{m,2u} < \Theta_1) \cap\right.$$

$$\left. \cdots \cap (\Theta_0 < a_{m,(k-1)u} < \Theta_1) \cap (a_{m,ku} \geq \Theta_1)\right]$$

$$= \sum_{D_m^k \in \Psi_{m,ku}^1} \sum_{D_m^{k-1} \in \Psi_{m,(k-1)u}^2} \cdots \sum_{d_m^1 \in \Psi_{m,u}^2}$$

$$\binom{u}{d_m^1}\binom{u}{d_m^2}\cdots\binom{u}{d_m^k}\left[(\epsilon_m)^{ku-D_m^k}(1-\epsilon_m)^{D_m^k}\right], \quad (17)$$

where $k = 1, \cdots, \bar{K}$ and $D_m^k = \sum_{i=1}^{k} d_m^i$. We will apply these results in Section 4.2 to derive the throughput of the CR network by the *law of total probability*.

Improved Sensing Policy . Under the memoryless sensing policy, some channels may not be sensed by any of the secondary users. Such an event occurs with probability $\Pr(U_m = 0) = \left(\frac{M-1}{M}\right)^N$, which is sufficiently large when M is large and/or the number of secondary users is close to the number of channels. Secondary users will not be able to estimate the state of a channel that nobody senses, and will neither access it in the transmission

phase. Therefore, the spectrum opportunities in that channel will be wasted when such events occur.

Motivated by this observation, we develop an *improved sensing* policy that attempts to reduce the chance that a channel is not sensed by any of the secondary users. The improved sensing policy incorporates a mechanism to spread secondary users to the channels. It also exploits channel state autocorrelation by considering sensing results and channel states in the previous time slot.

By the end of the sensing phase in a time slot t, the secondary users compute the channel availability $a_{m,k}$ for each channel m. During the following transmission phase, if a secondary user transmits on channel m, it can obtain more accurate channel state information: if its transmission is successful, then channel m is idle in time slot t; otherwise, channel m is busy in the time slot. Such channel information can be exchanged at the beginning of the sensing phase in the next time slot. Then, we can classify the M channels into three sets according to the channel states in time slot t, including

- The set of channels that are detected or believed to be idle, denoted by $\mathcal{B}_0(t)$.

- The set of channels that are detected or believed to be busy, denoted by $\mathcal{B}_1(t)$.

- The set of channels whose states are not identified due to lack of time or not sensed by any of the secondary users, denoted by $\mathcal{B}_2(t)$.

Let $|\mathcal{B}_0(t)|$, $|\mathcal{B}_1(t)|$ and $|\mathcal{B}_2(t)|$ be the cardinalities of $\mathcal{B}_0(t)$, $\mathcal{B}_1(t)$, and $\mathcal{B}_2(t)$, respectively.

If channel m is in set $\mathcal{B}_0(t)$ and the stop time on channel m is less than the maximum stop time $\bar{K}$, one user among those u_m users that are sensing this channel will be randomly chosen to switch to sense another channel in the set $m \cup \mathcal{B}_1(t) \cup \mathcal{B}_2(t)$ in time slot $(t+1)$. If channel m is in set $\mathcal{B}_1(t)$ and the stop time on channel m is less than the maximum stop time $\bar{K}$, the secondary users that are sensing this channel will randomly choose a channel in $m \cup \mathcal{B}_2(t)$ to sense in time slot $(t+1)$. With the above mechanism that reassigns secondary users to channels based on the sensing results in the previous time slot, we can reduce the chance that a licensed channel is not sensed by any of the users. This approach achieves the *load balancing* effect since it attempts to spread out secondary users to the channels.

3.2. Transmission Phase

We adopt the p-persistent CSMA protocol for data channel access for secondary users during the data transmission phase. Under this protocol, a secondary user delays its transmission when the channels are busy. Once one or more channels are detected idle, the secondary user will attempt to access the idle channel(s) for data transmission with probability p. We consider the heavy load domain, where each secondary user always has data to send to every other secondary user. The following two cases are investigated for opportunistic spectrum access for secondary users.

Case 1. Once the estimate of channel m, i.e., $a_{m,k}$, exceeds threshold Θ_1, each of the secondary users sensing channel m will send an RTS packet on channel m with probability p, to contend for the transmission opportunity on this channel. If there is only one secondary user that sends RTS, then it wins the channel; if there is no secondary user that sends RTS, then the channel will not be accessed and will be wasted; if there are more than one RTS packets sent on channel m, there is collision and none of the secondary users can use the channel.

We define P_m^{idle}, P_m^{succ} and P_m^{coll} as the probability that there is no RTS transmission on channel m, the probability that exactly one secondary user successfully transmits an RTS on channel m, and the probability that there is collision on channel m when multiple RTS packets are transmitted, respectively. Recall that U_m is the number of secondary users that choose channel m to sense. This set of secondary users also attempt to access channel m if it is found idle. With p-persistent CSMA, it follows that

$$P_m^{idle}(U_m) = (1-p)^{U_m} \tag{18}$$
$$P_m^{succ}(U_m) = U_m \times p \times (1-p)^{U_m-1} \tag{19}$$
$$P_m^{coll}(U_m) = 1 - P_m^{idle}(U_m) - P_m^{succ}(U_m)$$
$$= 1 - (1-p)^{U_m} - U_m \times p \times (1-p)^{U_m-1} \tag{20}$$

Case 2. We assume that the CR users can transmit data over more than one channels using the channel bonding/aggregation techniques [6, 9]. In this case, every secondary user keeps on sensing the channel until the channel state is identified or until the end of the sensing phase. At the beginning of the transmission phase, the set of idle channels are identified and are know to all the secondary users. Then every secondary user will transmit an RTS packet with probability p on the control channel, to contend for the entire set of idle channels. If there is only one secondary user that sends RTS on the control channel, it wins the entire set of idle channels. Otherwise, the idle channels will be wasted (i.e., when no RTS is sent, or more than one RTS are sent on the control channel).

We define P^{idle}, P^{succ} and P^{coll} as the probability of no RTS transmission on the control channel, the probability that exactly one RTS sent on the control channel, and the probability of collision on the control channel, respectively. For p-Persistent CSMA, we have

$$P^{idle}(N) = (1-p)^N \tag{21}$$
$$P^{succ}(N) = N \times (1-p)^{N-1} \tag{22}$$
$$P^{coll}(N) = 1 - P^{idle}(N) - P^{succ}(N)$$
$$= 1 - (1-p)^N - N \times p \times (1-p)^{N-1}. \tag{23}$$

4. Performance Analysis

4.1. Interference Analysis

One of the main challenges in designing a CR network MAC protocol is how to balance the tension between maximizing the capacity of secondary users and protecting primary users from harmful collisions. Let $\gamma_m \in [0,1]$ be the maximum tolerable collision probability to primary users on channel m: $\gamma_m = 0$ means that no secondary transmission is allowed, while $\gamma_m = 1$ means that secondary users have the same privilege as primary users when accessing the channels. The probability of collision caused by secondary users to primary users should be kept below γ_m.

We first derive the conditional probability that channel m is miss detected to be idle by u secondary users after mini-slot k, as follows.

$$\Pr(K_m = k \mid U_m = u, S_m = 1)$$
$$= \sum_{D_m^k \in \Psi_{m,ku}^1} \sum_{D_m^{k-1} \in \Psi_{m,(k-1)u}^2} \cdots \sum_{d_m^1 \in \Psi_{m,u}^2}$$
$$\binom{u}{d_m^1}\binom{u}{d_m^2}\cdots\binom{u}{d_m^k}(\delta_m)^{D_m^k}(1-\delta_m)^{ku-D_m^k}. \tag{24}$$

In Case 1, the idle channels are accessed by different secondary users. The probability that secondary users collide with primary users on channel m is

$$P_{m,1}^{intf} = \sum_{k=1}^{\bar{K}} \sum_{u=0}^{N} \Pr(K_m = k \mid U_m = u, S_m = 1) \times$$
$$\Pr(U_m = u) \times \left[P_m^{succ}(u) + P_m^{coll}(u)\right]. \tag{25}$$

In Case 2, a winning secondary user takes all the idle channels using the channel bonding/aggregation technique. The probability that secondary users collide with primary users on channel m is

$$P_{m,2}^{intf} = \sum_{k=1}^{\bar{K}} \sum_{u=0}^{N} \Pr(K_m = k \mid U_m = u, S_m = 1) \times$$
$$\Pr(U_m = u) \times P^{succ}(N). \tag{26}$$

For primary user protection, the probability of secondary users causing collision with primary users on channel m should be kept lower than or equal to γ_m, i.e.,

$$P_{m,i}^{intf} \leq \gamma_m, \text{ for } i = 1, 2. \tag{27}$$

This constraint is used to set the channel access probability p for the p-persistent CSMA protocol.

4.2. Throughput Analysis

Based on previous analysis, the expected throughput of the proposed CR MAC protocols adopting the two sensing policies, can be derived after the system attains steady state. Without loss of generality, we ignore the time spent on RTS/CTS exchanges, which can be approximated by a fixed amount of overhead.

In Case 1, the expected throughput of channel m that is sensed by u users, denoted by $\Lambda_m^1(u)$, can be derived as

$$\Lambda_m^1(u) = \sum_{k=1}^{\bar{K}} \Pr(K_m = k \mid U_m = u, S_m = 0) \times$$

$$R_m \times \frac{1}{T_s} \times \left[(\bar{K} - k)T_{ms} + T_{data}\right], \quad (28)$$

where R_m is the data rate of channel m, and T_s is the time slot duration given in (3).

Let $\vec{U} = [U_1, U_2, \cdots, U_M]$ denote the secondary user *sensing state vector*, where each element U_m represents the number of secondary users that choose channel m to sense and access. The aggregate throughput for the CR network, denoted by Ω_1, is

$$\Omega_1 = \sum_{\vec{U}} \Pr(\vec{U}) \sum_{\vec{S}} \Pr(\vec{S}) \sum_{m=1}^{M} \left[I_{[S_m=0]}\Lambda_m^1(u)P_m^{succ}(u)\right], \quad (29)$$

where $\vec{S}$ is the channel state vector defined in Section 2, $P_m^{succ}(u)$ is given in (19) and $I_{[S_m=0]}$ is an indicator that channel m is idle, i.e.,

$$I_{[S_m=0]} = \begin{cases} 1, & \text{if } S_m = 0 \\ 0, & \text{otherwise.} \end{cases} \quad (30)$$

In Case 1, the sensing process on channel m can stop early if the estimate of channel availability $a_{m,k}$ exceeds threshold Θ_1 or drops below the threshold Θ_0. In the former case, the remaining mini-slots can be used to transmit data. In Case 2, all CR users wait till the beginning of the transmission phase, and then contend for the idle channels by sending RTS packets on the control channel. The winning secondary user's data transmissions start at the beginning of the transmission phase (i.e., after $\bar{K}$ mini-slots). We can derive the throughput for channel m as follows.

$$\Lambda_m^2(u) = \sum_{k=1}^{\bar{K}} \Pr(K_m = k \mid U_m = u, S_m = 0) \times$$

$$R_m \times \frac{T_{data}}{T_s}, \quad (31)$$

The aggregate throughput for the CR network, denoted by Ω_2, is

$$\Omega_2 = \sum_{\vec{U}} \Pr(\vec{U}) \sum_{\vec{S}} \Pr(\vec{S}) \times$$

$$\sum_{m=1}^{M} \left[I_{[S_m=0]}\Lambda_m^2(u)P^{succ}(N)\right]. \quad (32)$$

5. Simulation Results

5.1. Simulation Settings

We evaluate the performance of the proposed CR MAC protocol using a customized simulator developed with MATLAB. We compare the following four schemes in the simulations:

- A simple random sensing scheme that each user chooses one channel to sense with equal probability, termed *Random* in the plots.

- The negotiate sensing scheme presented in [6], termed *Negotiate* in the plots.

- The memoryless sensing scheme as described in Section 3.1. In the figures, *Memoryless1* refers to transmission scheme Case 1 (i.e., idle channels are accessed by different secondary users, see Section 3.2), and *Memoryless2* refers to transmission scheme Case 2 (i.e., idle channels are accessed by a winning secondary user using channel bonding/aggregation techniques [9]).

- The improved sensing scheme presented in Section 3.1. In the figures, *Improved1* refers to transmission scheme Case 1, and *Improved2* refers to transmission scheme Case 2.

We choose the negotiate sensing scheme since it adopts a similar network model and assumptions. With this scheme, different secondary users attempt to select distinct channels to sense by overhearing the control packets on the control channel [6]. One of the major differences between negotiate sensing and the proposed schemes in this paper, is that negotiate sensing does not consider spectrum sensing errors in the MAC protocol design.

The simulation parameters are summarized in Table 2, which follow the typical values used in [6]. We run each simulation scenario for 10 times with different random seeds. Each point in the plots shown in this section is the average of 10 simulation runs. We plot 95% confidence intervals as error bars on the simulation curves, which are negligible in all the figures.

5.2. Simulation Results

We first verify our throughput analysis presented in Section 3. In Figs 5 and 6, we plot the throughputs

Table 2. Simulation Parameters

Symbol	Value	Definition
T_{ms}	9 μs	mini-slot interval
T_s	1.89 ms	time slot interval
M	5	number of licensed channels
N	8	number of secondary users
η	0.3	utilization of the licensed channels
ϵ	0.3	probability of false alarm
δ	0.3	probability of miss detection
R	1 Mb/s	data rate of each licensed channel
Θ_1	0.8	upper threshold for channel decision
Θ_0	0.2	lower threshold for channel decision
$\bar{K}$	5	maximum stop time for channel sensing

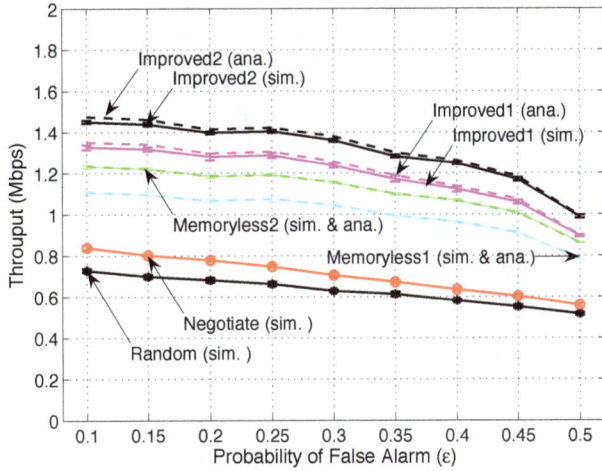

Figure 6. Throughput versus miss detection probability (with 95% confidence intervals for the simulation results).

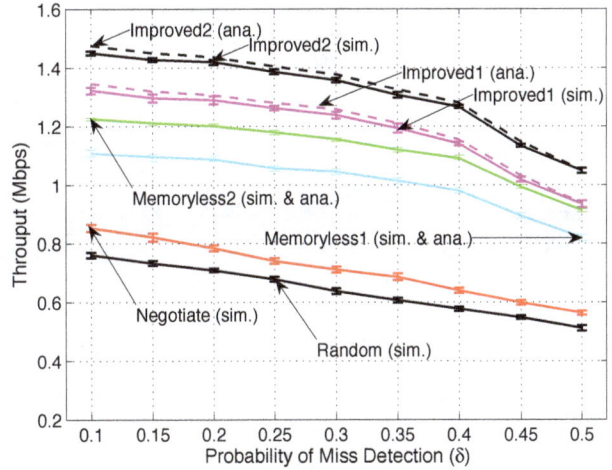

Figure 5. Throughput versus false alarm probability (with 95% confidence intervals for the simulation results).

for the CR MACs incorporating the memoryless sensing policy and the improved sensing policy, with both simulation and analysis curves (dashed curves). We observe that the simulation and analysis curves for the memoryless sensing CR MACs overlap completely with each other, indicating that our analysis is exact. Furthermore, although there is a gap between the simulation and analysis curves for the CR MACs with the improved sensing policy, the gap is generally very small. The gap is actually due to an approximation we used for the secondary user sensing state vector $\vec{U}$, for which deriving the exact form is non-trivial. In the analysis, we assume that the probability is 0 that a channel is not sensed by any secondary user. We find the analysis can serve as a tight upper bound for the CR MAC throughput performance when the improved sensing policy is incorporated.

We next investigate the impact of sensing errors on the CR MAC performance. We assume identical false alarm probabilities $\epsilon_m = \epsilon$, and identical miss detection probabilities $\delta_m = \delta$ for all the licensed channels. In

Fig. 5, we plot the throughputs obtained by the four schemes versus the false alarm probability ϵ. Specifically, we fix δ at 0.3 and increase ϵ from 0.1 to 0.5. Intuitively, a higher false alarm probability results in lower probability for secondary users to exploit the transmission opportunities in the licensed channels. This is illustrated in the figure, as all the four throughput curves decrease as ϵ is increased. The improved sensing MAC achieves the best performance, with about 10% gain over the memoryless sensing MAC and about 200% gain over the two existing approaches. The advantage of channel bonding/aggregation is also demonstrated in the figure, where Case 2 transmission scheme always achieves higher throughput than Case 1 scheme.

In Fig. 6, we examine the impact of miss detection probability δ on the CR network throughput. In these simulations, we fix ϵ at 0.3 and increase δ from 0.1 to 0.5. We find that the miss detection error has small impact on the throughputs of the random sensing and negotiate sensing protocols, since miss detection errors are not considered in the design of these protocols. However, both our proposed CR MAC schemes achieve considerable throughput gains over the random sensing and negotiate sensing schemes.

In Fig. 7, we plot the throughput of the four schemes under different channel utilization values ranging from 0.3 to 0.7. As the utilization of the licensed channels is increased, the transmission opportunities for secondary users are clearly reduced. Therefore the four curves are all decreasing function of η. The improved policy with transmission scheme Case 2 achieves the best performance among the four schemes, while random sensing has the poorest performance. When the channel utilization is $\eta = 0.3$, the improved policy achieves a 10% gain in throughput over the memoryless sensing

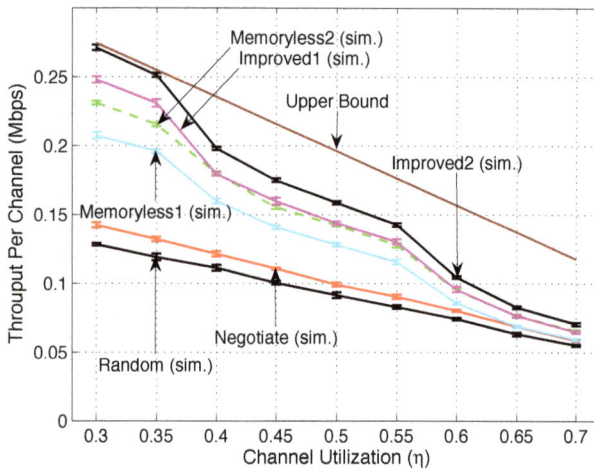

Figure 7. Throughput versus channel utilization (with 95% confidence intervals for the simulation results).

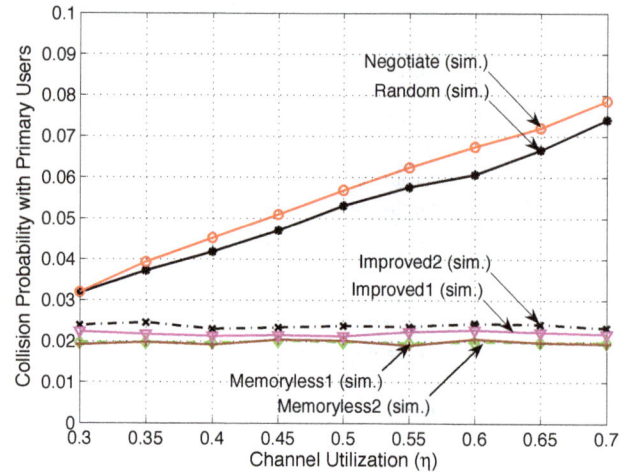

Figure 8. Collision probability with primary users when the maximum tolerable collision probability is $\gamma = 3.5\%$.

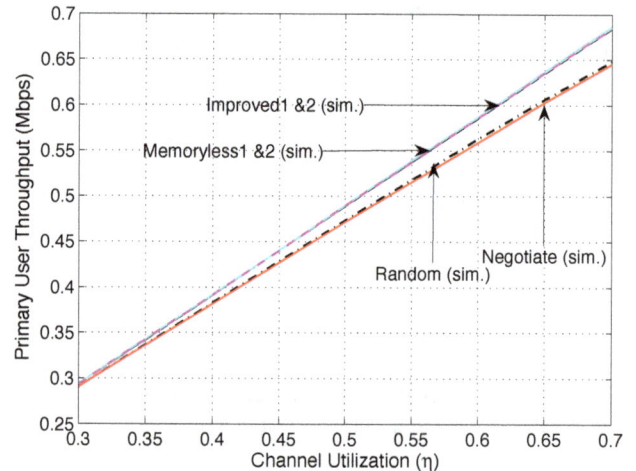

Figure 9. Total throughput of primary users when they become more active.

policy. We also plot the upper bound on the CR network throughput, as given by the channel idle probability in (2). When the channel utilization is low, the improved policy with transmission scheme Case 2 can achieve a throughput very close to the upper bound. The gap between the upper bound and the achievable throughput increase when the primary users get more busy.

In Fig. 8, we plot the collision probability caused by secondary transmissions to primary users, when the maximum allowable collision probability is set as $\gamma = 3.5\%$. We plot the measured collision probabilities in the simulations when the channel utilization is increased from 30% to 70%. It can be seen that the collision probabilities of random and negotiate sensing schemes increases along with η and soon exceed the 3.5% threshold. On the other hand, the collision probabilities of the proposed schemes are kept around 2.5% for the entire range of η examined.

Finally, we plot the throughput of the primary users in Fig. 9. The primary user throughput curves for all the four schemes increase when the channel utilization η is increased. The gap between the curves of the proposed schemes and those of random and negotiate sensing schemes, is due to the different collision rates secondary users introduce to primary users under these schemes (see Fig. 8). As η is increased, the proposed schemes introduces relatively constant collision rates to primary users (i.e., around 2.5%), while the random and negotiate sensing schemes introduce increasingly higher collision rates to primary users, which degrade the throughput of primary users.

6. Related Work

CR has been considered as a "spectrum agile radio" that enables dynamic spectrum access to exploit transmission opportunities in licensed spectrum bands [2, 3]. Several CR MAC protocols have been proposed in the literature. In [10], Le and Hossain propose a MAC protocol for opportunistic spectrum access in CR networks. Two channel selection schemes are proposed: uniform channel selection and spectrum opportunity-based channel selection. The latter considers the probability of spectrum availability and selects each channel with different probabilities based on the estimation of spectrum availability.

A decentralized cognitive MAC protocol is developed in [11] that allows secondary users to explore spectrum opportunities without a central coordinator or a dedicated control channel. However, a practical implementation would be complicated and hardware demanding. This is because each secondary user needs to be equipped with multiple sensors to detect the availability of each licensed channel in the proposed scheme.

In a piece of recent work [6], Su and Zhang propose a negotiation-based sensing policy (NSP), in which a secondary user knows which channels are already sensed and will choose a different channel to sense. In [12], the authors consider two types of hardware constraints: sensing constraint and transmission constraint. In [13], based on the information obtained by a delegate secondary user, each secondary user group selects and switches to the best data channel for data communication during the next period. In [14], the authors describe a policy such that a secondary user selects the channel that has the highest successful transmission probability to access. Many prior works [4, 6, 10, 12] assume perfect channel sensing, within which secondary users can always sense the channel correctly. Sensing errors are not considered.

The joint design of opportunistic spectrum access and sensing policies is studied in a recent work [15] in the presence of sensing errors. The authors develop a separation principle that decouples the designs of sensing and access policy. This interesting study is based on a constrained partially observable Markov decision process (POMDP) formulation and thus has an exponentially growing computational complexity [15].

The important issue of QoS provisioning in CR networks has been studied only in a few papers [16–24]. In [16], a game-theoretic framework is described for resource allocation for multimedia transmissions in spectrum agile wireless networks. In [17], the impact of system parameters residing in different network layers are jointly considered to achieve the best possible video quality for CR users. In [18], Ali and Yu jointly optimize video parameter with spectrum sensing and access strategy. In [19], video encoding rate, power control, relay selection and channel allocation are jointly considered for video over cooperative CR networks. The problem is formulated as a mixed-integer nonlinear problem and solved by a solution algorithm based on a combination of the branch and bound framework and convex relaxation techniques. In our prior work, we study the problem of scalable video multicast in an infrastructure-based CR network in [7, 24], multiuser scalable video streaming over a multi-hop CR network in [23], and multiuser downlink video streaming over a CR femtocell network in [20, 22], where effective algorithms that achieve optimality or with bounded performance are developed. We also investigate the problem of combing cooperative relay with CR for multiuser downlink video streaming, where interference alignment is incorporated to facilitate concurrent transmissions of multiple video packets.

7. Conclusion

We studied the problem of design and analysis of MAC protocols for CR networks in this paper. In particular, we proposed and analyzed two opportunistic multi-channel MAC protocols, adopting a memoryless sensing policy and an improved sensing policy, respectively. The impact of imperfect sensing (in the forms of miss detection and false alarm) are explicitly considered in the CR MAC design. We developed analytical models to evaluate the performance of the proposed protocols. Our simulation study demonstrates the accuracy of the analysis, as well as the superior throughput performance of the proposed CR MACs over existing approaches.

Acknowledgements. This work is supported in part by the US National Science Foundation (NSF) under Grants CNS-0953513, ECCS-0802113, DUE-1044021, IIP-1127952, and CNS-1145446, and through the NSF Wireless Internet Center for Advanced Technology (WICAT) at Auburn University. Any opinions, findings, and conclusions or recommendations expressed in this material are those of the author(s) and do not necessarily reflect the views of the Foundation.

References

[1] D. Hu and S. Mao, "Design and analysis of a sensing error-aware mac protocol for cognitive radio networks," in *Proc. IEEE GLOBECOM'09*, Honolulu, HI, Nov./Dec. 2009, pp. 5514–5519.

[2] Q. Zhao and B. Sadler, "A survey of dynamic spectrum access," *IEEE Signal Process. Mag.*, vol. 24, no. 3, pp. 79–89, May 2007.

[3] Y. Zhao, S. Mao, J. Neel, and J. Reed, "Performance evaluation of cognitive radios: metrics, utility functions, and methodologies," *IEEE Proc. IEEE*, vol. 97, no. 4, Apr. 2009.

[4] A. Motamedi and A. Bahai, "MAC protocol design for spectrum-agile wireless networks: Stochastic control approach," in *Proc. IEEE DySPAN'07*, Dublin, Ireland, Apr. 2007, pp. 448–451.

[5] S. Geirhofer, L. Tong, and B. Sadler, "Cognitive medium access: Constraining interference based on experimental models," *IEEE J. Sel. Areas Commun.*, vol. 26, no. 1, pp. 95–105, Jan. 2008.

[6] H. Su and X. Zhang, "Cross-layer based opportunistic MAC protocols for QoS provisionings over cognitive radio wireless networks," *IEEE J. Sel. Areas Commun.*, vol. 26, no. 1, pp. 118–129, Jan. 2008.

[7] D. Hu, S. Mao, and J. Reed, "On video multicast in cognitive radio networks," in *Proc. IEEE INFOCOM'09*, Rio de Janeiro, Brazil, Apr. 2009.

[8] H. Nan, T.-I. Hyon, and S.-J. Yoo, "Distributed coordinated spectrum sharing MAC protocol for cognitive radio," in *Proc. IEEE DySPAN'07*, Dublin, Ireland, Apr. 2007, pp. 240–249.

[9] C. Corderio, K. Challapali, D. Birru, and S. Shankar, "IEEE 802.22: An introduction to the first wireless standard based on cognitive radios," *J. Commun.*, vol. 1, no. 1, pp. 38–47, Apr. 2006.

[10] L. Le and E. Hossain, "OSA-MAC: A MAC protocol for opportunistic spectrum access in cognitive radio networks," in *Proc. IEEE WCNC'08*, Las Vegas, NV, Mar./Apr. 2008, pp. 1426–1430.

[11] Q. Zhao, L. Tong, A. Swami, and Y. Chen, "Decentralized cognitive MAC for opportunistic spectrum access in ad hoc networks: A POMDP framework," *IEEE J. Sel. Areas Commun.*, vol. 25, no. 3, pp. 589–600, Apr. 2007.

[12] J. Jia, Q. Zhang, and X. Shen, "HC-MAC: A hardware-constrained cognitive MAC for efficient spectrum management," *IEEE J. Sel. Areas Commun.*, vol. 26, no. 1, pp. 106–117, Jan. 2008.

[13] B. Hamdaoui and K. Shin, "OS-MAC: An efficient MAC protocol for spectrum-agile wireless networks," *IEEE Trans. Mobile Comput.*, vol. 7, no. 8, pp. 915–930, Aug. 2008.

[14] A.-C. Hsu, D. Weit, and C.-C. Kuo, "A cognitive MAC protocol using statistical channel allocation for wireless ad-hoc networks," in *Proc. IEEE WCNC'07*, Hong Kong, China, Mar. 2007, pp. 105–110.

[15] Y. Chen, Q. Zhao, and A. Swami, "Joint design and separation principle for opportunistic spectrum access in the presence of sensing errors," *IEEE Trans. Inf. Theory*, vol. 54, no. 5, pp. 2053–2071, May 2008.

[16] A. Fattahi, F. Fu, M. van der Schaar, and F. Paganni, "Mechanism-based resource allocation for multimedia transmission over spectrum agile wireless networks," *IEEE J. Sel. Areas Commun.*, vol. 25, no. 3, pp. 601–612, Apr. 2007.

[17] H. Luo, S. Ci, and D. Wu, "A cross-layer design for the performance improvement of real-time video transmission of secondary users over cognitive radio networks," *IEEE Trans. Circuits Syst. Video Technol.*, vol. 21, no. 8, pp. 1040–1048, Aug. 2011.

[18] S. Ali and F. R. Yu, "Cross-layer qos provisioning for multimedia transmissions in cognitive radio networks," in *Proc. IEEE WCNC'09*, Budapest, Hungary, Apr. 2009, pp. 1–5.

[19] Z. Guan, L. Ding, T. Melodia, and D. Yuan, "On the effect of cooperative relaying on the performance of video streaming applications in cognitive radio networks," in *Proc. IEEE ICC'11*, Kyoto, Japan, June 2011, pp. 1–6.

[20] D. Hu and S. Mao, "On medium grain scalable video streaming over cognitive radio femtocell networks," *IEEE J. Sel. Areas Commun.*, vol. 30, no. 4, Apr. 2012.

[21] ——, "Cooperative relay with interference alignment for video over cognitive radio networks," in *Proc. IEEE INFOCOM'12*, Orlando, FL, Mar. 2012.

[22] ——, "Resource allocation for medium grain scalable videos over femtocell cognitive radio networks," in *Proc. IEEE ICDCS'11*, Minneapolis, MN, June 2011, pp. 258–267.

[23] ——, "Streaming scalable videos over multi-hop cognitive radio networks," *IEEE Trans. Wireless Commun.*, vol. 9, no. 11, pp. 3501–3511, Nov. 2010.

[24] D. Hu, S. Mao, Y. Hou, and J. Reed, "Fine grained scalability video multicast in cognitive radio networks," *IEEE J. Sel. Areas Commun.*, vol. 29, no. 3, pp. 334–344, Apr. 2010.

8

Online Algorithms for Adaptive Optimization in Heterogeneous Delay Tolerant Networks

Wissam Chahin[1,*], Francesco De Pellegrini[2], Rachid El-Azouzi[1], Amar Prakash Azad[3]

[1]CERI/LIA, University of Avignon, 339, chemin des Meinajaries, Avignon, France
[2]CREATE-NET Via alla Cascata 56/D, Povo, Trento, Italy
[3]SOE, UCSC, USA

Abstract

Delay Tolerant Networks (DTNs) are an emerging type of networks which do not need a predefined infrastructure. In fact, data forwarding in DTNs relies on the contacts among nodes which may possess different features, radio range, battery consumption and radio interfaces. On the other hand, efficient message delivery under limited resources, e.g., battery or storage, requires to optimize forwarding policies. We tackle optimal forwarding control for a DTN composed of nodes of different types, forming a so-called heterogeneous network. Using our model, we characterize the optimal policies and provide a suitable framework to design a new class of multi-dimensional stochastic approximation algorithms working for heterogeneous DTNs. Crucially, our proposed algorithms drive online the source node to the optimal operating point without requiring explicit estimation of network parameters. A thorough analysis of the convergence properties and stability of our algorithms is presented.

Keywords: Delay tolerant networks, multi-dimensional optimal control, projected ODE, stochastic approximation.

1. Introduction

Delay Tolerant Networks (DTNs) are designed to sustain communications in networked systems, where persistent end-to-end connectivity cannot be guaranteed [1–3]. In DTNs, messages are carried from source to destination via relay nodes adopting the so-called "store and carry" forwarding, which leverages on nodes' mobility pattern. The key problem in DTNs is thus to efficiently route messages towards the intended destination. It is worth observing that traditional routing techniques would fail in this context due to frequent disruptions. Furthermore, mobile nodes rarely possess information on the upcoming encounters they are going to experience [4]. A number of schemes have been proposed for efficient message forwarding in DTNs [3, 5, 6].

Disseminating multiple copies of the message in the network is the straightforward routing solution to overcome disconnections. This ensures that at least some of them will reach the destination within some

deadline [7, 8] with high probability. This technique is named epidemic forwarding [9], in analogy to spread of infectious diseases.

In literature, several variants of epidemic forwarding exist, including *spray and wait* [7] and *two hop routing protocol* [10], implementing different trade-offs between delay and number of released copies. We confine our analysis to the *two hop routing protocol* because of two major technical advantages: first, compared to plain epidemic routing it performs natively a better trade-off between the number of released copies and the delivery probability [8]. Second, and most relevant with respect to the algorithmic design we propose later, forwarding control can be fully implemented on board of the source node. Under two hop routing, the source transmits a message copy to mobiles it encounters. A relay forwards the message copy it has to the destination only.

In some literature on DTNs, a common simplifying assumption (adopted for modeling reasons) is that DTN nodes have all similar physical characteristics, e.g., transmission range, mobility patterns, etc. Thus, the network is assumed to be *homogeneous*. Under this assumption, recent work provides insight into the performance of DTNs [8, 10, 11].

However, it is clear that DTN nodes may belong to different categories, e.g., mobiles, laptops, PDAs and thus differ in their transmission range, mobility, etc. A DTN with different types of nodes in turn is classified as *heterogeneous* [12–14]. To this respect, our starting assumption, as in [12] and [13], is that according to their physical characteristics, nodes group into classes homogeneous with respect to routing. More precisely, two nodes belong to the same class if they have *same intermeeting intensities with source and destination nodes*, i.e., the intensities by which they meet source and destination are the same.

In this context, the fundamental question that arises naturally when one models the trade off of network resources for delivery probability (namely, the number of messages or the energy expenditure), is how to exploit diversity of contact patterns to improve performance.

The structure of the paper is as follows. In the next section, we discuss related work and then identify the main contributions of the paper. Section 3 introduces the model. We identify the problem in section 4. In particular, we first aim at deriving the closed-form structure of the optimal forwarding policy for heterogeneous DTNs, both for static and dynamic control policies. Leveraging on the properties of the so determined optimal forwarding policy, in section 5 we design online algorithms which are able to converge to the optimal control policy over time. In particular, we first present the static algorithm then we make a step forward by introducing the dynamic algorithm that uses two-time scale stochastic approximations. It is worth noticing that one of our objectives is to show that online implementation of these algorithms can be made in such a way to depend only on local knowledge at source nodes. We provide extensive numerical results to validate our theoretical derivation in section 6. Section 7 concludes the paper.

In literature, some attempts to address optimal control for heterogeneous DTNs have been performed in [14] and [13]. The algorithmic formulation that we provide here introduces not only a closed form description of the optimal policy, but also an algorithmic distributed implementation suitable for disconnected operations.

2. Related Work and Contribution

With the aim of optimizing network performance, several previous works addressed the control of forwarding schemes in DTNs [4, 8, 10, 11]. The work [10] proposed to control two-hop forwarding and optimized the system performance by choosing the average duration of timers for message discarding.

Authors of [8] considered a homogeneous network and described a general framework for the optimal control of monotone relay policies. The optimal control was proved there to be of dynamic type; the first work claiming the optimality of dynamic policies was [11], limited to epidemic routing. In line with [14], our formulation builds also on multidimensional control, which results in existence of several thresholds. We go a step forward by describing the *closed form structure of the optimal control*. Also, we extend the approach in [15] leveraging stochastic approximation algorithms because they overcome the explicit estimation of network parameters. In fact, such an estimation is per se a difficult task in disconnected systems [16]. Furthermore, we observe that this operation becomes critical in the case of multiple classes of mobiles since a number of such estimates would be required.

In literature, the heterogeneity in mobile ad hoc DTNs is well documented [17, 18]. However, very few papers addressed this aspect from the modeling perspective. One such work is [13]; the authors assume that nodes may migrate from one class to another. In our framework nodes are fixed within one class. Also, in [19], the authors showed that the presence of heterogeneity has controversial effect on the delivery probability. I.e., it cannot be related in a straightforward manner to the performance of the system. In [14] a general setting for the optimality of controls of a DTN was presented. In [12], a routing scheme was proposed in heterogeneous DTNs based on the use of history information to identify the nodes of "highest utility" for routing. In algorithms we present here, source will forward the message according to the node's class and no other a priori information is needed for the source to take the forwarding decision.

Novel Contributions. The main contributions of this work are the following. First, we introduce a new perspective of the optimal forwarding problem in DTNs and characterize its structure. Second, based on this characterization we introduce a class of stochastic approximation algorithms that attain optimality when multiple classes of relays exist and can operate at runtime in spite of the lack of full information on the network state. Moreover, we deeply investigate these algorithms' performance and rigorously prove their convergence to some limit set of Ordinary Differential Equations (ODEs), then we use Lyapunov functions to confirm their stability.

Our work focuses on proposing a general algorithmic approach to optimize forwarding control in a distributive and energy efficient fashion when many classes of nodes co-exist. We believe this approach could be extended to account for protocols other than two hops. In order for this method to be operational in such scenarios, message replication control could be decentralized in a way that each relay node can control and keep track of the number of message copies it forwards.

3. System Model

Table 1. Glossary of Notations

Symbol	Meaning
$N + 2$	number of nodes
N_i	number of nodes of class i
λ_{si}	pairwise intermeeting intensity for node i with the source
λ_{id}	pairwise intermeeting intensity for node i with the destination
τ	timeout value
Δ	time slot
K	$\lfloor \tau/\Delta \rfloor$
$X_i(n)$	number of infected nodes of class i at time $n\Delta$, (x_i at n=0)
Ψ	energy constraint
Ψ_i	Maximum number of nodes of class i that can be infected
$F_D(t)$	delivery probability at (t)
$\mathbf{U}(.)$	transmission control vector
$U_i(n)$	forwarding probability to class i at time $n\Delta$
$\Pi_I(w)$	projection over I of the value w
h_i	the switch time of a dynamic policy of class i
θ_i	$= \sum_{k=0}^{K-1} U_i(k)$

Consider a network of $N + 2$ mobile nodes (composed of m classes), each equipped with some form of proximity wireless communications. One node, *source node*, has a message to be sent to a destination node. The network is assumed to be sparse, so that, at any time instant, nodes are isolated with high probability. Communication opportunities arise whenever, due to mobility, two nodes get within reciprocal communication range; such events are often named *contacts*. For the ease of reading, we collect all the main symbols used in the paper in Table 1.

The time between contacts of any two nodes, referred to as *intermeeting times*, is assumed to be exponentially distributed[1]. The contact rate is known to converge to a quantity that is independent of number of mobiles N in the network when the N grows large, under the fluid-approximations (see [12, 21, 22]). The validity of such a model has been discussed in [20], and its accuracy has been shown for a number of mobility models (Random

Walker, Random Direction, Random Waypoint) [8, 15]. There exist studies based on traces collected from real-life mobility [1] arguing that intermeeting times may follow a power-law distribution. In [12], it has been shown that the traces and many other exhibit exponential tails after a cutoff point. We choose the exponential intermeeting times model due to the mathematical tractability and the above reasons.

Heterogeneity. Mobiles can have different physical characteristics, such as transmission power, mobility, etc. In our model, nodes' heterogeneity is captured by the distribution of intermeeting times. In particular, intermeeting intensity parameters capture the physical characteristics of nodes [21], dictating the rate at which two nodes meet. More precisely, node i from class k is represented by the tuple $\lambda_i^k = \{\lambda_{si}, \lambda_{id}\}^k$ where λ_{si} (resp. λ_{id}) refers to the inter-meeting intensity of the node i from class k with the source node (resp. with the destination node). We denote the number of mobiles in class k by $N_k \geq 1$, therefore the total number of mobile nodes in m classes are $\sum_{k=1}^{m} N_k = N$. We refer to the m class system as the m-dimensional (mD) system for the sake of brevity.

There can be multiple source-destination pairs, but we assume that at a given time there is a single message, generated by a tagged source node. The message may eventually have many copies spread in the network.[2] For simplicity, we assume that a message is generated at time $t = 0$ and it remains relevant for some time τ. We do not assume any feedback that allows the source or other mobiles to know whether the message has made it successfully to destination within the allotted time τ.

We adopt a probabilistic *two-hop* routing protocol according to which the source passes the message to mobiles that do not have it with some probability $U \in [u_{min}, u_{max}]$. But, a relay node transmits the message only when it meets the destination node. Such relay policy is *monotone* [8] because the number of copies of the message increases over time. The message is called delivered when the destination receives the first copy of the message.

The problem we address in this paper is to *design online algorithms that drive the source node to an optimal operating point. An optimal operating point is such if the corresponding forwarding policy maximizes the probability to deliver the message by time τ under the constraint on the amount of energy spent by the source.*

---

[1] The *inter-meeting* rate is a function of number of nodes, reciprocal transmission power of nodes, mobility parameters, etc. [20]

[2] Results in the subsequent sections are valid for multiple simultaneous source-destination pairs. But, we should additionally assume that the bandwidth is large enough to ensure that the different forwarding processes are independent, though limited to one message per source. Source s, destination d and the N relays can be accordingly reindexed and the analysis in the subsequent sections remains valid for any such pairs.

The energy spent by the source node relates to the number of message copies transmitted. We assume that the energy spent by relay nodes is not a constraint. In fact, under two hop forwarding relays have to transmit at most once. In particular, we assume that i) each transmission consumes a constant amount of energy, and ii) all other activity requires negligible amount of energy. Under these assumptions the source node spends an energy amount that is proportional to the number of message copies.

A naive way to maximize the delivery probability is to maximize the number of infected nodes. However, under the constraint on the source energy budget we need to tackle an optimal forwarding control able to account for heterogeneity of the relay nodes.

We adopt a discrete time model where the time axis is divided into slots of small duration Δ. Time slot k is the interval $[k\Delta, (k+1)\Delta]$ and the number of slots is equal to $K = [\tau/\Delta]$. Moreover, the control during $[k\Delta, (k+1)\Delta]$ is a constant, denoted by $U(k)$.

Forwarding Control. The source node has the possibility to control dynamically the forwarding process to relay nodes: it will forward to nodes in class i with probability U_i. This will slow down the generation of message copies within class i, where $U_i : \{0, \ldots, K-1\} \to [u_{\min}, u_{\max}], i = 1, \ldots, m$. The control policy the source uses can be expressed by the m-dimensional control vector $\mathbf{U} = \{U_1, \ldots, U_m\}$.

Message Delivery Distribution. The source node aims at maximizing the fluid approximation for the cumulative distribution function (CDF) of the delay[3] $F_D(t) := P(T_d \leq t)$. It is based on a generalization of [8]:

$$F_D(t) = 1 - \exp\left(-\sum_j \lambda_{jd} \int_{s=0}^{t} X_j(s)ds\right). \quad (1)$$

Note that because of monotonicity, maximizing $F_D(t)$ in (1) is equivalent to maximizing $\sum_i \lambda_{is} \int_0^t X_i(s)ds$.

Energy Consumption. Let $\epsilon > 0$ be the energy consumed by the source for transmission of a single copy of the message. As explained in the previous section, energy consumed by the source is significant as compared to the energy spent by other relay nodes under two hop routing. Thus, the total energy consumed by the network to generate message copies during $[0, \tau]$ is

$$\mathcal{E}(\tau) = \epsilon \sum_i [X_i(\tau) - X_i(0)].$$

Notice that the total amount of energy spent by the source is proportional to the sum of messages (to all classes) and is the same for each message irrespectively of the mobiles' class.

Optimization Problem. The source's goal is to obtain the *multi class optimal policies* by optimizing over the m-dimensional control vector $\mathbf{U} = \{U_1, \ldots, U_m\}$, where $U_i : \{0, \ldots, K-1\} \to [u_{\min}, u_{\max}], i = 1, \ldots, m$, which solves

$$\max_{\mathbf{U}} F_D(\tau), \quad \text{subject to} \quad \mathbf{X}(\tau) \cdot \mathbf{1} \leq \Psi, \ \mathbf{X}(0) = \mathbf{x}, \quad (2)$$

where $\mathbf{X}(\tau) \cdot \mathbf{1} = \sum_i X_i(\tau)$. The initial condition $\mathbf{X}(0) = \{X_i(0), i \in \{1, \cdots, m\}\}$, constraint Ψ and total number of copies at time 0, i.e., $\mathbf{x}$ (where $\Psi \geq \mathbf{x} \mathbf{1}$) are input for the optimization problem. The control vector is $\mathbf{U}(\cdot) = \{U_i(\cdot) : i \in [1, \cdots, m]\}$ with the m-dimensional support $\mathbf{U}(\cdot) \in [u_{\min}, 1]^m$. Recall that maximizing $F_D(\tau)$ is equivalent to maximizing $\sum_j \lambda_{jd} \int_{s=0}^{\tau} X_j(s)ds$.

The optimal control problem reads

$$\max_{\mathbf{U}} \quad J = \sum_{t=1}^{K-1} \int_{k\Delta}^{(k+1)\Delta} \sum_{i=1}^{m} \lambda_{id} X_i(t)dt, \quad (3)$$

$$\text{subject to} \quad X(\tau) = \sum_{i=1}^{m} X_i(\tau) \leq \psi, \quad (4)$$

where $X_i(n)$ denotes number of mobiles, not including the destination for class i, that have a copy of the message at time $n\Delta$ (i.e. at the beginning of the n-th slot), $X_i(0) = x_i$. Under some standard assumptions, it forms a Markov chain with possible states $1, \ldots, N$ (refer to [15]). It is characterized by

$$X_i(n+1) = X_i(n) + (N_i - X_i(n))(1 - e^{-\lambda_{si}\Delta U_i(n)}), i = 1, \ldots, m. \quad (5)$$

The objective functional can be rewritten (after integrating (3)) as

$$J = \sum_{i=1}^{m} \tau N_i \lambda_{id} - \frac{\lambda_{id}}{\lambda_{si}} \sum_{n=0}^{K-1} (N_i - X_i(n)) \frac{1 - e^{-\lambda_{si}\Delta U_i(n)}}{U_i(n)}. \quad (6)$$

This model provides a useful framework for the subsequent algorithmic development that we detail later in the paper.

4. Optimal Control

Our goal is to find the optimal transmission policy for the problem stated in (2) for each class of mobiles. In the following, we first obtain the optimal policy from the static class of policies in which the *transmission control* is time-invariant for each class of mobile nodes. To avoid cumbersome notation, we derive the optimal static control for two classes, which can be easily extended for general m classes (we

---

[3]The controlled version reported in (1) derives from the separable differential equation in the form $\frac{d}{dt} F_D(t) = \lim_{h \to 0} \frac{\mathbb{P}[T_d > t+h] - \mathbb{P}[T_d > t]}{h} = [1 - F_D(t)] \sum_j \lambda_{jd} X_j(s)$.

describe this extension later). Note that static control is suboptimal with respect to dynamic optimal control policies [8]. Yet, it is rather convenient for the sake of implementation simplicity.

In the case of dynamic control policies, we will also obtain the optimal policies, which turn out to be of threshold type. Various methodologies have been developed to establish the threshold structure of optimal transmission policies in DTNs: one based on the Pontryagin maximum principle [8], another based on some sample path comparisons[15], some on stochastic ordering, etc. These approaches, developed in the context of DTNs with one type of population, are not applicable to our problem since the model is no longer scalar. Accordingly, we develop a new approach that establishes the optimality of threshold type policies for each class.

Denote $\overline{X}_i(\tau)$ as the number of message copies the source can transmit to nodes of class i by time τ without any control, i.e. $U_i = 1$. Such transmissions are also referred to as uncontrolled transmission [8]. Indeed, a controlled dynamics simply refers to the slowed down transmission [8, eq. 1] in which the number of message under a control U_i is simply $\overline{X}_i(U_i\tau)$. In our multi-class structure, denoting $\mathbf{U} = \{U_1, \cdots, U_m\}$ as the control vector, we refer the total number of message copies in the network by $\overline{\mathcal{X}}(\{\mathbf{U}\tau\})$ which is the sum of message copies over all classes, i.e. $\overline{\mathcal{X}}(\{U_1\tau, \cdots, U_m\tau\}) = \sum_{i=1}^m \overline{X}_i(U_i\tau)$.

4.1. Optimal Static Control Policy

Our goal is to maximize $F_D(\tau)$ while keeping $\mathcal{E}(\tau)$ low so as to satisfy $\sum_{i=1}^m X_i(\tau) \leq \psi$. In view of (1), this is equivalent to maximizing $\sum_i \lambda_{is} \int_0^t X_i(s)ds$. The control for class i is denoted by U_i under the *static* control policy, a control that is constant over time. In what follows, we derive the optimal static policy for 2D (2 node classes) for the sake of exposition clarity. In the subsequent we illustrate that the method can be directly extended for mD (m node classes).

Theorem 1. (Static Optimal Control-2D): Consider the problem of maximizing $F_D(\tau)$ subject to the energy constraint $\mathcal{E}(\tau) \leq \Psi$.

  i. If $\overline{\mathcal{X}}(\{u_{\min}\tau, u_{\min}\tau\}) > \psi$, then there is no feasible solution.

  ii. If $\overline{\mathcal{X}}(\{\tau, \tau\}) < \psi$, then the policy $\mathbf{U}$ is optimal if and only if $\mathbf{U} = \{1, 1\}$ for $t \in [0, \tau]$ a.e., otherwise,

  iii. If $\overline{\mathcal{X}}(\{\tau, u_{\min}\tau\}) \leq \psi$ then the best static policy is $\mathbf{U} = \{1, v_2\}$, where v_2 is a constant.

  iv. If $\overline{\mathcal{X}}(\{\tau, u_{\min}\tau\}) > \psi$, it is sufficient to swap the class indexes in the above statement (for the case iii)).

Static policy v_2 is given as

$$v_2 = \frac{\overline{X}_1^{-1}(\psi - \overline{\mathcal{X}}_{-1}(u_{\min}\tau))}{\tau}, \text{ and} \qquad (7)$$

$$\overline{\mathcal{X}}_{-1}(u_{\min}\tau) = \overline{\mathcal{X}}(u_{\min}\tau) - \overline{X}_1(U_1\tau). \qquad (8)$$

Proof: Parts (i) and (ii) are obvious. We show part (iii) and (iv) in the following. From (5), the number of nodes with message copies for each class is simply $X_i(t) = N_i - (N_i - X_i(0))e^{-\lambda_{si}U_it}$, $0 \leq t \leq \tau$. Hence, we can consider ϕ the bijection such that $\phi_i(U_i) = X_i(\tau)$, $i = 1, 2$. From (1), maximizing $F_D(\cdot)$ is equivalent to maximizing the following (a function of $(X_1(\tau), X_2(\tau))$), and it reads in particular

$$J(\cdot) = \sum_{i=1,2} \lambda_{id}\left[N_i\tau - \frac{X_i(\tau) - X_i(0)}{\lambda_{si}U_i(X_i(\tau))}\right], \qquad (9)$$

where $\phi^{-1}(X_i(\tau)) = U_i(X_i(\tau))$. Using (5), we have $U_i(X_i(\tau)) = -\frac{1}{\lambda_{si}\tau}\log\left(\frac{N_i - X_i(\tau)}{N_i - X_i(0)}\right)$. This (bijection) allows the equivalence of maxima with respect to $X_i(\tau)$ to u_i.

Denote $L = \{(X_1(\tau), X_2(\tau))|\overline{X}_i(u\tau) \leq X_i(\tau) \leq \overline{X}_i(\tau), i = 1, 2,$ such that $X_1(\tau) + X_2(\tau) \leq \psi\}$. Indeed, since $\frac{\partial U_i}{\partial X_i(\tau)} > 0$, it is easy to note that $\nabla J \neq 0$ for all points that fall in the interior of L. Moreover, we have

$$\frac{\partial J}{\partial X_i(\tau)} = -\frac{\lambda_{id}\tau}{\lambda_{si}U_i(X_i(\tau))}\left[1 + \frac{X_i(\tau) - X_i(0)}{U_i(X_i(\tau))}\frac{\partial U_i}{\partial X_i(\tau)}\right] < 0,$$

where we used the fact that $X_i(\tau) > X_i(0)$, $U_i(\cdot) > 0$ and $\frac{dX_i(\tau)}{dU_i} > 0$ (from the rule of the derivative of the inverse function) to determine the sign of the bracketed right end product. Therefore, $J(\cdot)$ cannot attain its maximum in the interior of L, so it does in the ∂L (on the boundary). This is also depicted geometrically for the cases ii) and iii) in Fig. 1.

Note, the term $\frac{X_i(\tau) - X_i(0)}{U_i(X_i(\tau))}$ has a negative gradient with respect to $X_i(\tau)$, so that the maximum can only be attained on the intersection of L with the line $X_1(\tau) + X_2(\tau) = \Psi$. In the following, concavity of $J(\cdot)$ ensures unique maxima on one of the boundaries.

The vectorial maximization of $J(\cdot)$ reduces to single variable maximization due to restricting $J(\cdot)$ in the

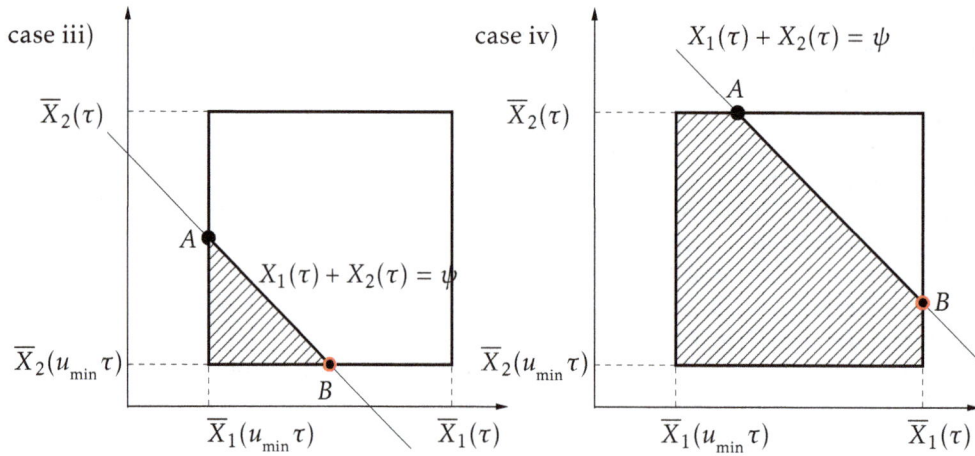

Figure 1. Geometric interpretation of Theorem. 1: case ii) and iii). The feasible region of $X_1(\tau)$ and $X_2(\tau)$ is depicted by horizontal and vertical axis respectively. The shaded area represents the set of points satisfying $X_1(\tau) + X_2(\tau) \leq \Psi$.

region ($X_1(\tau) + X_2(\tau) = \Psi$), e.g., $X_i(\tau)$,

$$
\begin{aligned}
J(X_1(\tau)) &= \lambda_{1d}\tau\left[N_1 + \frac{X_1(\tau) - X_1(0)}{\log\left(\frac{N_1 - X_1(\tau)}{N_1 - X_1(0)}\right)}\right] \\
&+ \lambda_{2d}\tau\left[N_2 + \frac{\Psi - X_1(\tau) - X_2(0)}{\log\left(\frac{N_2 - \Psi + X_1(\tau)}{N_2 - X_2(0)}\right)}\right] \\
&= \lambda_{1d}\tau N_1\left[1 + \frac{N_1 - X_1(0)}{N_1}f\left(\frac{X_1(\tau) - X_1(0)}{N_1 - X_1(0)}\right)\right] \\
&+ \lambda_{2d}\tau N_2\left[1 + \frac{N_2 - X_2(0)}{N_2}f\left(\frac{\Psi - X_1(\tau) - X_2(0)}{N_2 - X_2(0)}\right)\right],
\end{aligned}
$$

where $f(x) = x/\log(1-x)$. To see that $J(X_1(\tau))$ is concave, firstly $f(x) = x/\log(1-x)$ is a concave function in $[0,1]$, since it is differentiable and $\ddot{f}(x) > 0$ in $(0,1)$. Furthermore, $f(Ax + B)$ is still concave for any linear combination of the argument $Ax + B$. Finally, any linear combination of convex functions through nonnegative coefficients is still a concave function.

Now, since $J(X_1(\tau))$ is a concave function in $[\overline{X}_1(u_{\min}\tau), \overline{X}_2(\tau)]$, it attains its maximum at the extrema of the segment, the points corresponding to A and B in Fig. 1. The explicit form for parts (iii) and (iv) is obtained calculating $\overline{X}_1(v_1\tau)$, $\overline{X}_2(v_2\tau)$ under the assumption that $U_2 = u_{\min}$ and $U_1 = 1$ is optimal, respectively. This concludes the proof. ◇

Remark 1. The explicit closed form of the optimal static policy in parts (iii) and (iv) of theorem 1 can be easily computed as

- The optimal policy for part (iii) is $\mathbf{U} = \{1, v_2\} = \{1, \frac{-1}{\lambda_{s2}\tau}\log(\frac{N-\psi}{N_2-X_2(0)} - \frac{N_1-X_1(0)}{N_2-X_2(0)}e^{-\lambda_{s1}\tau})\}$.

- The optimal policy for part (iv) is $\mathbf{U} = \{v_1, u_{\min}\} = \{\frac{-1}{\lambda_{s1}\tau}\log(\frac{N-\psi}{N_1-X_1(0)} - \frac{N_2-X_2(0)}{N_1-X_1(0)}e^{-\lambda_{s2}\tau u_{\min}}), u_{\min}\}$.

Having illustrated the 2D optimal policy for static control, we note that it can be directly extended to m-dimensions. Without loss of generality, we assume that the indexing of nodes is in the increasing order of the value of λ_i for the following theorem. We denote by $e_k(c)$ for the m-dimensional vector which has 0 on all its components but for the k-th component has a value c.

Theorem 2. (Static Optimal Control-mD): Consider the problem of maximizing $\bar{F}_D(\tau)$, in a static policy for the case of m classes (mD case), the following holds

i. If $\overline{\mathcal{X}}(\{1u_{\min}\tau\}) > \psi$ there is no feasible solution.

ii. If $\overline{\mathcal{X}}(\{1\tau\}) < \psi$ then the optimal control policy is $\mathbf{U} = \mathbf{1}$, otherwise,

iii. Counting down k from m to 1, for every k:

    a. If $\overline{\mathcal{X}}(\{(\mathbf{1} - e_k(1 - u_{\min}))\tau\}) < \psi$, then the optimal control policy is $\mathbf{U} = \{\mathbf{1} - e_k(1 - v_k)\}$, otherwise,

    b. If $\overline{\mathcal{X}}(\{(\mathbf{1} - e_{k-1}(1 - u_{\min}) - e_k(1 - u_{\min}))\tau\}) < \psi$, then the optimal control policy is $\mathbf{U} = \{\mathbf{1} - e_{k-1}(1 - v_{k-1}) - e_k(1 - u_{\min})\}$.

iv. If $\overline{\mathcal{X}}(\{(\mathbf{1} - \sum_{k=1}^{m} e_k(1 - u_{\min}))\tau\}) < \psi$, then the optimal control policy is $\mathbf{U} = \{\mathbf{1} - e_1(1 - v_1) - \sum_{k=2}^{m} e_k(1 - u_{\min})\}$.

The constant v_k is given by

$$
v_k = \frac{\overline{X}_k^{-1}(\psi - \sum_{i=1}^{k-1}\overline{X}_i(\tau) - \sum_{i=k+1}^{m}\overline{X}_i(u_{\min}\tau))}{\tau}.
$$

Proof: The proof is a direct extension of the 2D case. ◇

4.2. Optimal Dynamic Control Policy

In what follows, we consider the problem of maximizing $F_D(\tau)$ under the control $\mathbf{U}(k) \in [u_{min}, 1]^m$. The solution to the dynamic control problem will be shown to consist of polices involving thresholds. With no loss of generality, we consider the case of two classes (2D) in the following analysis. It can be easily extended to the mD case. This subsection provides the suitable framework for the two-time scale dynamic algorithm introduced later in the paper.

Definition 1 (Threshold policy). A mD threshold policy is a control policy $\mathbf{U} : [0, \tau] \to \{u_{min}, u_{max}\}^m$ with the related switching parameter h_i for the i-th class if $U_i(k) = u_{max}$ for $k \le h_i$ a.e. and $U_i(k) = u_{min}$ for $k > h_i$ i.e. component switches from u_{max} to u_{min} at most once. We denote h_i the switch time (the *threshold*) of a dynamic policy with respect to the i–th component.

Let ψ_1 be the maximum number of nodes of class 1 that can be infected: then the source can infect ($\psi_2 = \psi - \psi_1$) nodes of class 2. Then, when the constraint is saturated, the optimal threshold policy (U_1^*, U_2^*) satisfies the following relation

$$S_i(\psi_i) \overset{\text{def}}{=} \sum_{k=0}^{K-1} U_i^*(k, \psi_i) = -\frac{1}{\lambda_{si}\Delta} \log\left(\frac{N_i - \psi_i}{N_i - X_i(0)}\right).$$

Let $h_i(\psi_i)$ the threshold for the optimal policy U_i^*. Then we have

$$S_i(\psi_i) = h_i(\psi_i) \cdot u_{max} + (K - 1 - h_i(\psi_i)) \cdot u_{min} + g(\psi_i),$$
$$(10)$$

where $g(\psi_i) = -\frac{1}{\lambda_{si}\Delta} \log(\frac{N_i - \psi_i}{N_i - X_i(0)}) - (h_i(\psi_i) \cdot u_{max} + (K - 1 - h_i(\psi_i)) \cdot u_{min})$. Notice that (10) defines a bijection, so that $(U_1^*, U_2^*) \sim (\psi_1^*, \psi - \psi_1^*)$: we can now consider the problem from slightly different perspective. The source goal is to find the optimal ψ_1^* that maximizes the delivery probability. The advantage of the new formulation is that the joint constraint is replaced by two separate constraints, one per class. In turn we can express our initial optimization as

$$\max_{\psi_1} \quad \bar{J}(\psi_1) = \max_{\psi_1}(\bar{J}_1(\psi_1) + \bar{J}_2(\psi - \psi_1))$$
$$\text{s.t.} \quad X_1(\tau) \le \psi_1, X_2(\tau) \le \psi - \psi_1, \quad (11)$$

where $\bar{J}_i(\cdot)$ is

$$\bar{J}_i(\psi_i) = \tau N_i \lambda_{id} - \frac{\lambda_{id}}{\lambda_{si}} \sum_{k=0}^{K-1} (N_i - X_i(k, \psi_i))\frac{1 - e^{-\lambda_{si}\Delta . U_i(k, \psi_i)}}{U_i(k, \psi_i)},$$

for $i = 1, 2$. Clearly, the solution of this new problem solves directly the original optimization problem (3).

Theorem 3. There exists an unique optimal value (ψ_1^*) that maximizes the delivery probability in (11).

Proof:

In order to prove the uniqueness, it is sufficient to prove that J is concave in ψ_1. We start by proving that $\bar{J}_1(\psi_1)$ is concave. From equation (10), it follows that $\frac{dg(\psi_1)}{d\psi_1} = \frac{1}{\lambda_{s1}\Delta}$, then the derivative of function $\bar{J}_1$ can be expressed as follows: $\frac{d\bar{J}_1(\psi_1)}{d\psi_1} =$

$$-\frac{\lambda_{1d}}{\lambda_{s1}}(N - X_1^{h_1}(\psi_1)).\frac{\lambda_{s1}\Delta.g(\psi_1)e^{-\lambda_{s1}\Delta.g(\psi_1)} - 1 + e^{-\lambda_{s1}\Delta.g(\psi_1)}}{U_1^{h_1}(\psi_1)^2}$$
$$+\frac{\lambda_{1d}}{\lambda_{s1}}\frac{1 - e^{-\lambda_{si}\Delta.u_{min}}}{u_{min}} \cdot \sum_{k=h_1+1}^{K-1} \frac{dX_1^k(\psi_1)}{d\psi_1} \quad (12)$$

The first term is clearly decreasing in ψ_1, since $g(\psi_1)$ is an increasing function in ψ_1. Let us now determine the derivative of second term as follows:

• For $k = h_1 + 1$, we have

$$X_1^{h_1+1}(\psi_1) = X_1^{h_1} + (N_1 - X_1^{h_1})(1 - e^{-\lambda_{s1}\Delta.g(\psi_1)}),$$

then

$$\frac{dX_1^{h_1+1}(\psi_1)}{d\psi_1} = (N_1 - X_1^{h_1})e^{-\lambda_{s1}\Delta.g(\psi_1)}.$$

• For $k = h_1 + 2$, we have

$$X_1^{h_1+2}(\psi_1) = X_1^{h_1+1}(\psi_1) + (N_1 - X_1^{h_1+1}(\psi_1))(1 - e^{-\lambda_{s1}\Delta.u_{min}}),$$

then

$$\frac{dX_1^{h_1+2}(\psi_1)}{d\psi_1} = \frac{dX_1^{h_1+1}(\psi_1)}{d\psi_1}e^{-\lambda_{s1}\Delta.u_{min}}.$$

• Making all the steps up to $k = K - 1$ we can derive the useful formula of the derivative of the second term in (12):

$$\sum_{k=h_1+1}^{K-1} \frac{dX_1^k(\psi_1)}{d\psi_1} = \frac{dX_1^{h_1+1}(\psi_1)}{d\psi_1}e^{-\lambda_{s1}\Delta.\sum_{k=h_1+2}^{K-1} u_{min}}, \quad (13)$$

Since $g(\psi_1)$ is increasing function in ψ_1, it is easy to check that the derivative of function $X_1^{h_1+1}(\psi_1)$ is a decreasing function, It follows from (13) that the second term is decreasing function in ψ_1. Hence the function $\bar{J}_1$ is a concave function. Using the same steps we can show that $\bar{J}_2(\psi - \psi_1)$ is a concave function in ψ_1, and since the sum of two concave functions is a concave function, $\bar{J}(\psi_1)$ is a concave function, hence proved. ◇

Having characterized the optimal forwarding policies for the m-class DTN, in the following we focus on algorithms that the source can adapt to achieve the optimal performance. In particular, we propose algorithms in which source node does not require a prior knowledge of system parameters, rather has the ability to estimate online and adjust the forwarding accordingly, we call *blind online algorithms*. We resort on stochastic approximation based approach to design the algorithms.

5. Blind Online Algorithms for Adaptive Optimization

In this section, we propose two online algorithms to attain optimal control of forwarding by the source node: i) Algorithm 1 applies to the static control case , and ii) Algorithm 2 applies to the dynamic control case. Both algorithms are *blind*, do not require a-priori knowledge of network parameters which is intermeeting intensities and number of mobiles in our context. Observe that in the heterogeneous case, each class of nodes has its own (unknown) parameters: intermeeting intensities (λ_{si} and λ_{id}) and the number of nodes (N_i) for each class i. These algorithms will only depend on the source ability to distinguish between classes, e.g., according to node's type (whether it is a throwbox, a smartphone, etc): leveraging on the structure of the optimal solution, this will be enough to find the optimal forwarding control that the source should adopt without explicit estimation of such parameters.

Remark 2. Let $\theta_i = \sum_{k=0}^{K-1} U_i(k)$, $i = 1, 2$, then θ_i identifies both the static and dynamic policies: the static policy is $U_i = \theta_i/K$, while for the dynamic policy threshold writes $h_i = \max\{h \in \mathbb{N} : v(h) = h \cdot u_{max} + (K - h) \cdot u_{min} \leq \theta_i\}$, and $U_i(h) = \theta_i - v(h)$.

5.1. Blind Online Algorithm for Static Control

Our static algorithm is an extension of [15] to the multi-dimensional case. Each step of the algorithm corresponds to a round of duration τ. For the sake of notation, let $\widehat{X}(\theta_i^k, t)$, $i = 1, 2$ the number of nodes of the ith class that are potentially infected by the source in the current round up to time t, by averaging over several consecutive rounds, using interpolation, the source node is able to obtain an estimate of the average number of copies ($\widehat{X}(\theta_i^k)$) that could potentially be attained using the current set of K policies. $\widehat{X}(\theta_i^k)$ is used to update θ_i according to the formula showed in Algorithm 1, in which, for $I = [\theta_{min}, \theta_{max}]$, the projection function Π_I is defined as follows:

$$\Pi_I(\theta_i) = \begin{cases} \theta_{max} & \text{if} & \theta_i \geq \theta_{max}, \\ \theta_i & \text{if} & \theta_{min} \leq \theta_i \leq \theta_{max}, \\ \theta_{max} & \text{if} & \theta_i \leq \theta_{min}. \end{cases}$$

This stochastic approximation algorithm will implicitly discover the fastest class. This permits to adjust the values of (θ_1, θ_2) for the next round in order to, eventually, estimate the optimal (θ_1, θ_2) in I, which uniquely

---

Algorithm 1 Stochastic approximation of the optimal policy using online estimation (2D static case)

1: **input:** $I = [0, K - 1]$, $\theta_1^0 = \theta_2^0 = K/2$, $k = 0$
2: **while** $max(|\theta_1^{k+1} - \theta_1^k|, |\theta_2^{k+1} - \theta_2^k|) > \epsilon$ **do**
3:     $\widehat{X}_1(\theta_1^k) = \text{interp}\big(\widehat{X}_1(\theta_1^k), \theta_1^k\big)$
4:     $\widehat{X}_2(\theta_2^k) = \text{interp}\big(\widehat{X}_2(\theta_2^k), \theta_2^k\big)$
5:     **if** $\widehat{X}_1(\theta_1^k) >= \widehat{X}_2(\theta_2^k)$ **then**
6:       $\theta_1^{k+1} = \Pi_I\big(\theta_1^k + a_k(\Psi - \widehat{X}_1(\theta_1^k))\big)$
7:       $\theta_2^{k+1} = \Pi_I\big(\theta_2^k + a_k(\Psi - \widehat{X}_1(\theta_1^k)) - \widehat{X}_2(\theta_2^k))\big)$
8:     **else**
9:       $\theta_1^{k+1} = \Pi_I\big(\theta_1^k + a_k(\Psi - \widehat{X}_1(\theta_1^k)) - \widehat{X}_2(\theta_2^k))\big)$
10:      $\theta_2^{k+1} = \Pi_I\big(\theta_2^k + a_k(\Psi - \widehat{X}_2(\theta_2^k))\big)$
11:    **end if**
12:    $k \leftarrow k + 1$
13: **end while**

---

determine the optimal static policy $\mathbf{U} = (U_1^*, U_2^*)$. In the following theorem we discuss the convergence of our algorithm.

Theorem 4. If the sequence $\{a_k\}$ verifies that $a_k > 0$, $\forall k$, $\sum_{k=0}^{+\infty} = +\infty$ and $\sum_{k=0}^{+\infty} a_k^2 < +\infty$, then the sequence (θ_1^k, θ_2^k) converges to the optimal solution (θ_1^*, θ_2^*).

Proof: First, we show that the sequence (θ_1^k, θ_2^k) converges to some limit set of the following Ordinary Differential Equation (ODE)

$$\dot{\theta}^1 = G_1(\theta^1) + z_1 = \Psi - E[X_1 + X_2|(\theta^1, \theta_{min})] + z_1, \quad (14)$$
$$\dot{\theta}^2 = G_2(\theta^1, \theta^2) + z_2 = \Psi - E[X_1 + X_2|(\theta^1, \theta^2)] + z_2, \quad (15)$$
$$z = (z_1, z_2) \in -C((\theta^1, \theta^2)),$$

where z_i, $i = 1, 2$, is the projection or constrain term, the minimum force needed to keep the trajectory of the ODEs in $[\theta_{min}, \theta_{max}]$ and the set $C(\vec{\theta})$ is defined as follows [23]: for $\vec{\theta} = (\theta_1, \theta_2) \in (\theta_{min}, \theta_{max})^2$, we have $C(\vec{\theta}) = \{(0, 0)\}$; and for (θ_1, θ_2) in the boundary of $[\theta_{min}, \theta_{max}]^2$, we let $C(\vec{\theta})$ be the infinite convex cone generated by the outer normals at $\vec{\theta}$ of the faces on which $\vec{\theta}$ lies. Put simply, $C(\vec{\theta})$ contains the possible values needed to keep $\vec{\theta}$ in $[\theta_{min}, \theta_{max}]$. For example, if $\theta_1 = \theta_{max}$ and $G_1(\theta_1)$ point out of $[\theta_{min}, \theta_{max}]$ then $z_1(t) = G_1(\theta_1)$. Hence the function $z(\cdot)$ is determined by $[\theta_{min}, \theta_{max}]^2$ and the functions $G_i(\cdot)$, $i = 1, 2$.

Since $G_1(\theta_1)$ (resp. $G_2(\theta_1, \theta_2)$) is decreasing function in θ_1 (resp. θ_1 and θ_2), then the equilibrium is unique. Moreover, it is easy to check that the optimal solution (θ_1^*, θ_2^*) (see theorem 1) is the unique equilibrium of (14)-(15). As discussed in [23], the convergence of such stochastic algorithm is guaranteed when the sequence (a_k) verifies, $a_k > 0$, $\forall k$, $\sum_{k=0}^{+\infty} a_k = +\infty$ and $\sum_{k=0}^{+\infty} a_k^2 < +\infty$. We now need to show that

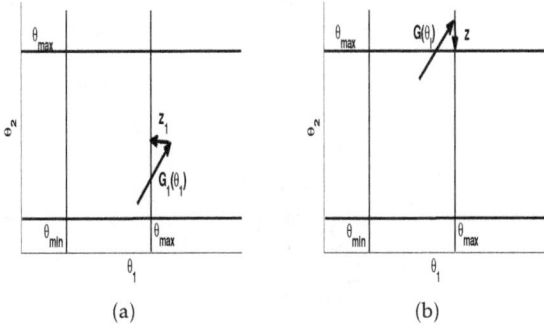

Figure 2. Projection of the value θ_i over I

(θ_1^*, θ_2^*) is globally asymptotically stable of the system (14)-(15). We use the Lyapunov function $V(\theta_1, \theta_2) = (\theta_1 - \theta_1^*)^2 + (\theta_2 - \theta_2^*)^2$. Then we have

$$
\begin{aligned}
\dot{V}(\vec{\theta}) &= 2\dot{\theta}_1(\theta_1 - \theta_1^*) + 2\dot{\theta}_1(\theta_2 - \theta_2^*) \\
&= 2(G_1(\theta_1) + z_1)(\theta_1 - \theta_1^*) \\
&\quad + 2(G_2(\theta_1, \theta_2) + z_2)(\theta_2 - \theta_2^*).
\end{aligned} \tag{16}
$$

Since there is at most one active constraint, we have two cases (see Fig. 2):

- If $\theta_{min} < \theta_i < \theta^{max}$ for $i = 1, 2$, then $z_1 = 0$ and $z_2 = 0$.

- If $\theta_i = \theta_{max}$ (resp. θ_{min}) and $G_i(\theta_{max}) > 0$ (resp. $G_i(\theta_{min}) < 0$) for only one class i, then $z_i = -G_i(\theta_{max})$ (resp. $z_i = -G_i(\theta_{min})$) and $z_{-i} = 0$ for other class.

And because $G_1(\theta_1)$ and $G_2(\theta_1, \theta_2)$ are strictly decreasing functions, it is easy to check that $\dot{V}(\vec{\theta})$ is decreasing. Hence the optimal solution is asymptotically stable.

5.2. Blind Online Algorithm for Dynamic Control

We apply the *two-time-scale stochastic approximation algorithm*, which is a stochastic recursive algorithm. Compared to standard stochastic approximation techniques in literature [23], here some of the components are updated using a step-size much smaller than those of the remaining components. For further insight into the convergence properties of this class of algorithms, the reader is referred to [24].

The algorithm introduced here drives the source to the optimal (ψ_1) that maximizes $J(\psi_1)$, i.e. maximizes the probability of success $F_D(\tau) = 1 - \exp\Big(-\sum_j \lambda_{jd} \int_{s=0}^{t} X_j(\psi_1, s)ds\Big)$. Since the system parameters (such as N_i and λ_i, $i = 1, 2$) are unknown, the value of $J(\psi_1)$ is also unknown, but a noisy estimate of it is known, namely $f(\psi_1)$ such that $\mathbb{E}[f|\psi_1] = J(\psi_1)$.

The two-time-scale stochastic approximation algorithm is then formulated as follows:

$$
\begin{aligned}
\theta_i^{k+1} &= \Pi_I\big(\theta_i^k + a_k(\psi_i^k - X_i(\theta_i^k))\big), \\
\psi_2^k &= \psi - \psi_1^k, i = 1, 2,
\end{aligned} \tag{17}
$$

$$
\psi_1^{k+1} = \Pi_H\Big(\psi_1^k + b_k \frac{f(\psi_1^k + c_k) - f(\psi_1^k - c_k)}{c_k}\Big), \tag{18}
$$

where $I = [0, K - 1]$, $H = [0, \psi]$ and $\{a_k\}, \{b_k\}$ are sequences of non-increasing positive constants satisfying $\sum_{k=0}^{+\infty} a_k = +\infty, \sum_{k=0}^{+\infty} a_k^2 < +\infty$, $\sum_{k=0}^{+\infty} b_k = +\infty, \sum_{k=0}^{+\infty} b_k^2 < +\infty$, and $\lim_{k \to +\infty} \frac{b_k}{a_k} = 0$. The last condition implies that $b_k \to 0$ at a faster rate than a_k, implying that (18) moves on a slower timescale than (17). An example of such stepsizes are $a_k = \frac{1}{k}$, $b_k = \frac{1}{1+k \log k}$ and so on. Further requirements are imposed on sequence (c_k), which we defer to Thm. 6 for the sake of clearness.

More in detail, at each round k the following steps are executed:

1: Fix (ψ_1) at the value (ψ_1^k) and learn the optimal values of θ_1^k, θ_2^k for $(\psi_1^k + c_k), (\psi_1^k - c_k)$ using the following algorithm:

$$
\begin{aligned}
\theta_1^{k+1} &= \Pi_I\big(\theta_1^k + a_k(\psi_1^k - X_1(\theta_1^k))\big), \\
\theta_2^{k+1} &= \Pi_I\big(\theta_2^k + a_k(\psi_2^k - X_2(\theta_2^k))\big).
\end{aligned} \tag{19}
$$

2: Measure the noisy estimate of the success probability, $f(\psi_1)$, at $(\psi_1^k + c_k), (\psi_1^k - c_k)$ when θ_1^k, θ_2^k, obtained at the first step, are applied.

3: Use $f(\psi_i^k + c_k), f(\psi_i^k - c_k)$ to update the value of (ψ_1^k) according to Kiefer-Wolfowitz algorithm as shown at step (11) of algorithm 2.

In the following theorem we proof the convergence of the Kiefer-Wolfowitz part of the algorithm that appears in (18); this serves as an introduction to theorem 6.

Theorem 5. If the sequence (b_k) verifies: $b_k > 0$, $\forall k$, and $\sum_{k=0}^{+\infty} b_k = +\infty, \sum_{k=0}^{+\infty} b_k^2 < +\infty$ and $c_k \to 0$, then (ψ_1^k) converges to the optimal solution (ψ_1^*).

Proof: Consider two sequences $\{b_k, c_k, k \geq 1\}$ satisfying $c_k \to 0$, $\sum_k b_k = \infty$, $\sum_k b_k c_k < \infty$, $\sum_k (b_k/c_k)^2 < \infty$, and the recursive updates of ψ_1:

$$
\psi_1^{k+1} = \psi_1^k + b_k \frac{f(\psi_1^k + c_k) - f(\psi_1^k - c_k)}{c_k}. \tag{20}
$$

This recursive schema converges stochastically to the optimal value (ψ_1^*) that maximizes $J(\psi_1)$ provided that $J(\psi_1)$ satisfies the following conditions:

1. $J(\psi_1)$ is a strictly quasi-concave function.

2. There exists β and B such that $|\psi_1^a - \psi_1^*| + |\psi_1^b - \psi_1^*| < \beta$ implies $|J(\psi_1^a) - J(\psi_1^b)| < B|\psi_1^a - \psi_1^b|$.

3. There exists ρ and R such that $|\psi_1^a - \psi_1^b| < \rho$ implies $|J(\psi_1^a) - J(\psi_1^b)| < R$.

4. For every $\delta > 0$ there exists a positive $\pi(\delta)$ such that $|\psi_1 - \psi_1^*| > \delta$ implies

$$\inf_{0<\epsilon<\frac{\delta}{2}} \frac{|J(\psi_1 + \epsilon) - J(\psi_1 - \epsilon)|}{\epsilon} > \pi(\delta).$$

In order to prove that $J(\psi_1)$ satisfies these conditions, we only need to prove that $J(\psi_1)$ is concave and has a unique maximum solution ψ_1^* ([25]-lemma 2, theorem 1), which is already proved in theorem 3, hence proved. ◇

Theorem 6. The sequence (θ_i^k, ψ_1^k) defined in the iteration (17) and (18) converges a.s. to $(\theta_i(\psi_1^*), \psi_1^*)$.

Proof: This can be proved directly by the results of ([26], Chapter 6) together with the sure convergence of the two related single-time-scale stochastic approximation algorithms – those defined in (17) and (18), respectively – and that each algorithm has a globally asymptotically stable equilibrium, appearing in the following.

The first algorithm is the aforementioned algorithm (Alg. 1) taking as entries the fixed values of $\widehat{X}_1(\theta_1^k), \widehat{X}_2(\theta_2^k)$ (which are ψ_1^k, ψ_2^k for the current round k). In fact, (17) sees ψ_1 as quasi-static (i.e., 'almost a constant') and it is easy to prove that the sequence (θ_1^k, θ_2^k) converges to some limit set of the following Ordinary Differential Equation (ODE)

$$\begin{aligned}
\dot{\theta}_1 &= G_1(\theta_1) + z_1 \\
&= \psi_1 - E[X_1 | \theta_1] + z_1, \quad z_1 \in -C_1(\theta_1), \quad (21) \\
\dot{\theta}_2 &= G_2(\theta_2) + z_2 \\
&= (\psi - \psi_1) - E[X_2 | \theta_2] + z_2, \quad z_2 \in -C_2(\theta_2), (22)
\end{aligned}$$

where z_i, $i = 1, 2$, is the minimum force needed to keep the solution θ_i in $I = [\theta_{min}, \theta_{max}]$.

If θ_i is in I on some time interval, then $z_i(\cdot)$ is zero on that interval ($C_i(\theta_i)$ contains only the zero element). If θ_i is on the interior of a boundary of I (i.e., θ_i equals either θ_{min} or θ_{max}) and $G_i(\theta_i)$ points out of I, then $z_i(\cdot)$ points backward inside I, i.e. $C_i(\theta_i)$ is the infinite convex cone generated by the outer normals at θ_i of the faces on which θ_i lies. For example, let $\theta_i = \theta_{max}$, with $G_i(\theta_i) > 0$, then, $z_i(t) = -G_i(\theta_i)$. ◇

Following the same reasoning in the proof of theorem 4, it is easy to verify that each of the ODEs (21) and (22) has a globally asymptotically stable equilibrium $\theta_i^*(\psi_1)$.

The second algorithm is the Kiefer-Wolfowitz whose convergence is proved in theorem 5 and its asymptotic behavior is characterized [23] by the ODE

Algorithm 2 Stochastic approximation of the optimal policy using online estimation (2D Dynamic case)

1: **input:** $I = [0, K - 1], H = [0, \psi]$, $\theta_1^0 = \theta_2^0 = K/2$, $\psi_1^0 = \psi$, $c_k = \frac{1}{k^{0.001}}$
2: **while** $|\psi_1^{k+1} - \psi_1^k| > \epsilon$ **do**
3:     set $\psi_1^+ = \min(\psi_1^k + c_k, \psi)$, $\psi_2^+ = \psi - \psi_1^+$
4:     $\theta_1^k = \Pi_I\big(\theta_1^k + a_k(\psi_1^+ - \widehat{X}_1(\theta_1^k))\big)$
5:     $\theta_2^k = \Pi_I\big(\theta_2^k + a_k(\psi_2^+ - \widehat{X}_2(\theta_2^k))\big)$
6:     Measure $f(\psi_1^+) = func(\theta_1^k, \theta_2^k)$
7:     set $\psi_1^- = \max(\psi_1^k - c_k, 0)$, $\psi_2^- = \psi - \psi_1^-$
8:     $\theta_1^k = \Pi_I\big(\theta_1^k + a_k(\psi_1^- - \widehat{X}_1(\theta_1^k))\big)$
9:     $\theta_2^k = \Pi_I\big(\theta_2^k + a_k(\psi_2^- - \widehat{X}_2(\theta_2^k))\big)$
10:    Measure $f(\psi_1^-) = func(\theta_1^k, \theta_2^k)$
11:    $\psi_1^{k+1} = \Pi_H\big(\psi_1^k + b_k(\frac{f(\psi_1^+) - f(\psi_1^-)}{c_k})\big)$
12:    $k \leftarrow k + 1$
13: **end while**

$$\begin{aligned}
\dot{\psi}_1 &= G(\theta_1(\psi_1), \theta_2(\psi_1), \psi_1) + z_3 \\
&= \frac{\partial J(\psi_1)}{\partial \psi_1} + z_3, z_3 \in -C_3(\psi_1), \quad (23)
\end{aligned}$$

where z_3, is the minimum force needed to keep the solution ψ_1 in $H = [0, \psi]$. In order to prove that (ψ_1^*) is globally asymptotically stable of the ODE (23), we use the Lyapunov function $V(\psi_1) = (\psi_1 - \psi_1^*)^2$. Then we have

$$\begin{aligned}
\dot{V}(\vec{\psi}_1) &= 2\dot{\psi}_1(\psi_1 - \psi_1^*) \\
&= 2\frac{\partial J(\psi_1)}{\partial \psi_1}(\psi_1 - \psi_1^*) < 0. \quad (24)
\end{aligned}$$

Asymptotic global stability follows from Lyapunov's theorem. ◇

5.3. Discussion on Implementation Issues

Since the proposed algorithms do not require a-priori knowledge of network parameters like intermeeting intensities and number of mobiles of each class, they can be easily implemented in real scenario. In other words, for relay nodes to apply those algorithms, a simple coded version of the proposed algorithms can be loaded on each device's memory. Each time a node needs to transmit a message, i.e., becomes a source, it executes this code. Moreover, as the forwarding protocol employed here is two hop, source can keep track of number of message copies in the network. After few rounds, as we will show in next section, the source will be driven to apply the optimal policies characterized in the previous sections. Certainly timers should be set for algorithm rounds, but those timers will operate distributively on each node.

6. Numerical Investigation

In this section we illustrate numerical experiments that validate our results. Using Matlab scripts, we have studied the impact of the inter-meeting intensity and the energy constraint (ψ) on the 2-D system optimal dynamic policies. We are also interested in the performance of the stochastic approximation algorithms described earlier; we will investigate their ability to drive the two hop forwarding, both in the static as well as dynamic scenario, to the optimal operating point.

In Fig. 3, we reported the optimal control in the case when intermeeting intensities with both the source and the destination are the same ($\lambda_{si} = \lambda_{id}$ for i=1,2), we can temporarily refer to both intensities by mentioning one of them, let us say λ_{s1}.

The first observation on Fig. 3(a) is that when λ_{s1} is small, the control tends to infect both classes over a larger interval. However, as long as $\lambda_{s1} < \lambda_{s2}$, the switch time for class 2 is higher and the source infects more nodes of class 2 (Fig. 3(c)). As λ_{s1} increases, an interesting case is when $\lambda_{s1} = \lambda_{s2}$: the switch times are equal and the source infects the same number of nodes per class.

The situation is flipped after equality is reached: when λ_{s1} exceeds λ_{s2}, the source tends to depend more on class 1 and infects more nodes of that class. This can be explained by the fact that class 1 nodes meet the source and destination with higher frequency ($\lambda_{1d} = \lambda_{s1} > \lambda_{s2}$). Intuitively, the probability of success improves as λ_{s1} increases as in Fig. 3(b).

In Fig. 4, we depict the switch times for both classes when each class's nodes meet the source at a rate that is different from the rate at which they meet the destination ($\lambda_{si} \neq \lambda_{id}$ for i=1,2). Fig. 4(a) shows that the switch time of class 1 increases as λ_{1d} increases but still is smaller than the switch time for class 2, as long as ($\lambda_{1d} < \lambda_{2d}$), after this point the source will infect class 1 nodes for longer time because they have higher chance now to meet the destination. The same observation is captured in Fig. 4(b). In Fig. 4(c) we observe that at the beginning when ($\lambda_{1d} < \lambda_{2d}$), the source infects more nodes of class 2 than class 1, then X_1 keeps increasing until it becomes larger than X_2 when ($\lambda_{1d} > \lambda_{2d}$). Intuitively, the probability of success improves as λ_{1d} increases as in Fig. 4(d).

Another important aspect is the impact of the energy constraint on the dynamic policies adopted by the source (see Fig. 5). We notice that the larger the energy constraint is, the higher the switch times are (Fig. 5(a) and 5(c)). This is because increasing ψ allows the source to infect more nodes by augmenting the optimal switch times while satisfying the constraint (Fig. 5(b) and 5(d)). An interesting observation is captured in Fig. 5(b) (Fig.5(d)): when ψ has a small value ($\psi = 10$), the source

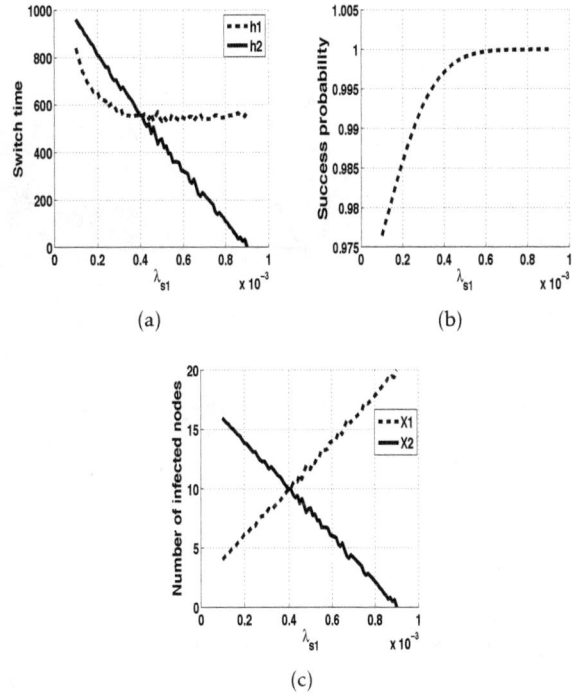

Figure 3. The impact of the inter-meeting intensity ($\lambda_{si} = \lambda_{id}$ for i=1,2) on switching times, delivery probability and number of infected nodes in the 2-D case, where $\psi = 20, N_1 = N_2 = 50, \lambda_{s2} = 0.4 \times 10^{-3}, \tau = 1000$ s.

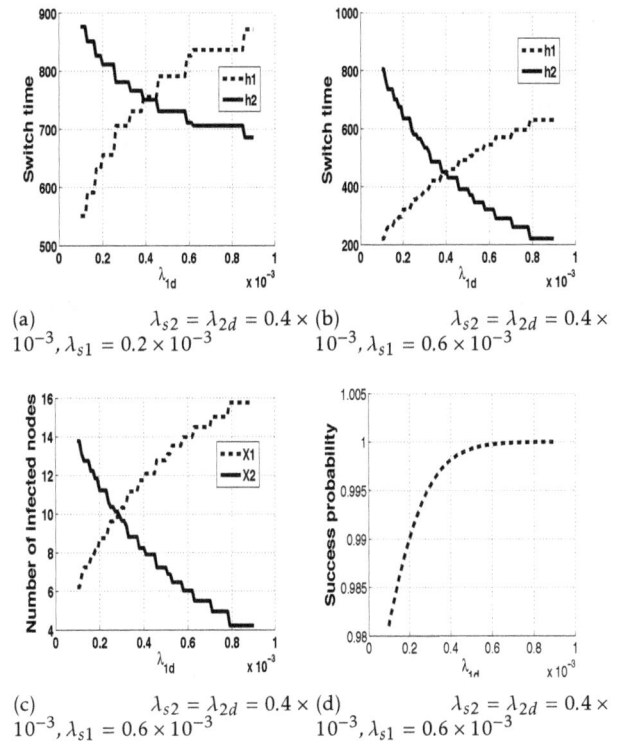

(a) $\lambda_{s2} = \lambda_{2d} = 0.4 \times 10^{-3}, \lambda_{s1} = 0.2 \times 10^{-3}$ (b) $\lambda_{s2} = \lambda_{2d} = 0.4 \times 10^{-3}, \lambda_{s1} = 0.6 \times 10^{-3}$

(c) $\lambda_{s2} = \lambda_{2d} = 0.4 \times 10^{-3}, \lambda_{s1} = 0.6 \times 10^{-3}$ (d) $\lambda_{s2} = \lambda_{2d} = 0.4 \times 10^{-3}, \lambda_{s1} = 0.6 \times 10^{-3}$

Figure 4. The impact of the inter-meeting intensity ($\lambda_{s1} \neq \lambda_{1d}$) on switching times and number of infected nodes in the 2-D case, where $\psi = 20, N_1 = N_2 = 50, \tau = 1000$ s.

(a) $\lambda_{s1} = 0.6 \times 10^{-3}, \lambda_{s2} = 0.4 \times$ (b) $\lambda_{s1} = 0.6 \times 10^{-3}, \lambda_{s2} = 0.4 \times$
10^{-3} 10^{-3}

(c) $\lambda_{s1} = 0.4 \times 10^{-3}, \lambda_{s2} = 0.6 \times$ (d) $\lambda_{s1} = 0.4 \times 10^{-3}, \lambda_{s2} = 0.6 \times$
10^{-3} 10^{-3}

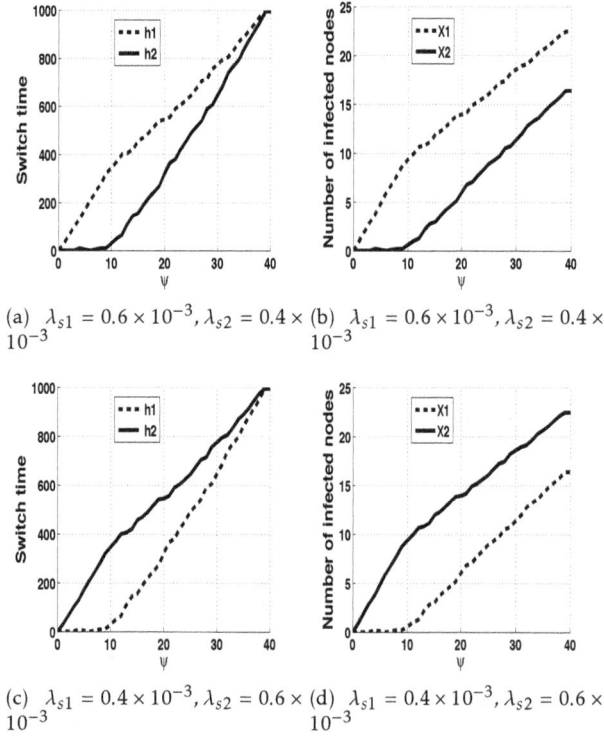

Figure 5. The impact of the energy constraint (ψ) on switching times and number of infected nodes in the 2-D case, where $N_1 = N_2 = 50$.

(a) $\psi = 30$ (b) $\psi = 60$

(c) $\psi = 90$

Figure 6. The stochastic approximation using online estimation ($\lambda_{s1} > \lambda_{s2}$) in the 2-D case (static algorithm 1), where $\tau = 1000, N_1 = N_2 = 1000, \lambda_{s1} = \lambda_{1d} = 0.34 \times 10^{-4}, \lambda_{s2} = \lambda_{2d} = 0.14 \times 10^{-4}$.

forwards the message only to class 1 (class 2) nodes, respectively. This indicates that under a tight energy constraint, the source can rely on one class only (the one with higher λ_s) to have its message delivered. For the case where $\psi = 40$, the source will forward the message all the time to both classes in order to saturate the constraint.

Fig. 6 shows sample paths of our learning algorithm for the problem of static control. In Fig. 6(a) the algorithm is showed to converge to $(U_1^*, U_2^*) = (u_1^*, u_{min})$ with $0 < U_1^* < 1$. Indeed, since the source may infect at most 30 nodes, from theorem (1, case (iv)) and remark (1), the optimal solution is (v_1, u_{min}) with $v_1 = \frac{-1}{\lambda_{s1}\tau} \log\left(\frac{N-\psi}{N_1-X_1(0)} - \frac{N_2-X_2(0)}{N_1-X_1(0)} e^{-\lambda_{s2}\tau u}\right)$, we can see that U_1 converges to v_1 and U_2 to u_{min}.

In Fig. 6(b) we plot case (iii) of theorem 1 where the energy constraint is 60 and the algorithm converges to $(U_1^*, U_2^*) = (1, v_2)$ with $v_2 = \frac{-1}{\lambda_{s2}\tau} \log\left(\frac{N-\psi}{N_2-X_2(0)} - \frac{N_1-X_1(0)}{N_2-X_2(0)} e^{-\lambda_{s1}\tau}\right)$. In this case the source will infect nodes of both classes in order to saturate the constraint by giving the message with probability 1 to the class with larger intermeeting intensities $\{\lambda_{si}, \lambda_{id}\}$ (class 1) and with some probability v_2 to the other class.

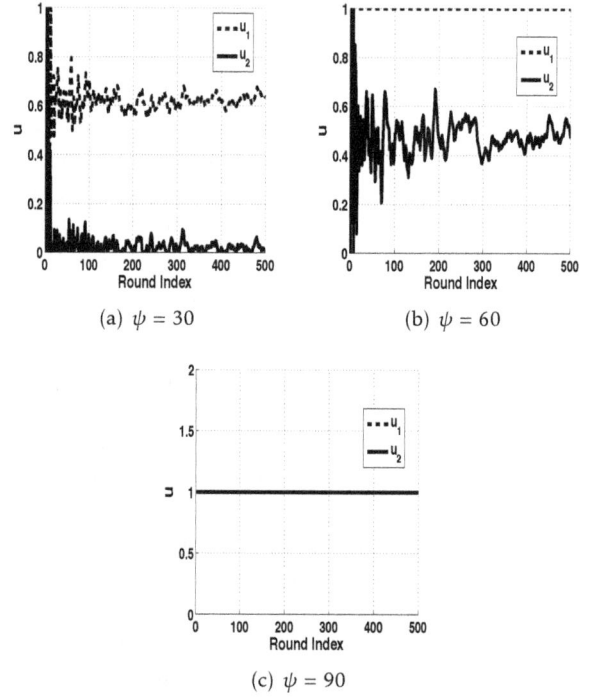

Fig. 6(c) shows the case where the constraint has a large value ($\psi = 90$) and the source can not reach the constraint even when forwarding to both classes with probability 1.

The performance of our learning algorithm in the dynamic case is shown in Fig. 7. In Figs. 7(a) and 7(b) we depict the convergence of ψ_1 (recall that ψ_1 is the maximum number of nodes of class 1 that can be infected) to the optimal value that maximizes the delivery probability under two different values of the energy constraint ($\psi = 20, \psi = 50$), and in Figs. 7(c) and 7(d) we show the corresponding switch times for both cases.

To better understand Fig. 7, let us observe Figs. 7(a) and 7(c) where the energy constraint has a small value ($\Psi = 20$), we notice that since class 1 has bigger intermeeting intensities ($\lambda_{s1}, \lambda_{1d}$), the algorithm drives the source to forward the message to class 1 for longer time resulting in more infected nodes in class 1 (19 infected nodes out of 20), which is the value at which the probability of success attains its maximum Fig.7(e). While in Figs. 7(b) and 7(d) where ψ is large ($\psi = 50$), the algorithm converges to ($\psi_1^* = 32$) the value that maximizes the success probability in Fig.7(f). Note that since it meets nodes of class 1 more often ($\lambda_{s1} > \lambda_{s2}$), the source can infect more nodes of class 1 with smaller switching time for this class while it has to forward

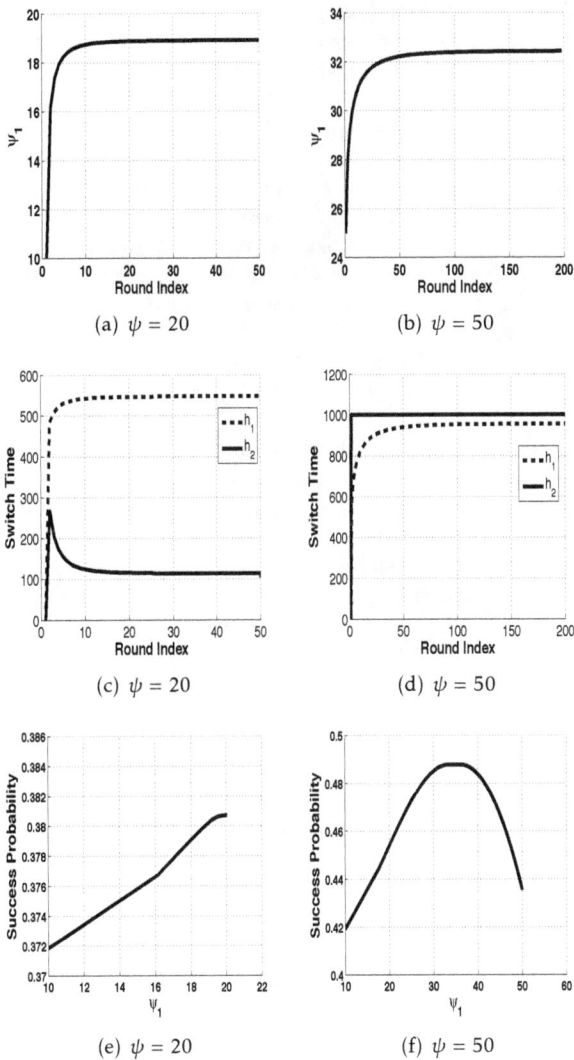

(a) $\psi = 20$ (b) $\psi = 50$

(c) $\psi = 20$ (d) $\psi = 50$

(e) $\psi = 20$ (f) $\psi = 50$

Figure 7. The stochastic approximation using online estimation $(\lambda_{s1} > \lambda_{s2})$ in the 2-D case (Dynamic algorithm 2), where $\tau = 1000, N_1 = N_2 = 1000, \lambda_{s1} = \lambda_{1d} = 0.34 \times 10^{-4}, \lambda_{s2} = \lambda_{2d} = 0.14 \times 10^{-4}$.

all the time to class 2 in order to saturate the energy constraint.

7. Conclusion

In this paper we studied the problem of optimal relaying for DTNs consisting of *heterogeneous* mobile nodes. We have considered controlled two hop forwarding policies for source-destination message delivery using multiple classes of relays. Using our model we characterized the optimal forwarding policy in the family of the multi-dimensional DTNs for both static and dynamic control cases.

Finally, we have designed novel algorithms based on the theory of stochastic approximations. Those algorithms enable nodes in a heterogeneous DTN to

tune up independently and optimally the parameters for both static and dynamic optimal forwarding policies. Thus, nodes autonomously adapt to the operating point of a system comprising multiple classes of nodes. The distinctive feature of our so-called *blind online algorithms* is that the source node does not need to know explicitly the system parameters a priory. Instead, our novel implementation works at runtime permitting online adaptation to the specific existing conditions.

References

[1] CHAINTREAU, A., HUI, P., CROWCROFT, J., DIOT, C., GASS, R. and SCOTT, J. (2007) Impact of human mobility on opportunistic forwarding algorithms. *IEEE Transactions on Mobile Computing* 6: 606–620.

[2] BURLEIGH, S., TORGERSON, L., FALL, K., CERF, V., DURST, B., SCOTT, K. and WEISS, H. (2003) Delay-tolerant networking: an approach to interplanetary Internet. *IEEE Comm. Magazine* 41: 128–136.

[3] PELUSI, L., PASSARELLA, A. and CONTI, M. (2006) Opportunistic networking: data forwarding in disconnected mobile ad hoc networks. *IEEE Communications Magazine* 44(11): 134–141.

[4] ZHAO, W., AMMAR, M. and ZEGURA, E. (2005) Controlling the mobility of multiple data transport ferries in a delay-tolerant network. In *Proc. of IEEE INFOCOM* (Miami USA).

[5] ZHANG, Z. (2006) Routing in intermittently connected mobile ad hoc networks and delay tolerant networks: overview and challenges. *Communications Surveys & Tutorials, IEEE* 8(1): 24–37.

[6] JONES, E.P. and WARD, P.A. (2006) Routing strategies for delay-tolerant networks. *Submitted to ACM Computer Communication Review (CCR)* .

[7] SPYROPOULOS, T., PSOUNIS, K. and RAGHAVENDRA, C. (2008) Efficient routing in intermittently connected mobile networks: the multi-copy case. *ACM/IEEE Transactions on Networking* 16: 77–90.

[8] ALTMAN, E., BAŞAR, T. and DE PELLEGRINI, F. (2010) Optimal monotone forwarding policies in delay tolerant mobile ad-hoc networks. *Performance Evaluation* 67(4): 299 – 317.

[9] VAHDAT, A. and BECKER, D. (2000) *Epidemic Routing for Partially Connected Ad Hoc Networks*. Tech. Rep. CS-2000-06, Duke University.

[10] HANBALI, A.A., NAIN, P. and ALTMAN, E. (2006) Performance of ad hoc networks with two-hop relay routing and limited packet lifetime. In *Proc. of Valuetools* (New York NY USA: ACM): 49.

[11] NEGLIA, G. and ZHANG, X. (2006) Optimal delay-power tradeoff in sparse delay tolerant networks: a preliminary study. In *Proc. of ACM SIGCOMM CHANTS 2006*: 237–244.

[12] SPYROPOULOS, T., TURLETTI, T. and OBRACZKA, K. (2009) Routing in delay-tolerant networks comprising heterogeneous node populations. *Mobile Computing, IEEE Transactions on* 8(8): 1132 –1147.

[13] CHAINTREAU, A., BOUDEC, J.Y.L. and RISTANOVIC, N. (2009) The age of gossip: Spatial mean-field regime. In

Proc. of ACM SIGMETRICS (Seattle, Washington, USA).

[14] DE PELLEGRINI, F., ALTMAN, E. and BASAR, T. (2010) Optimal monotone forwarding policies in delay tolerant mobile ad hoc networks with multiple classes of nodes. In *proc. of WiOpt WDM Workshop* (Avignon, France).

[15] ALTMAN, E., NEGLIA, G., DE PELLEGRINI, F. and MIORANDI, D. (2009) Decentralized stochastic control of delay tolerant networks. In *Proc. of INFOCOM* (Rio de Janeiro, Brazil).

[16] GUERRIERI, A., CARRERAS, I., PELLEGRINI, F.D., MIORANDI, D. and MONTRESOR, A. (2010) Distributed estimation of global parameters in delay-tolerant networks. *Elsevier Comput. Commun.* **33**(13): 1472–1482.

[17] HUI, P., CROWCROFT, J. and YONEKI, E. (2008) Bubble rap: social-based forwarding in delay tolerant networks. In *Proc of ACM MobiHoc* (New York, NY, USA: ACM): 241–250. doi:http://doi.acm.org/10.1145/1374618.1374652.

[18] CONAN, V., LEGUAY, J. and FRIEDMAN, T. (2007) Characterizing pairwise inter-contact patterns in delay tolerant networks. In *Proc. of ACM Autonomics*: 1–9.

[19] LEE, C.H. and EUNT, D.Y. (2009) Heterogeneity in contact dynamics: helpful or harmful to forwarding algorithms in DTNs? In *Proc. of WiOPT* (Seoul, Korea): 72–81.

[20] GROENEVELT, R. and NAIN, P. (2005) Message delay in MANETs. In *Proc. of Sigmetrics* (Banff, Canada: ACM): 412–413. See also R. Groenevelt, Stochastic Models for Mobile Ad Hoc Networks. PhD thesis, University of Nice-Sophia Antipolis, April 2005.

[21] ZHANG, X., NEGLIA, G., KUROSE, J. and TOWSLEY, D. (2007) Performance modeling of epidemic routing. *Elsevier Computer Networks* **51**: 2867–2891.

[22] BAKHSHI, R., CLOTH, L., FOKKINK, W. and HAVERKORT, B. (2008) Meanfield analysis for the evaluation of gossip protocols. *Sigmetrics Performance Evaluation Review archive* **36**: 31–39.

[23] KUSHNER, H.J. and YIN, G.G. (2003) *Stochastic Approximation and Recursive Algorithms and Applications* (Springer, 2nd Edition).

[24] BORKAR, V.S. (1997) Stochastic approximation with two time scales (Elsevier).

[25] (2010) *Stochastic approximation to optimize the performance of human operators.*

[26] BORKAR, V.S. (2008) *Stochastic Approximation: A Dynamical Systems Viewpoint* (Cambridge University Press).

9

Logical Link Control and Channel Scheduling for Multichannel Underwater Sensor Networks

Jun Li *, Mylène Toulgoat, Yifeng Zhou, and Louise Lamont

Communications Research Centre Canada, 3701 Carling Avenue, Ottawa, ON. K2H 8S2 Canada

Abstract

With recent developments in terrestrial wireless networks and advances in acoustic communications, multichannel technologies have been proposed to be used in underwater networks to increase data transmission rate over bandwidth-limited underwater channels. Due to high bit error rates in underwater networks, an efficient error control technique is critical in the logical link control (LLC) sublayer to establish reliable data communications over intrinsically unreliable underwater channels. In this paper, we propose a novel protocol stack architecture featuring cross-layer design of LLC sublayer and more efficient packet-to-channel scheduling for multichannel underwater sensor networks. In the proposed stack architecture, a selective-repeat automatic repeat request (SR-ARQ) based error control protocol is combined with a dynamic channel scheduling policy at the LLC sublayer. The dynamic channel scheduling policy uses the channel state information provided via cross-layer design. It is demonstrated that the proposed protocol stack architecture leads to more efficient transmission of multiple packets over parallel channels. Simulation studies are conducted to evaluate the packet delay performance of the proposed cross-layer protocol stack architecture with two different scheduling policies: the proposed dynamic channel scheduling and a static channel scheduling. Simulation results show that the dynamic channel scheduling used in the cross-layer protocol stack outperforms the static channel scheduling. It is observed that, when the dynamic channel scheduling is used, the number of parallel channels has only an insignificant impact on the average packet delay. This confirms that underwater sensor networks will benefit from the use of multichannel communications.

Keywords: Underwater sensor networks, multichannel communications, cross-layer design, logical link control (LLC), channel scheduling, modeling and simulation, packet delay.

1. Introduction

Underwater sensor networks will play an important role in fulfilling enhanced capability of maritime situational awareness and response in coastal waters. Compared with terrestrial wireless sensor networks, underwater sensor networks experience slower propagation speed, lower transmission rates, and poorer quality of communication links [1–3]. In seawater, for instance, the speed of acoustic signals is in the order of 10^3 meters per second; the data transmission rate for an acoustic modem can be up to a few *kbps* with a transmission range up to several kilometers; high bit error rates (BER) are expected in underwater sensor networks due to multipath interferences and Doppler distortions.

Multichannel technologies have been adopted in next-generation terrestrial wireless communications to increase data transmission rate. For instance, a multiple-input multiple-output antennas (MIMO) system uses multiple channels consisting of distinct antenna pairs [4], while orthogonal frequency division multiplexing (OFDM) applies disjoint frequency bands to form multiple channels [5, 6]. Both technologies have been used in wireless network standards such as WiMax (IEEE 802.16) [7] and LTE (3GPP Long Term Evolution) [8]. Thanks to these developments in terrestrial wireless networks and recent advances in acoustic communications [9–12], the multichannel technologies are used in underwater networks to

---

*Corresponding author. Email: jun.li@crc.gc.ca

increase data transmission rate over more bandwidth-limited underwater channels.

Given the exceptionally high bit error rates in underwater networks, an efficient error control technique is critical in the logical link control (LLC) sublayer to establish reliable data communications over intrinsically unreliable underwater channels. In comparison to multichannel medium access control (MAC) schemes (e.g., [13–15]) and routing protocols (e.g., [16–18]) reported in the literature for long-delay underwater sensor networks, little research on the LLC design in multichannel underwater sensor networks has been conducted. Indeed, several studies in the context of single-channel underwater communications have been reported [3, 19, 20]. All LLC schemes reported in these studies are modified versions of the stop-and-wait automatic-repeat-request protocol (SW-ARQ), which has long been known to be less efficient than the selective-repeat ARQ scheme (SR-ARQ) in both throughput and delay performance.

In multichannel terrestrial wireless networks, LLC schemes, which are often designed based on these classical single-channel ARQ protocols, (i.e., SW-ARQ, the go-back-N ARQ protocol (GBN-ARQ), and SR-ARQ) and thus referred to as multichannel ARQ, have become an integral part of the LLC sublayer for high-speed multimedia services [21, 22]. In the literature, several studies on multichannel ARQ protocols for terrestrial wireless networks have been reported. For instance, system throughput performance in multichannel ARQ protocols was studied in [23–25]. Chang and Yang [26] analyzed the average packet delay for the three classical ARQ protocols over multiple identical channels (i.e., all channels have the same transmission rate and the same error rate). Fujii and Hayashida and Komatu [27] derived the probability distribution function of the packet delay for GBN-ARQ over multiple channels that have the same transmission rate but possibly different error rates. Ding [28] considered ARQ protocols for parallel channels that possibly have both different transmission rates and different error rates, and derived approximate expressions for their mean packet delay. The resequencing issue in multichannel ARQ protocols was addressed by Shacham and Chin [29], and recently by Li and Zhao [30]. The packet delay distribution function for SW-ARQ over multiple channels was studied in [31] using an end-to-end analytical approach.

Motivated by the approaches applied to LLC designs for terrestrial wireless networks, in this paper we describe a multichannel underwater sensor network system, where each transmitter-receiver pair will be connected by a generic number of forward channels. Given the physical multichannel system, we propose a novel cross-layer protocol stack architecture for multichannel underwater sensor networks. In the proposed cross-layer design, a dynamic packet-to-channel scheduling policy takes advantage of the channel state information and is combined with a SR-ARQ based error control scheme to provide improved network performance. A simplified version of the proposed protocol stack is implemented and the packet delay performance is evaluated using computer simulations.

The main contributions of this paper include a novel cross-layer protocol stack architecture used for multichannel underwater sensor networks. The proposed protocol stack architecture uses a SR-ARQ based design at the logical link control sublayer, can make the channel state information available at the sublayer via cross-layer design. Using the channel state information, a dynamic channel scheduling can be used for simultaneously and more efficiently transmitting multiple packets over parallel channels. Simulation results show that the dynamic channel scheduling approach enhances the performance of multichannel underwater sensor networks over a static channel scheduling case. With the dynamic channel scheduling, the average packet delay increases with the average of error rates of the parallel channels, but decreases with the variance in the error rates; with the static channel scheduling, the average packet delay increases with both the average error rate and the variance in the error rates. In addition, if the average error rate among parallel channels remains fixed, the number of parallel channels has an insignificant impact on the average packet delay when the dynamic channel scheduling is applied. However, the average packet delay is severely affected by the number of parallel channels when the static channel scheduling is used.

The rest of this paper is organized as follows. Section 2 describes a multichannel underwater sensor network and proposes a protocol stack architecture featuring cross-layer design of the logical link control sublayer protocols. The SR-ARQ based logical link control design is proposed in Section 3, followed by a dynamic channel scheduling in Section 4. Simulation results are presented and discussed in Section 5, followed by the final section concluding this study.

2. Multichannel Underwater Sensor Networks

In this section, we describe a multichannel communication model for underwater sensor networks and propose a protocol stack architecture for the sensor node.

2.1. Multichannel Communication Model

Multiple underwater sensors are deployed on the seabed in a choke point where surveillance and reconnaissance of surface vessels and submarines are required. Each sensor is capable of collecting, processing and communicating sensory data acquired from its surroundings to another sensor node or to a more capable communication unit, such as an autonomous underwater vehicle (AUV). As a result, these multiple sensors form an underwater surveillance network. The physical communication device in each sensor node is a multichannel system, *e.g.*, a CDMA system in [32] or a MIMO-OFDM system in [12]. In addition, the physical device can conduct full-duplex data communication, which has been proven feasible in underwater networks using CDMA techniques [32]. Then, for a connected pair of sensor nodes, a multichannel communication is illustrated in Figure 1, where node A will transmit data packets and receive acknowledgements, while node B will receive data packets and transmit acknowledgements. We assume that the forward link (from A to B) consists of M ($M \geq$ 2) parallel channels that can transmit data packets simultaneously. Each channel is characterized by a data transmission rate, which is used to characterize the effective channel bandwidth and defined as the number of bits of data transmitted over the channel during a unit of time, and a packet error rate, which characterizes the lossy property of the underwater channel caused by multipath interferences and Doppler distortions. A feedback channel (from B to A) is used for transmitting acknowledgement frames (see Section 3.2) and is assumed to be error-free. The following assumptions are also used in this study.

- The M forward channels have the same transmission rate, and they are slotted in time with one unit (or slot) equal to the transmission time of a packet over a channel, *i.e.*, the transmission rate of each channel is one packet per slot. Meanwhile, it is equal to the transmission time of an acknowledgement frame over the feedback channel.

- The propagation delay of a data packet on a forward channel and the propagation delay of an acknowledgement frame on the feedback channel are the same as given by τ slots. Then the packet round trip time (RTT) equals $2(\tau + 1)$.

- A high-rate cyclic redundancy check (CRC) error-detection code (*e.g.*, a 16-bit polynomial in [33]) is used so that an erroneous packet received over a forward channel can always be detected.

- Packet errors that occur on different channels are assumed to be mutually independent.

2.2. Cross–Layer Protocol Stack Architecture

Each individual sensor node is configured with the protocol stack shown in Figure 2. In a multi-hop underwater sensor network, the upper layer

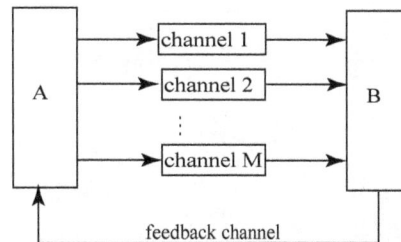

Figure 1. Multichannel Communication Model

corresponds to the network layer where a route for each source-destination pair needs to be determined. In the simulation study conducted in this paper, which involves only two one-hop neighbors, the upper layer in Figure 2 will act as either a packet generator (in the transmitter) or a data sink (in the receiver). Packets generated at the upper layer are sequentially assigned to integer numbers, referred to as their sequence numbers. All channels share the same set of packet sequence numbers. Below the upper layer is the data link layer, which is composed of the LLC sublayer and the MAC sublayer, and is next to the physical layer. The focus of this work is design of the logical link control sublayer, which is responsible for correcting corrupted packets caused by poor transmission conditions on channels. To that end, a multichannel MAC protocol (*e.g.*, the one in [14]) is assumed in the MAC sublayer responsible for solving the packet transmission problem due to collisions. The physical layer involves channel coding and modulation at the transmitter, and demodulation and decoding at the receiver. Meanwhile, the channel state information, *e.g.*, the bit error rate of each channel, can be assessed in the physical layer. The LLC design to be elaborated in Section 3 provides services to the upper layer and relies on its immediate lower layer (*i.e.*, the MAC sublayer) to perform required functions. Moreover, as a cross-layer design of the LLC sublayer, our proposed LLC design is allowed to access the bit error rate information of physical channels at the physical layer and use this information for packet transmission scheduling, which will be discussed in detail in Section 4.

3. Logical Link Control Design

In this section, we give an overview of the SR-ARQ scheme for single-channel communications and propose a SR-ARQ based logical link control design for multichannel underwater sensor networks.

For a logical link control sublayer design, each sensor node has a buffer for storing packets. When the sensor node acts as the transmitter, its buffer is referred to as the transmission queue, where packets can wait for transmission and retransmission based on the first-in-first-out service discipline, *i.e.*, packets with

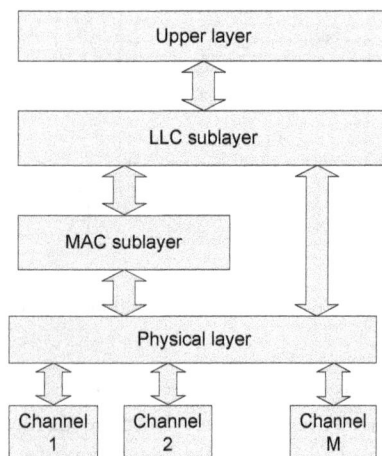

Figure 2. Protocol Stack Architecture

smaller sequence numbers have higher priority to be (re)transmitted than a packet with a larger sequence number. When the sensor node acts as the receiver, its buffer is denoted as the resequencing queue, where out-of-sequenced packets waiting for delivery to the upper layer (e.g., network layer) are temporarily stored.

3.1. SR-ARQ in Single-Channel Communications

In single-channel SR-ARQ, the transmitter sends packets continuously, while the receiver generates either a negative acknowledgement (NACK) for an erroneously received packet or a positive acknowledgement (ACK) for a correctly received packet. These control packets are sent over the feedback channel. Once a NACK arrives at the transmitter, the transmitter retransmits the negatively acknowledged packet without retransmitting the packets following it. To preserve the same order of packets as they arrived from the upper layer, the resequencing queue at the receiver is used to store mis-ordered packets, which are the correctly received packets with the condition that at least one packet with a smaller sequence number has not been correctly received. Via the resequencing queue of the receiver, packets are sequentially delivered from the LLC sublayer to the network layer.

3.2. SR-ARQ Based Multichannel LLC

The M channels are numbered by $i = 1, 2, \cdots, M$. At the beginning of a slot, the transmitter starts transmitting a block of M packets, one packet per channel, and completes transmission at the end of the slot. The receiver receives the block of M packets, which were transmitted in slot t for $t = 1, 2, \cdots$, in slot $t + \tau + 1$ (see Figure 3). The receiver responds to an erroneously received or lost packet by generating a NACK and a correctly received packet by generating an ACK. Then the receiver sends an acknowledgement frame

containing the M acknowledgements (ACKs/NACKs) corresponding to the most recently received block of M packets to the transmitter. Transmission of the acknowledgement frame starts at the beginning of slot $t + \tau + 1$ and completes at the end of the slot. After sending the acknowledgement frame, the receiver discards erroneously received packets, delivers the packets in sequence, and stores the out-of-sequenced packets in the resequencing queue.

The transmitter receives the acknowledgement frame, which is associated with the block of M packets transmitted in slot t, in slot $t + 2\tau + 1$. It checks each acknowledgement in the acknowledgement frame, and prepares the next block of M packets to transmit in slot $t + 2(\tau + 1)$ according to the following rule: If there is no NACK in the acknowledgement frame, the next block to be transmitted is composed of M new packets (never transmitted before); if the acknowledgement frame contains k NACKs, the next block of M packets consist of those k negatively acknowledged old packets (transmitted before), and $M - k$ new packets (see Figure 3). Meanwhile, the transmitter removes these positively acknowledged packets from the transmission queue. These selected M packets are to be transmitted in slot $t + 2(\tau + 1)$ according to the channel scheduling policy elaborated in the next section.

4. Dynamic Channel Scheduling Policy

As shown in Figure 2, the proposed LLC design has the knowledge about the current bit error rate of each channel, from which the packet error rate (PER) of the channel can be obtained.

In fact, packet error rate information of channels can be used by the LLC design for scheduling transmission of the block of M packets over the M channels in each slot. We denote the following packet-to-channel scheduling policy by the dynamic channel scheduling. To transmit the block of M packets in a slot, the best channel (i.e., a channel with the smallest error rate) is assigned to the packet associated with the smallest sequence number in the block; the second best channel is assigned to the packet associated with the second smallest sequence number; and so forth. It is noted that, if the communication system uses the same modulation scheme (e.g., M-ary Phase-Shift Keying (MPSK)) for all M channels, which is often true in practice, the dynamic packet-to-channel scheduling policy can be implemented based on the signal-to-interference-plus-noise ratio (SINR) value of each channel. That is, the channel with the largest SINR value is assigned to the packet associated with the smallest sequence number in the block; the channel with the second largest SINR value is assigned to the packet associated with the second smallest sequence number; and so forth. The dynamic channel scheduling is illustrated in Figure 3, where the PER of channel 1 is not greater than that of channel 2 which is not greater than that of channel 3.

With this approach, the number of out-of-sequenced packets in the resequencing queue is reduced as the number of out-of-sequenced packets incurred by the loss of a packet having a smaller sequence number is always greater than or equal to the number of out-of-sequenced packets incurred by the loss of a larger sequence number packet. For instance, in Figure 3 at the beginning of slot 9, four blocks (packet $1 - 12$) have been received. In the resequencing queue of the receiver, five packets are out-of-sequenced and waiting. Among them, five packets (i.e., packet $5; 7; 8; 10; 12$) are queued because of the loss of packet 4, four packets (i.e., packet $7; 8; 10; 12$) are queued because of the loss of packet 6, and one (i.e., packet 12) is due to the loss of packet 11. Since correctly receiving the packet associated with the smallest sequence number among the block to be transmitted can lead to immediate delivery of some out-of-sequenced packets in the resequencing queue, this packet should be arranged for transmission with the least possibility for transmission error. As shown in the next section, this dynamic channel assignment will significantly improve the average packet delay performance.

Nevertheless, if the channel bit error rate information is not available in the LLC sublayer, a static channel scheduling can be used to simultaneously transmit a block of M packets over the M channels. The static channel scheduling is illustrated in Figure 4 and works as follows. To transmit the block of M packets in a slot, an old packet (i.e., a packet to be retransmitted) is always assigned to the same channel for retransmission as the originally assigned one, while a new packet (i.e., a packet to be transmitted for the first time) is assigned to a uniformly chosen channel among those available for transmitting new packets. In a real-world multichannel communication environment, a packet to be retransmitted is often assigned to a different channel for retransmission from the previously assigned one due to the time-correlation property of the channel error process. For the time-uncorrelated channels, which are assumed in this study, these two static channel scheduling methods actually have the same effect on the system performance.

As will be shown from simulation results in the next section, the dynamic channel scheduling using the cross-layer design approach outperforms the static channel scheduling.

5. Performance Evaluation

In this section, we conduct a simulation study on the performance of the SR-ARQ based LLC design for multichannel underwater sensor networks. The performance metric that we consider is the average

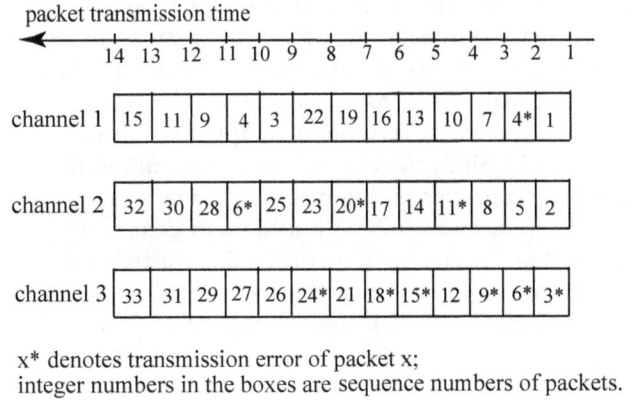

x* denotes transmission error of packet x;
integer numbers in the boxes are sequence numbers of packets.

Figure 3. Dynamic Channel Scheduling ($M = 3; \tau = 3$)

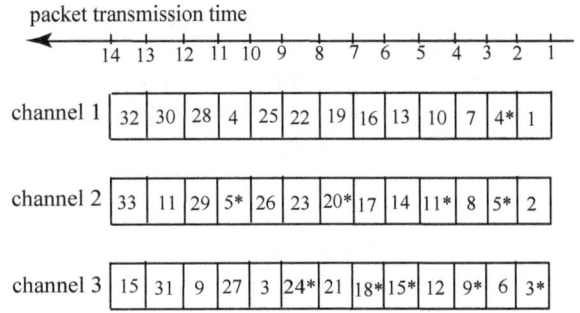

Figure 4. Static Channel Scheduling ($M = 3; \tau = 3$)

packet delay. The delay of a packet is defined as the amount of time (i.e., the number of slots) between the instant at which the packet is transmitted for the first time and the instant at which it leaves the resequencing queue in the receiver. We investigate the impact of the two channel scheduling policies and the system parameters on the average packet delay performance through simulations.

5.1. Simulation Environment

We use the SimPy simulator [34], which is an object-oriented, process based discrete-event simulation platform based on the standard programming language Python. SR-ARQ based LLC design is first implemented with SimPy. Then two individual processes, one considered as the transmitter and the other as the receiver, form an M-channel underwater sensor network. Each process independently operates an object of SR-ARQ based LLC. The transmitter continuously sends data packets and receives acknowledgement frames, and the receiver receives data packets and sends out acknowledgement frames. Data packets are transmitted over M parallel channels, while acknowledgement frames are transmitted via a separate feedback channel with no errors.

In the following simulation study, the round trip time of a packet is fixed to be 8 slots, $\tau = 2$. We assume that the packet lossy property of a channel is time-invariant. That is, the probability p_i that a packet transmitted over channel i is erroneously received or simply lost is a real number in $(0, 1)$. Since the channels may have different packet lossy properties, p_i might be different from p_j, for $i, j = 1, \cdots, M$ and $i \neq j$. Without loss of generality, we assume that the channels are ordered according the their packet error rates, i.e., $p_1 \leq p_2 \leq \cdots \leq p_M$. Then, we use Δ_i to represent the ratio of p_{i+1} to p_i for $i = 1, \cdots, M-1$, i.e.,

$$\Delta_i = \frac{p_{i+1}}{p_i}, \qquad i = 1, \cdots, M-1. \tag{1}$$

It is clear that, the larger the value of Δ_i, the greater the difference between the error rates of channels i and $i + 1$. For this study we let $\Delta = \Delta_1 = \cdots = \Delta_{M-1}$. After letting p denote the average of the error rates for the M channels, i.e.,

$$p = \frac{1}{M} \sum_{i=1}^{M} p_i, \tag{2}$$

the triad (M, Δ, p) uniquely determines the packet error rate sequence $(p_1, p_2, \cdots, p_M)$.

5.2. Simulation Results

We plot the simulation results of the average packet delay for the SR-ARQ based LLC with the dynamic and static channel scheduling in Figure 5, Figure 6, and Figure 7. From these plots we observe that the dynamic channel scheduling improves the packet delay performance in multichannel underwater sensor network environments over the static channel scheduling. For instance, for $M = 16$, the average packet delay is reduced by as much as 70% when the packet-to-channel scheduling policy changes from the static channel scheduling to the dynamic channel scheduling. When $\Delta = 1.5$, the average packet delay for the dynamic channel scheduling is only one third of that for the static channel scheduling.

The average packet delay is plotted in Figure 5 for $\Delta = 1.2$, $p = 0.25$, and M varying from 2 to 16. As expected, the average delay difference between the two channel scheduling policies becomes larger with M. Meanwhile, as M increases, the average packet delay with the dynamic channel scheduling slightly increases at first and then slightly decreases. This shows that, under the saturated traffic condition, the overall impact of the number of parallel channels on the packet delay performance is insignificant for this scheduling policy. Since the number of channels has only an insignificant impact on the average packet delay, the use of parallel channels will be a favorable option for packet error control in a multichannel underwater communication

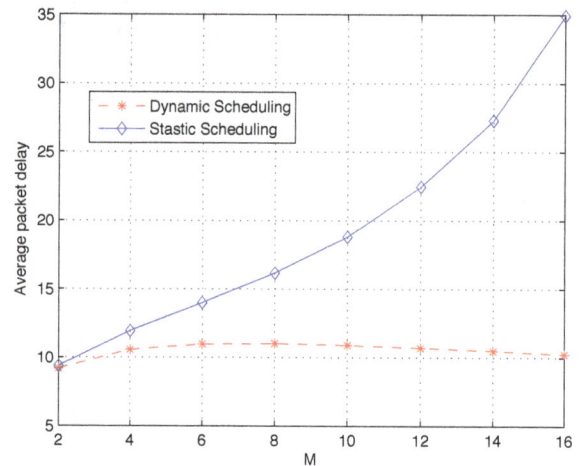

Figure 5. Average Packet Delay vs. M ($\Delta = 1.2$, $p = 0.25$)

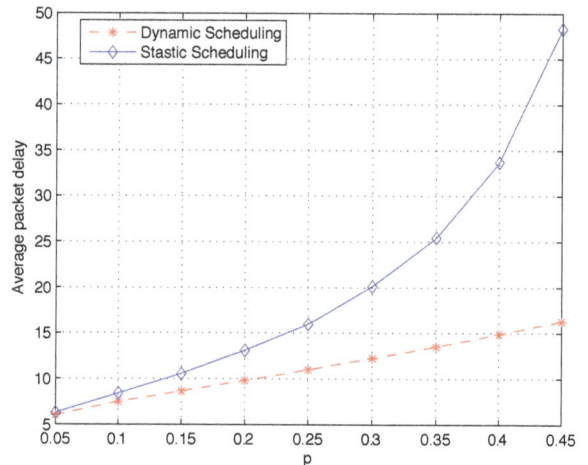

Figure 6. Average Packet Delay vs. p ($\Delta = 1.2$, $M = 8$)

system with the SR-ARQ based LLC. It is noted that, for the multichannel LLC design under non-saturated traffic conditions, packet end-to-end delay includes another delay component, the packet waiting time at the transmitter, in addition to the packet delay defined in this study. Under a non-saturated traffic condition, it is clear that the increase of the transmission rate mainly results in the reduction of the packet waiting time at the transmitter, and hence the packet end-to-end delay. So the above observation corroborates the fact that the increase of the number of parallel channels leads to the increase of the transmission rate hence the decrease of the overall packet delay for multichannel underwater communication systems with non-saturated traffic.

In Figure 6, we plot the average packet delay when $M = 8$, $\Delta = 1.2$, and p is varying from 0.05 to 0.45. The average packet delay increases as p does, while the increasing rate with the dynamic channel scheduling is

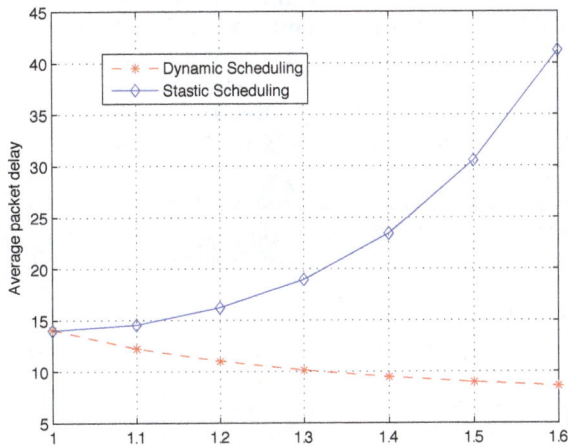

Figure 7. Average Packet Delay vs. Δ (*M* = 8, *p* = 0.25)

smaller than that with the static channel scheduling. The average packet delay is shown in Figure 7 when $M = 8$, $p = 0.25$, and Δ is varying from 1.1 to 1.7. As Δ increases, the average packet delay decreases when the dynamic channel scheduling is applied, but it increases when the static channel scheduling is used. For example, when Δ increases from 1.1 to 1.5, the average packet delay with the dynamic channel scheduling decreases almost by 50%, but the average packet delay with the static channel scheduling increases by 100%. This is explained by the fact that for greater variance in the error rates, the error rates of the first few channels is smaller. For instance, in Figure 7, the error rates of channels 1 to 4 when Δ = 1.2 are smaller than the corresponding ones when Δ = 1.1. Intuitively, the packets transmitted over the first few channels have a higher probability of being correctly received (and delivered to the upper layer). This results in a smaller possibility for the other packets to be queued in the resequencing queue. Therefore, the average waiting time of a packet queued in the resequencing queue is reduced, and so is the total average packet delay.

6. Conclusion

In this paper, we proposed a SR-ARQ based logical link control design for multichannel underwater sensor networks and a cross-layer design of packet-to-channel scheduling policy for more efficient transmission of multiple packets over parallel channels. The dynamic channel scheduling is proposed when channel state information is obtained at the LLC sublayer through the cross-layer design approach, while the static channel scheduling is the option when no channel state information is available at the LLC sublayer. We performed a simulation study on the average packet

delay for the proposed LLC design with the two channel scheduling policies. From simulation results, we observed that the dynamic channel scheduling always achieves a better packet delay performance than the static channel scheduling. The average packet delay with the dynamic channel scheduling increases with the average error rate of all channels, but decreases with the variance in the error rates of the parallel channels. More interestingly, we observed that the number of parallel channels has an insignificant impact on the average packet delay, when the dynamic channel scheduling is applied in multichannel underwater communications, and hence the use of parallel channels is a favorable option for multichannel underwater networks.

Acknowledgement. This work reported herein was supported by Defence Research and Development Canada (DRDC).

References

[1] Che, X., Wells, I., Dickers, G., Kear, P. and Gong, X. (2010) Re-Evaluation of RF Electromagnetic Communication in Underwater Sensor Networks. *IEEE Communications Magazine*, 48(12): 143-151.

[2] Jiang, Z. (2008) Underwater Acoustic Networks ï£¡ Issues and Solutions. *International Journal of Intelligent Control and Systems*, 13(3): 152-161.

[3] Stojanovic, M. (2005) Optimization of a Data Link Protocol for an Underwater Acoustic Channel. In *Proceedings of Oceans 2005 - Europe (Brest, France)*, 1: 68-73.

[4] Winters, J.H. (1987) On the Capacity of Radio Communication Systems with Diversity in a Rayleigh Fading Environment. *IEEE Journal on Selected Areas in Communications*, 5(5): 871-878.

[5] Bahai, A.R.S., Saltzberg, B.R. and Ergen, M. (2004) Multi Carrier Digital Communications: Theory and Applications of OFDM. *New York, Springer*.

[6] Chang, R.W. (1966) Synthesis of Band-Limited Orthogonal Signals for Multichannel Data Transmission. *Bell System Technical Journal*, 45: 1775-1796.

[7] IEEE Standard (2004) *802.16: Air Interface for Fixed Broadband Wireless Access Systems*.

[8] 3GPP Technical Specification Group (2008) *3GPP TS 36.201: Evolved Universal Terrestrial Radio Access (E-UTRA): Long Term Evolution (LTE) Physical Layer*, source: *www.3gpp.org*.

[9] Abdi, A. and Guo, H. (2009) A New Compact Multichannel Receiver for Underwater Wireless Communication Networks. *IEEE Transactions on Wireless Communications*, 8(7): 3326-3329.

[10] *AquaNetwork: Underwater wireless modem with networking capability*. Source: http://www.patentstorm.us/patents/7065068.html.

[11] Li, B., Zhou, S., Stojanovic, M., Freitag, L. and Willett, P. (2008) Multicarrier Communication Over Underwater Acoustic Channels With Nonuniform Doppler Shifts. *IEEE Journal of Oceanic Engineering*, 33(2): 198-209.

[12] LI, B., HUANG, J., ZHOU, S., BALL, K., STOJANOVIC, M., FREITAG, L. and WILLETT, P. (2009) MIMO-OFDM for High-Rate Underwater Acoustic Communications. *IEEE Journal of Oceanic Engineering*, 34(4): 634-644.

[13] CHAO, C.-M. and WANG, Y.-Z. (2010) A Multiple Rendezvous Multichannel MAC Protocol for Underwater Sensor Networks. In *Proceedings of IEEE Wireless Communications and Networking Conference (WCNC 2010) (Sydney, Australia)*, 1-6.

[14] SHAHABUDEEN, S., CHITRE, M. and MOTANI, M. (2007) A Multi-channel MAC Protocol for AUV Networks. In *Proceedings of Oceans 2007 - Europe (Aberdeen)*, 1-6.

[15] ZHOU, Z., PENGT, Z., CUI, J.-H. and JIANG, Z. (2010) Handling Triple Hidden Terminal Problems for Multi-Channel MAC in Long-Delay Underwater Sensor Networks. In *Proceedings of IEEE INFOCOM 2010 (San Diego, USA)*, 1-5.

[16] LEE, U., WANG, P., HOH, Y., VIEIRA, L.F.M., GERLA, M., and CUI, J.-H. (2010) Pressure Routing for Underwater Sensor Networks. In *Proceedings of IEEE INFOCOM (INFOCOM 2010) (San Diego, CA, USA)*.

[17] XU, M. and LIU, G. (2011) Fault Tolerant Routing in Three-dimensional Underwater Acoustic Sensor Networks. In *Proceedings of International Conference on Wireless Communications and Signal Processing (WCSP 2011) (Nanjing, China)*, 1-5.

[18] ZHENG, J. and JAMALIPOUR, A. (2009) Wireless Sensor Networks: A Networking Perspective, Wiley-IEEE Press, ISBN: 9780470167632.

[19] DALADIER, J.M. and LABRADOR, M.A. (2009) An Adaptive Logical Link Layer Protocol for Underwater Acoustic Communication Channels. In *Proceedings of Oceans 2009 (Biloxi, MS)*, 1-8.

[20] GAO, M., SOH, W.-S. and TAO, M. (2009) A Transmission Scheme for Continuous ARQ Protocols over Underwater Acoustic Channels. In *Proceedings of IEEE International Conference on Communications (ICC 2009) (Dresden, Germany)*, 1-5.

[21] FORKEL, I., KLENNER, H. and KEMPER, A. (2005) High Speed Downlink Packet Access (HSDPA) – Enhanced Data Rates for UMTS Evolution. *Computer Networks: The International Journal of Computer and Telecommunications Networking*, 49(3): 325-340.

[22] GHOSH, A., WOLTER, D.R., ANDREWS, J.G. and CHEN, R. (2005) Broadband Wireless Access with WiMax/8O2.16: Current Performance Benchmarks and Future Potential. *IEEE Communication Magazine*, 43(2): 129-136.

[23] DING, Z. and RICE, M. (2006) ARQ Error Control for Parallel Multichannel Communications. *IEEE Transactions on Wireless Communications*, 5(11): 3039-3044.

[24] HU, T., AFSHARTOUS, D. and YOUNG, G. (2004) Parallel Stop and Wait ARQ in UMTS - Peformance and Modeling. In *Proceedings of the 2004 World Wireless Congress (San Francisco, CA)*.

[25] WU, W.-C., VASSILIADIS, S. and CHUNG, T.-Y. (1993) Performance Analysis of Multi-Channel ARQ Protocols. In *Proceedings of the 36th Midwest Symposium on Circuits and Systems*, 2: 1328-1331.

[26] CHANG, J.-F. and YANG, T.-H. (1993) Multichannel ARQ Protocols. *IEEE Transactions on Communications*, 41(4): 592-598.

[27] FUJII, S., HAYASHIDA, Y. and KOMATU, M. (2001) Exact Analysis of Delay Performance of Go-Back-N ARQ Scheme over Multiple Parallel Channels. *Electronics and Communications in Japan, part 1*, 84(9): 27-41.

[28] DING, Z. (2006) ARQ Techniques for MIMO Communication Systems. PH.D. THESIS, Department of Electrical and Computer Engineering, Brigham Yong University Provo.

[29] SHACHAM, N. and SHIN, B.C. (1992) A Selective-Repeat-ARQ Protocol for Parallel Channels and Its Resequencing Analysis. *IEEE Transactions on Communications*, 40(4): 773-782.

[30] LI, J. and ZHAO, Y.Q. (2009) Resequencing Analysis of Stop-and-Wait ARQ for Parallel Multichannel Communications. *IEEE/ACM Transactions on Networking*, 17(3): 817-830.

[31] LI, J. and ZHAO, Y.Q. (2009) Packet Delay Analysis for Multichannel Communication Systems with MSW-ARQ. *Performance Evaluation*, 66(7): 380-394.

[32] XIE, G., GIBSON, J. and BEKTAS, K. (2004) Evaluating the Feasibility of Establishing Full-Duplex Underwater Acoustic Channels. In *Proceedings of the Third Annual Mediterranean Ad Hoc Networking Workshop (Med-Hoc-Net 2004) (Bordrum, Turkey)*.

[33] KOOPMAN, P. and CHAKRAVARTY, T. (2004) Cyclic Redundancy Code (CRC) Polynomial Selection for Embedded Networks. In *Proceedings of the International Conference on Dependable Systems and Networks (Florence, Italy)*, 145-154.

[34] MATLOFF, N. (2008) Introduction to Discrete-Event Simulation and the SimPy Language. Source: `http://heather.cs.ucdavis.edu/~matloff/156/PLN/DESimIntro.pdf`.

10

The Prohibitive Link between Position-based Routing and Planarity

David Cairns[1], Marwan Fayed[1*], Hussein T. Mouftah[2]

[1]Computing Science and Math, University of Stirling, Stirling, FK9 4LA, UK; [2]SITE, University of Ottawa, Ottawa, ON, K1N 6N5, Canada

Abstract

Position-based routing is touted as an ideal routing strategy for resource-constrained wireless networks. One persistent barrier to adoption is due to its recovery phase, where messages are forwarded according to left- or right-hand rule (LHR). This is often referred to as face-routing. In this paper we investigate the limits of LHR with respect to planarity. We show that the gap between non-planarity and successful delivery is a single link within a single configuration. Our work begins with an analysis to enumerate all node configurations that cause intersections in the unit-disc graph. We find that left-hand rule is able to recover from all but a single case, the 'umbrella' configuration so named for its appearance. We use this information to propose the Prohibitive Link Detection Protocol (PLDP) that can guarantee delivery over non-planar graphs using standard face-routing techniques. As the name implies, the protocol detects and circumvents the 'bad' links that hamper LHR. The goal of this work is to maintain routing guarantees while disturbing the network graph as little as possible. In doing so, a new starting point emerges from which to build rich distributed protocols in the spirit of CLDP and GDSTR.

Keywords: position-based routing, geographic routing, face routing, wireless routing

1. Introduction

The construction of network subgraphs appropriate for position-based (or geographic) routing protocols has, to date, remained a complex problem. These subgraphs are needed to recover from the local minima problem (see [3]) that prevents delivery and plagues position-based protocols. Network subgraphs constructed for recovery using only 1-hop information risk inaccuracies that cause routing failures [15, 24]. One remedy is to allow nodes to cooperate. If permitted, cooperating nodes may construct a network subgraph that remedies any inaccuracies [14, 19, 24, 25]. Yet the resources needed to power the many rounds of communication between nodes, risks being prohibitive in such a resource-constrained environment. The ideal wireless network subgraph would guarantee successful delivery while a) needing only 1-hop information and b) be able to acquire such information passively.

Traditionally, position-based routing protocols construct subgraphs (herein referred to as just 'graph') from available links in somewhat of a bottom-up fashion. Generally the idea is to extract a specific type of graph from the available nodes and links in the network. During the setup of such graphs each node evaluates available links to find those that preserve some global properties. Planar graphs [6] and k-spanners [23] are two such examples. The analogous question would be to ask, "what is the set of edges that must be preserved to guarantee a given feature in the graph?"

Our work is motivated by the opposite question, "What is the set of edges that must be *deleted* while still providing guarantees?" This work is a step in that direction. Without sacrificing the scalability and success of position-based routing, the goal of this work is to disturb the network as little as possible. To this end it is necessary to understand the causes for a position-based routing protocol to fail to recover from local minima and deal with those causes directly.

*Corresponding author. mmf@cs.stir.ac.uk

In this paper we investigate routing according to left- or right-hand rule (LHR). Using LHR, a node, upon receipt of a message, will forward to the neighbour that sits next in counter-clockwise order in the network graph. When used to recover from greedy routing failures, LHR guarantees success if implemented over planar graphs; for this reason it is often called 'face-routing'. We note, however, that if planarity is violated then LHR is only guaranteed to eventually return to the point of origin. Our work seeks to understand and correct the underlying causes of these failures.

We have chosen LHR for three reasons. First, it is most prevalent in position-based routing literature and hence well-studied. Second, it is a simple rule requiring little-to-no overhead. Finally, the ideal network graph remains elusive. To re-iterate, we envision the ideal graph as overcoming the inaccuracies that lead to routing failures; as one that results from knowledge of the 1-hop neighbourhood; and as one where each node transmits a constant number of messages.

We build on a provable enumeration of the possible types of intersections in a unit-disc graph (UDG), within which any two nodes are neighbours if separated by a maximum distance of one unit. Our analysis reveals that only three types of intersections are possible. A trace of face-routing over each intersecting neighbourhood further reveals that in only one of these configurations does LHR fail to recover: We call this the 'umbrella' configuration, so named for its appearance. The umbrella configuration naturally hides links and nodes from a face-routing traversal, partitioning the graph with respect to the traversal. Unless there appears some other non-local route to join these partitions, potential routes will be unavailable to any face-routing technique.

We use this information to propose the Prohibitive Link Detection Protocol (PLDP). PLDP identifies the umbrella configuration and removes from it a single link. In doing so, PLDP provides a graph over which any face-based method may provably guarantee delivery using standard geographic and face-routing rules.

In our evaluation we compare the setup and quality of PLDP graphs with CLDP and GDSTR using Netsim2 [19, 20]. Our simulation results demonstrate that PLDP performance is similar to current face-routing schemes. Where PLDP separates itself is in the number of messages required to setup the network: Most nodes will need to generate no setup packets and from those that do, a very small number of packets is needed. Our evaluation will show that the small number of setup packets are indicative of the infrequency of the umbrella configuration, the consequence of which is that PLDP is able to preserve most of the links in the original network graph. In a manner of speaking our evaluations suggest that

traditional face-routing schemes may be 'over-solving" the problem by planarizing networks.

On the surface this seems an unfair comparison since both GDSTR and CLDP operate without the unit disc assumption. We emphasize that our goal is to provide a better understanding of the underlying motivations for such distributed protocols and, in doing so, provide a new starting point for distributed protocols that may out-perform those of the current generation. PLDP may be unable to compete directly due to the unit disc assumption, but we will show that it provides a novel direction from which to build.

In summary, this paper seeks to provide a basis that is a lateral shift away from planarity so that better cooperative position-based protocols may be built. By investigating the underlying causes for failure, planarity is shown to be unnecessary in the majority of cases. We propose PLDP as a means to relax constraints on the network graph while preserving the promise of local face-routing techniques.

2. Related Work

Recovery algorithms in Euclidean position-based routing are equivalently known as *face routing* [1, 2] and *perimeter routing* [13]. Face routing was first proposed by Bose et al. in [1] with some theoretical bounds. Karp et al. independently proposed an identical mechanism in [13] but with work on a MAC-compatible implementation. Variants have since emerged addressing, for example, theoretical bounds in [16–18]. In [11], face-routing is augmented into a "select-and-protest" reactive protocol in order to reduce the information required to planarize the graph.

Wireless network graphs may consist of intersecting edges so it is necessary for planar subgraph methods to prune edges from the network graph so that it is planar while remaining connected. Gabriel Graphs (*GG*) and Relative Neighbourhood Graphs (*RNG*) are planar graphs whose constructions are localised, a characteristic particularly suitable to sensor environments. Intersecting edges are eliminated by connecting pairs of nodes through *witness* nodes, if such a node exists in a common region. It has since been shown that 'Hello' messages may hinder network performance [10]. This is addressed in face-routing directly by [4] and more generally in [7, 9, 22]. Further work in [26] reduces the path length during the recovery phase.

These distributed constructions are unable to resolve links broken by obstacles or interference [12, 21]. Recent breakthroughs have begun to surmount the impracticalities of face-routing while maintaining delivery guarantees [14, 19].

The Greedy Distributed Spanning Tree Protocol (GDSTR) algorithm in [19] builds on the fact that any message can be successfully delivered via depth-first search if the network is connected via a spanning tree. This fact alone does not solve the problem: GDSTR provides optimizations to reduce the otherwise inefficient delivery requiring up to $2n - 3$ hops. The authors in [19] describe a new type of spanning tree, the *hull tree*, to route more efficiently. A hull tree is a spanning tree with one added piece of information: each node records the convex hull that contains all of its descendants in the tree. In GDSTR forwarding occurs greedily, as do most position-based protocols. If a message reaches a void, a recovery mode is initiated where convex hulls are used to determine the regions of the network that contain unreachable destinations. This information is used by GDSTR to route along the spanning tree to forward to the appropriate convex hull. If a node is found en route that is closer to the destination than the node where the message was stuck, then GDSTR returns to greedy forwarding. GDSTR is known to scale well as the neighbourhood size grows. Furthermore, the use of multiple hull trees adds fault-tolerance to the network and if multiple trees are rooted at opposite ends of the network, routing efficiency improves.

The Cross-Link Detection Protocol (CLDP) proposed in [14], and later improved in [8], circumvents voids by face-routing. It uses left-hand rule over a planar subgraph of the network; its design however, is motivated by the observation that routing difficulties in planar subgraph methods arise, in part, due to the constructions themselves. (Recall from previous that successful local planar subgraph constructions rely on the unit disc graph.) For this reason, CLDP proposes an alternate construction of planar subgraphs that assumes only that links are bidirectional. CLDP operates in a distributed fashion, exchanging some localised operation for accurate information. The idea behind CLDP is that each node is able to probe the vicinity for intersecting links. A probe packet is initialised with the endpoints of the first link to be probed and forwarded according to left-hand rule. The probe eventually returns to its point of origin with a vector of the path taken. This information is shared with nearby nodes to prune links appropriately. To avoid the slow process of scheduling serial probes by neighbouring nodes, a system for concurrent probing is proposed. Concurrent probing is achieved by implementing a mechanism to 'lock' links so that no more than one link is removed at a time from any vicinity. CLDP is one of very few protocols to have been implemented on testbeds [14]. The associated communication complexities and storage costs revealed in this process (see [15, 19]) are motivation to develop alternative approaches to guarantee delivery.

A more recent approach is to think about how the network might embed onto a different physical space.

One such work appears in the FaceTrace project [25] which imagines that nodes in the network sit on a high-genus topological surface, such as a torus. It is a novel technique that extracts onto these surfaces faces from the network itself, rather than faces associated with local minima in the network. In doing so, planarity emerges naturally. In simulation FaceTrace exhibits routing quality of a very high order but the setup cost is reported to be similar to those of GDSTR, numbering many orders of magnitude.

Protocols such as CLDP and GDSTR, in order to be feasible for physical networks, sacrifice efficiency for accuracy. CLDP requires high-complexity negotiations within each neighbourhood in order to prune appropriate links. GDSTR reduces the messaging complexity but must broadcast information to construct and maintain its hull trees. It remains an open question whether such trade-offs are a necessity. The work in [11] is a step in the right direction. Its recognition that there are available short-cuts when routing according to LHR is further evidence that the planarity assumption may be excessive.

By contrast, in this work we show that there exists a locally constructed, non-planar graph construction over which face-based protocols guarantee success.

3. Links that Prohibit Routing Success

In the previous section we noted that routing according to left-hand rule (LHR) alone, fails to provide a guarantee of success. Though this fact is well known, the reasons and circumstances under which delivery may fail are poorly understood. In this section we investigate the limits imposed by intersections on face-based recovery.

Our investigation begins with an enumeration of all of the types of intersections that may appear in the UDG. We focus this work on the unit disc graph (UDG), where all communication ranges are normalised. The UDG is appropriate since it limits potential routing options yet still poses a challenge to LHR routing.

3.1. An Analysis of Intersection Types

Consider any two intersecting edges. We provide the edges ac and bd in Figure 1 for reference. The nodes a, b, c, d at the end points of these edges form a 4-gon (shown in Figure 1 using dashed lines). The question we ask is, which of the edges of the 4-gon may or may not be communicating links in the unit-disc graph? In order for at least one such edge to exist, we need to show that *all four sides cannot be greater than both diagonals.* Using cosine rule we know,

$$(ac)^2 = (ad)^2 + (dc)^2 - 2(ad)(dc)\cos D. \qquad (1)$$

If $|ac|$ is less than or equal to 1, then

$$(ad)^2 + (dc)^2 - 2(ad)(dc)\cos D \leq 1. \qquad (2)$$

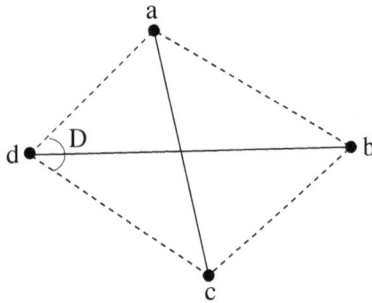

Figure 1. Intersecting links between two pairs of nodes may impose edges in a 4-gon.

When $D \geq \frac{\pi}{2}$, then $\cos D \leq 0$. In this case, $(ad)^2 + (dc)^2 \leq 1$, which means $(ad) \leq 1$ and $(dc) \leq 1$. Thus, if an angle of the 4gon is right or obtuse, then both incident edges must exist in the UDG. (By contrast, incident edges when $D < \frac{\pi}{2}$ may or may not exist.)

This implies and restricts the possible configurations that allow intersections to three in number, all shown in Figure 2. The two cases where the nodes of intersecting edges produce a 4-gon with two obtuse angles is shown in Figures 2a and 2b, while the 4-gon containing a single obtuse angle is shown in Figure 2c. (It is impossible for a 4-gon to be constructed with three obtuse edges; and that edges incident to an acute angle may or may not appear in the unit-disc graph.)

3.2. The Prohibitive Link

The finite and small number of possible intersections allows us to carefully examine the behaviour of a left-hand traversal over all possible cases. A left-hand traversal is deemed successful when it can identify a single unique face in an intersecting environment.

We show in Figure 3 the traces corresponding to the three intersections in Figure 2. Traversals are shown using a dotted line. In the first two cases an LHR traversal succeeds in identifying a single face irrespective of the point of entry into the intersecting environment. (We show via inductive proof in Section 4.2 that the same holds true when intersections are composed together.)

The 'bad' configuration occurs during a traversal of the intersection shown in Figure 3c. Here the different points of entry reveal that there are two faces with respect to LHR. This means there are two ways in which LHR may fail. The first is demonstrated by the dashed-dot-dash line originating at node d. (Entry at nodes a and b are analogous.) A traversal using left- or right-hand rule will never traverse edge ac while travelling through this intersection. Supposing c must be traversed in order to reach the destination, LHR will fail. The second possible failure occurs when an LHR traversal encounters this intersection first via node c

in Figure 3c using the dashed line. LHR traverses the inside of the triangle $\triangle abd$ and exits without ever seeing edges that protrude from the outside of the triangle. As before, any such edges leading to the destination may be overlooked by an LHR traversal.

This represents the case where network node a communicates with b, c, d, and b with d; node c communicates only with a. We call this case the *umbrella* configuration for its appearance.

The cause of both failures lies in the relationship between $\triangle abd$ and ac in Figure 3c: There exists an edge from the triangle that is accessible only from inside the triangle. In other words, a traversal around the inside or the outside of the triangle fails to encounter all edges leading to the triangle. Both failures are solved by removing any of the edges that form the triangle. The easiest of these to identify and remove from the network graph is the edge of the triangle that forms the intersection in the umbrella configuration. In Figure 3c this link is represented by $\overline{bd}$. We call this the *prohibitive link*.

The prohibitive link is most easily identified by the only node that is able to see all four nodes in the umbrella configuration. In Figure 3c this responsibility falls on node a. It looks for intersections consisting of pairs links. The first link is formed by itself and an immediate neighour, with subsequent links formed between two immediate neighbours.

We revisit this subject and build a networking protocol in the next section. Before closing this section the outcome following a removal of the prohibitive link from the umbrella configuration is demonstrated in Figure 4. The intersection that was the umbrella configuration is reduced to a planar set of edges easily navigated by left- or right-hand rule.

3.3. A Note on the Sufficiency of Prohibitive Links

We note that it is sufficient to delete the set of prohibitive links while preserving delivery guarantees using face-routing schemes, yet it may be unnecessary to delete all of the links in the set. The minimal set of deletions remains an open problem.

An example of where it is sufficient but unnecessary to delete a link is shown in Figure 5. Node a recognises prohibitive link $\overline{bd}$ in its neighbourhood. Removal of $\overline{bd}$ guarantees that all links in and out of the intersecting nodes will eventually be encountered. With respect to the local neighbourhood it is necessary to remove the prohibitive link. However the exitence of alternate paths outside of a neighbourhood, as shown in Figure 5 through nodes p and q, suggest that it may be unnecessary to remove the prohibitive link.

We will see in our evaluations in Section 5 that the occurrence of umbrella configurations is so infrequent that this trade-off between global knowledge and local

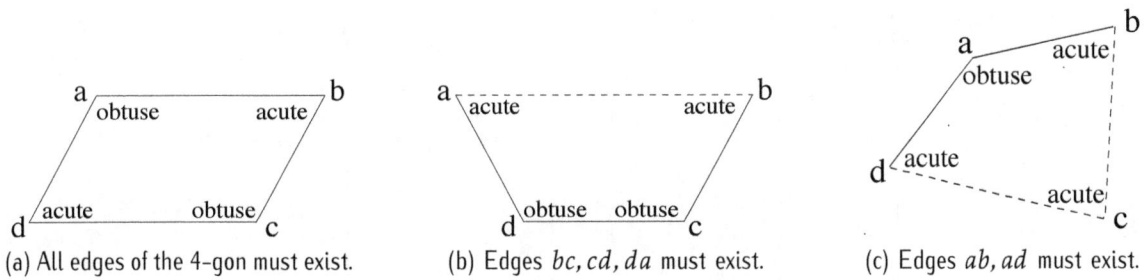

Figure 2. Possible 4-gons when edges intersect in the UDG; dashed lines indicating edges that may or may not appear.

(a) All edges of the 4-gon must exist.   (b) Edges bc, cd, da must exist.   (c) Edges ab, ad must exist.

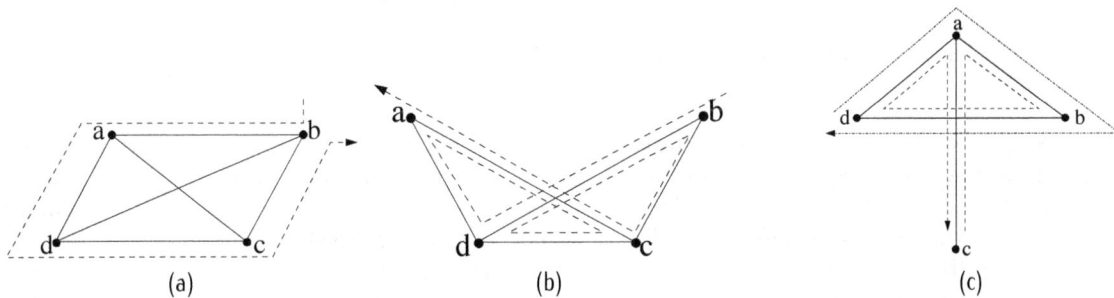

(a)   (b)   (c)

Figure 3. A unique face emerges from all but the 'umbrella' shape, shown in (c).

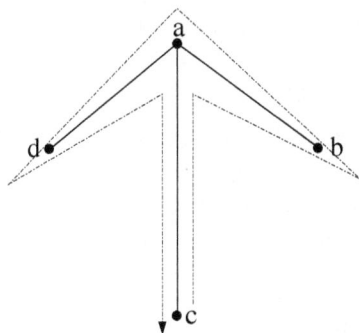

Figure 4. Removing prohibitive link bd allows LHR to traverse all edges.

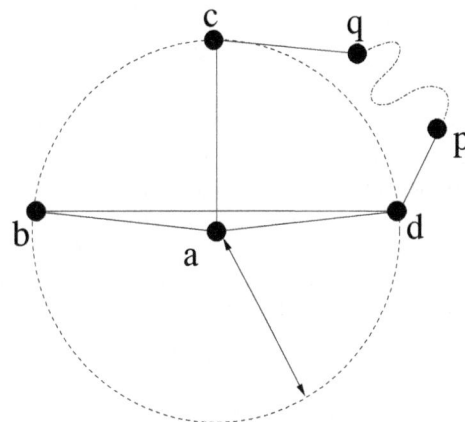

Figure 5. Removal of the prohibitive link is only sufficient.

decisions may not even be worth considering. In the next section we use our knowledge of the prohibitive link to construct the prohibitive-link detection and routing protocol.

4. Prohibitive Link Detection Protocol

We have enumerated all possible intersections in the unit-disc graph and identified the type of intersection with the link that prohibits successful delivery when routing according to right- or left-hand rule. In this section we present a Prohibitive Link Detection Protocol (PLDP). Proofs of correctness may be found in [5].

4.1. PLDP Overview

We assume a static graph where each node is assigned a coordinate in a 2-dimensional Euclidean space. We assume that the graph is connected and that all links are bi-directional. PLDP functions adequately in a mobile space provided that changes in position occur over a greater time-frame than is required to re-evaluate local prohibitive links and transmit local updates. In this work communication range is fixed and uniform across all nodes.

The face-routing family of protocols preserve their delivery guarantees in PLDP graphs. During their normal operation nodes route in a greedy fashion

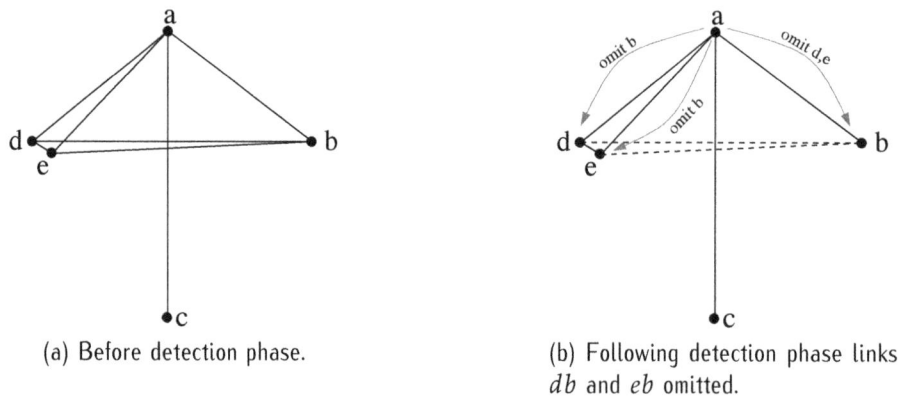

(a) Before detection phase.

(b) Following detection phase links
db and *eb* omitted.

Figure 6. Local neighbourhood before and after the PLDP detection phase.

and forward messages to the neighbour that most reduces the distance to the destination. Where no such neighbour exists a message is deemed 'stuck' in a local minima and is forwarded according to left- (or right-)hand rule. The node initially selected is the first to appear left (or right) of the line segment from the current location to the destination. The first node found that sits closer to the destination than the 'stuck' location returns to the greedy forwarding phase. Proofs of correctness and the ability to guarantee delivery to the destination appears in Section 4.2.

During the PLDP detection phase each node inspects its neighbourhood using neighbour positions reported in ordinary 'hello' packets. Each node evaluates intersections within range and flags any three neighbours that compose an umbrella configuration, as described in Section 3.2. Once sufficient information is compiled a node sends a notification packet to the neighbours that anchor prohibitive links.

Notification messages exchanged between nodes consist of either a delete or an insert instruction. As is suggested by its label, a delete deactivates a prohibitive link at the anchors of the link. Similarly an insert instruction reactivates a link previously deemed prohibitive. This allows for corrections as the network state changes.

We emphasize that PLDP takes a passive approach when looking for the recovery subgraph: In contrast to the 'active' approach taken by protocols such as CLDP and GDSTR, PLDP sends instruction messages only upon witnessing an umbrella configuration, and only to the neighbours that anchor the prohibitive link. The reduction in overall messaging is evaulated in Section 5.3.

The detection phase is demonstrated in Figure 6. In Figure 6a, node *a* determines that two intersections in its vicinity contain prohibitive links, those links being *bd* and *eb*. Nodes *b, d,* and *e* have no knowledge of node *c*'s existence. The responsibility falls on node *a*

to inform neighbours of their prohibitive links. Moving to Figure 6b, node *a* instructs each of *d* and *e* to ignore their links to *b* during recovery; similarly node *a* instructs *b* to omit links to *d* and *e*.

Alternatively, notifications may be avoided entirely by producing and sending 'hello' notification packets that include neighbour information. Having been provided a 2-hop view of its neighbourhood a node can see all of the information it needs to identify prohibitive links (albeit at the cost of a larger 'hello' packet).

4.2. Statement of Correctness

Having identified and removed prohibitive links in umbrella configurations, we show in this section that PLDP will successfully route a message between two nodes if a path exists. We remind our reader that during the routing phase of PLDP, any standard position-based routing technique consisting of greedy + face-routing recovery may be implemented.

The following argument progresses first by defining the graph embedding so that we might state our claim. We first establish connectivity of the network embedding, and then show correctness by tracing a face-routing traversal within its intersections.

Definition 4.1. Let G be an embedding of a graph, and $UDG(G)$ be the unit-disc graph over G. We define $Umb(UDG(G))$ as the subgraph of G where umbrella intersections are removed.

Proposition 4.2. If $UDG(G)$ is connected, then so too is $Umb(UDG(G))$.

Proof. We begin with an umbrella configuration in a unit disc graph. The prohibitive link is only removed when there is an alternate path remaining between all nodes in the neighbourhood. Thus, removal of the prohibitive link cannot disconnect the nodes in the umbrella configuration. Any connected component in $UDG(G)$ remains connected in G'.   □

Having shown connectivity we can now inspect traversals in the network embedding.

Proposition 4.3. We claim that in G', a traversal T, consisting of left-hand rule with memory, will find and traverse a unique face.

Proof. We prove by induction on the neighbourhoods witnessed by T. Consider the first neighbourhood, k_0, visible to starting node v. If no intersection is visible to v then the next edge in T is trivial. If, however, an intersection exists in k_0 then it must be in the form depicted in either of Figures 7a or 7b (see Section 3.1 for proof). We evaluate both cases below.

1. Consider the intersection in Figure 7a. For any $v \in \{m, n, o, p\}$ and destination d, if $\overline{vd}$ intersects with no local edges (ie. $\overline{vd}$ does not pass through quadrilateral $(mnop)$) then the next left edge - and thus first edge in the current face - is trivial. If, however, $\overline{vd}$ does pass through $(mnop)$ as shown in Figure 7a then there are two cases:

   $v = n$ The starting vertex is situated in the quadrilateral such that a single vertex sits left of $\overrightarrow{vd}$ and two vertices sit on the right. In Figure 7a this case is represented by $v = n$. Node n forwards to m. Both mo and mp intersect with nd, the line segment from the destination to the point where T started, so T will escape this neighbourhood when m chooses the next CCW edge from $\overrightarrow{mp}$.

   $v = o$ The starting vertex is situated in the quadrilateral such that a single vertex sits right of $\overrightarrow{vd}$ and two vertices sit on the left. In Figure 7a this case is represented by $v = o$. Here, too, o forwards to m and m chooses the next CCW edge from $\overline{mp}$.

   In either case, the face of interest begins at vertex m where a cycle, if traversed, will be declared.

2. Consider the intersection in Figure 7b. Let starting node be $v \in q, r, s, t$ and destination d sit such that $\overline{rd}$ intersects $\overline{qs}$ and $\overline{sd}$ intersects $\overline{rt}$. The trivial case is $v = q$. Three cases remain:

   $v = r$ The starting vertex is situated in the quadrilateral such that a single vertex sits left of $\overrightarrow{vd}$ and two vertices sit on the right. In Figure 7b this case is represented by $v = r$. r forwards to q where T will escape the neighbourhood. (Recall that when T intersects with $\overrightarrow{vd}$, T switches faces.) In this case the cycle will be detected at r.

   $v = s$ The starting vertex is situated in the quadrilateral such that a single vertex sits

right of $\overrightarrow{vd}$ and two vertices sit on the left. In Figure 7b this case is represented by $v = s$. s forwards along $\overline{sq}$ where T escapes the neighbourhood. On its return, T will detect a cycle at s since node t will avoid the edge $\overline{sq}$ since it was traversed previously.

$v = t$ The starting vertex is situated in the quadrilateral such that all three vertices sit left of $\overrightarrow{vd}$. In Figure 7b this case is represented by $v = t$. Here, T traverses $\overline{tr}, \overline{rq}$ before its escape from q. (Note that $\overline{qs}$ is not a valid edge since it intersects the edge previously traversed, $\overline{tr}$. T will detect this cycle at node t.

Assume now that for any neighbourhood, k_i, traversal T exits on the same face on which it enters. We show that for neighbourhood k_{i+1} traversal T exits on the same unique face on which it enters.

Referring once more to Figure 7, there are two types of neighbourhoods to consider. Those intersections whose endpoints join into a quadrilateral such as in Figure 7a require little consideration. For any entry point m, n, o, p on the quadrilateral, T will exit on the outside of this neighbourhood.

Similarly in Figure 7b, traversals entering on $\{q, r, s\}$ are trivial. We focus on traversals of T that reach node t. From t the next CCW edges in T are $\{\overline{tr}, \overline{rq}\}$ since $\overline{qs}$ intersects $\overline{tr}$. From q, T is forwarded along the next CCW edge. $\square$

Corollary 4.4. For any G', a traversal, T, consisting of left-hand rule with memory, guarantees a path will be found provided a path exists, or complete the face if no path exists.

Proof. We know that for a set of unique faces (ie. no intersections) in an embedding, that a left-hand traversal from source to destination is guaranteed to find a path provided one exists. Thus T, which finds sets of unique (non-intersecting) faces will find a path if it exists, or complete the face where no path exists. $\square$

Finally, we note that traversal T requires no memory to succeed. A trace with no memory through all examples reveals that T will escape from any intersecting neighbourhood via the same egress links as above. T will achieve this by traversing the remaining links in the quadrilateral formed by the intersection.

5. Simulation Results

The previous section describes the PLDP protocol. We demonstrate the practical performance of PLDP via simulation in the sections that follow.

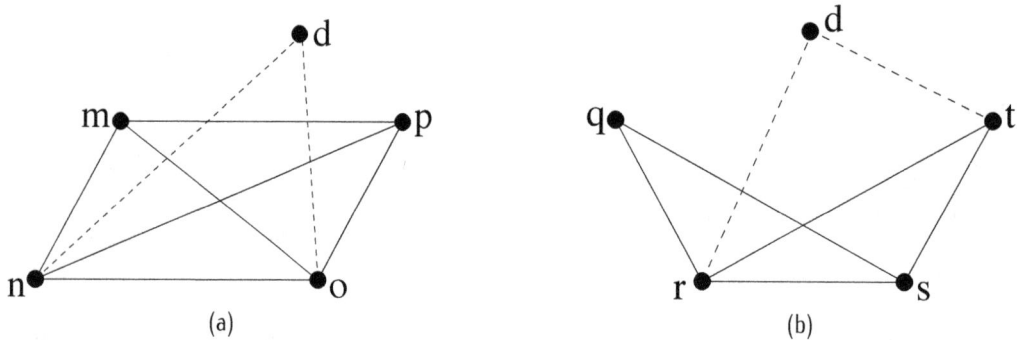

Figure 7. Once prohibitive links are omitted, two possible contentious configurations remain.

5.1. Experimental Design

So that we might better evaluate the performance of PLDP we have implemented PLDP into the netsim platform for geographic routing simulations used to evaluate GDSTR in [19] and GSpring in [20]. We compare PLDP primarily against two protocols. The first is CLDP [14, 15], a novel distributed planarization protocol that corrects for real-world events that violate the unit disc assumption. The second is GDSTR [19], also a distributed algorithm that reduces the high communication cost of CLDP but forces the network to cooperate as a whole.

The comparison of PLDP against CLDP and GDSTR may seem somewhat unfair given that CLDP and GDSTR operate outside of the UDG model. Our intention, rather than to 'compete' directly is to question the need for the one assumption on which all face-based routing techniques are based, that planarity is required for correct operation.

For completeness we set our experiments against a backdrop that includes more conventional face-routing schemes. Evaluations of the PLDP and CLDP network graph are made using GPSR [13] and GOAFR [16, 18]. GPSR design and accomplishments have served as the foundation on which later efforts have been built; it has long been considered the baseline for benchmark performance, while GOAFR provides some optimal theoretical bounds. Finally, GDSTR is implemented with two trees.

Simulation networks are composed of nodes placed uniformly in a space that is 1000 units squared; each node having a communication range of 10 units. Node density varies by increasing network size; neighbourhoods ranged from 4 to 16 nodes, on average. Each set of tests consisted of 5 runs, using the same five networks drawn from non-overlapping streams in each set of tests. A sample network with 250 nodes and an average degree of ~ 8 appears in Figure 8.

We test the validity of the CLDP and PLDP subraphs by routing with GPSR and with GOAFR.

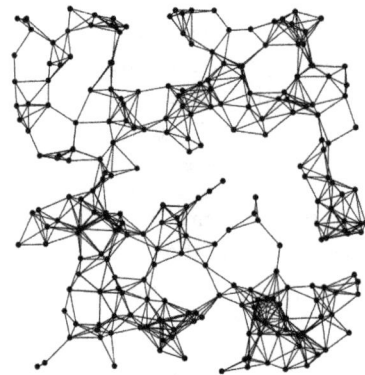

Figure 8. A sample network containing 250 nodes with an average degree of ~ 8.

Our primary performance metrics are hop stretch and message overhead. The latter takes our discussion into an investigation of the frequency of the umbrella configuration.

5.2. Hop Stretch

Hop stretch is defined as the ratio of hops taken vs hops of the minimum path. We consider only those paths over which packets were routed during face-based recovery; this is to avoid the distortion of results that would otherwise occur during the dominant greedy phase.

Observations for hop stretch are shown in Figure 9. In this figure we plot for PLDP with no changes to the routing protocols overtop, namely GOAFR and GPSR. The performance of both routing schemes is noticeably worse during the recovery phases over PLDP graphs than it is over planar graphs. In the best case scenario, routing over PLDP graphs during recovery takes 1/3rd greater number of hops than the next best scheme tested. Why is this, and is there anything that can be done?

To understand the cause we refer back to Figure 3b, in which we trace an LHR traversal. In this configuration

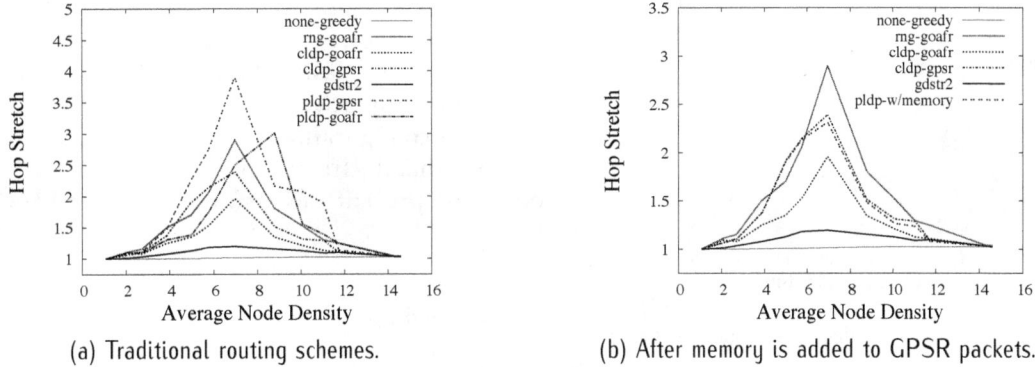

(a) Traditional routing schemes.

(b) After memory is added to GPSR packets.

Figure 9. Hop stretch of face-routing schemes on PLDP-induced graphs.

an LHR traversal that begins at node b and exits at node a requires 7 hops to escape. By contrast, were this region planarized then an LHR traversal requires 3 hops[1].

It is possible to resolve this issue by injecting memory into the routed packet. This memory consists of a record of the links traversed during face-routing so that LHR can avoid next hops that intersect with previous hops. In our evaluations up to this point we have tested unaltered routing protocols over the PLDP graph. Specifically this means that packets have no memory of where they have been. Thus many intersecting links will be traversed, only to return later along the same link in the opposite direction.

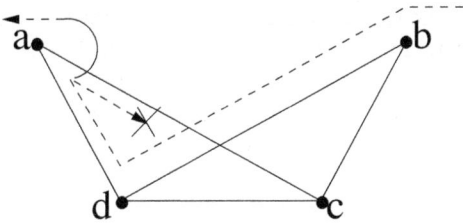

Figure 10. When packets record traversed links, nodes can substantially reduce hop counts.

So that our readers may clearly understand what is happening we point to Figure 10. A packet enters the intersecting region at node b who, according to LHR forwards to node d, who then forwards to a. Node a sees that the next link in counter-clockwise order $\overline{ac}$ intersects with link $\overline{bd}$, previously seen by this packet. In this example a packet escapes the intersecting region using 5 fewer hops than if the region had been planarized.

A noticeably different picture emerges if we record the recovery path and allow the routing protocol to

skip past those links that would intersect previously traversed links. The effect of this "with memory" approach is demonstrated in Figure 9b, in which we plot GPSR with memory over a PLDP graph. Noting the change in range along the y-axis, we can see that the hop stretch along the PLDP graph has been diminished by roughly 1/2.

In the next we evaluate the messaging cost associated with the setup and maintenance of the PLDP graph.

5.3. Message Cost

It is difficult, though necessary, to compare the setup of PLDP graphs with those setup by CLDP and GDSTR. The difficulty arises because of the difference in assumptions and goals: PLDP in its current form relies on the unit-disc assumption, whereas CLDP and GDSTR make none.

The comparison is necessary since, for all of their achievements, face-based protocols rely on the underlying assumption that some form of planarity is required for guaranteed delivery. PLDP graphs recognise that this assumption is stronger than necessary. Presumably, if we can reduce the set of undesired events then we can create more efficient real-world protocols.

The total number of messages sent in a network, on average, appears on log scale in Figure 11. The results obtained for CLDP and GDSTR are consistent with previous results: CLDP's relatively expensive messaging cost is reduced by an order of magnitude using the GDSTR approach. By contrast the number of PLDP packets produced is three orders of magnitude smaller than GDSTR. In our simulated networks PLDP produced close to zero packets until the average node density reached about 8 nodes. In the densest networks of approximately 16 nodes, 60 PLDP packets are sent.

In our trials these small numbers suggested that the number of prohibitive links is much smaller than we expected. In the next section we validate the small

---

[1] The overall cost to a path stretch is much lower than would appear since the portion of time a packet sends in recovery is much lower than the portion of time that a packet sends in greedy mode.

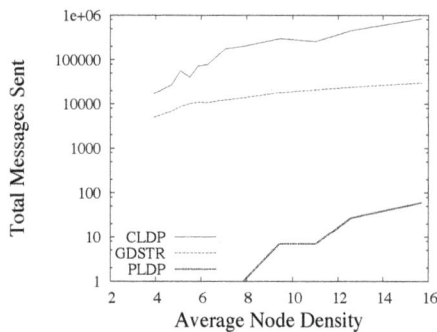

Figure 11. Packets sent or fowarded per node to achieve stability.

number of messages by investigating the frequency of prohibitive links.

5.4. Umbrella Observations

The very small number of PLDP packets produced implies that the number of prohibitive links is very small. To evaluate this hypothesis we generate large network graphs of varying density, distribution, and topology. Network nodes are distributed in a 200x200 unit space, each node with a fixed range of 8 units. We vary node density by changing the network size. Note that by changing size instead of communication range we can vary the density without affecting the maximum network diameter. Network sizes are 1500, 2500, and 3500 nodes. (In the uniform networks this results in average neighbourhood sizes of ~7, 12, and 17 nodes.) To obtain results unbiased by isolated nodes we tabulate and experiment over the largest connected component of each network as described by Table 1.

Table 1. Largest connected components in tested networks with 99% confidence intervals.

Initial	Size of largest connected component		
Network Size	Uniform	Normal	Skewed
3500	3499.9 ± 0.2	3450.8 ± 5.3	3403.8 ± 12.3
2500	2499.8 ± 0.4	2433.8 ± 4.9	2382.2 ± 10.9
1500	1490.0 ± 7.5	1406.7 ± 7.7	1359.0 ± 12.9

Nodes locations are chosen from a normal or skewed (Pareto) distribution in addition to the uniform distribution traditionally used to generate wireless network topologies. Uniformly distributed networks may be sufficient to provide insight yet are poor representations of many real deployments. Normal coordinates are generated with an average of 100 (the center) and a standard deviation of 40. Skewed coordinates are chosen from the Pareto distribution with scale parameter 1.0 and shape parameter 100.5. Example topologies appear in Figure 12.

In each network we count the number of intersecting links. From those we count the number of intersections

that form the umbrella configuration. The results are tabulated and averaged in Table 2 with 99% confidence intervals. The ratio of the two numbers appears in the last column, indicating that in all cases the proportion of intersections that form the umbrella configuration is slightly more than 1%. This suggests that the frequency of configurations that might otherwise prevent successful delivery via LHR is quite small.

6. Conclusions

In this paper we have explored an new approach to graph construction for successful forwarding in position-based routing. It is instructive to compare this approach with previous work.

Traditionally, the success of face-routing schemes relies on the assumption that the underlying graph is planar. This is restrictive; local constructions of planar graphs risk inaccuracies, while co-operative (or global) constructions are resource intensive. In either case there has yet to appear an examination of the challenges that face left-hand rule in the presence of intersections.

By contrast, the approach taken in this work was to enumerate the configurations that form an intersection in the network graph. We then scrutinised each with a left-hand rule traversal so as to isolate the 'bad' configurations from which left-hand rule is unable to recover. In doing so we recognised the existence of a prohibitive link that has the potential to conceal other viable links from a left-hand rule traversal. We then presented PLDP, a protocol that detects and avoids the prohibitive link to successfully deliver packets. It operates locally and, unlike planarization methods, omits only essential links.

Our simulation results demonstrate that routing performance over PLDP graphs is similar to current face-routing schemes. Success rates over all graphs for all schemes is 100%, while the path stretch in PLDP graphs is competitive with other methods. Where PLDP separates itself is in its messaging cost. Messages in PLDP are associated with the removal and maintenance of prohibitive links, which are shown to appear rarely. This suggests that traditional planar schemes may be 'over-solving' the problem.

We are working to release our source code as part of the Netsim2 package. We are pleased to make it available upon request in the meantime. Currently we are working to remove the unit-disc assumption. Then, using the approach presented in this paper we expect to augment PLDP for general case networks where communication error and non-uniform range is commonplace.

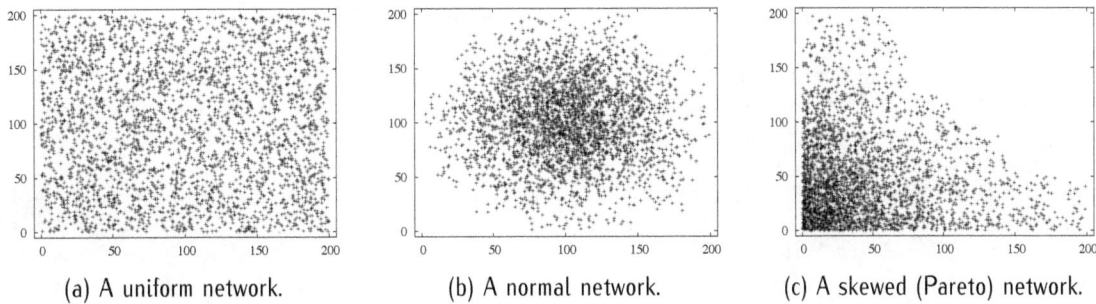

(a) A uniform network.　　　(b) A normal network.　　　(c) A skewed (Pareto) network.

Figure 12. Example networks of 3000 (density $\simeq$ 7) nodes with varying topologies.

Table 2. The number of umbrellas in tested networks with 99% confidence intervals.

Size (Density)	Node Distribution	Intersections	Umbrellas	Ratio U/I
	uniform	119536.4 ± 9563.5	1586.4 ± 96.1	0.013
1500 (7.5)	norm	2275283.8 ± 226415.4	30521.2 ± 3360.0	0.013
	skew	11577261.2 ± 8833878.2	130028.6 ± 97577.0	0.011
	uniform	939384.8 ± 38816.3	12454.6 ± 802.6	0.013
2500 (12.5)	norm	16631429.2 ± 1775319.2	225528.4 ± 22547.8	0.014
	skew	90216371.2 ± 51567741.3	1007242.4 ± 546546.8	0.011
	uniform	3692688.2 ± 124431.4	49094.8 ± 1984.6	0.013
3500 (17.5)	norm	67160718.8 ± 4266681.4	900722.2 ± 58436.6	0.014
	skew	280116248.2 ± 97476260.3	3199356.0 ± 1024083.1	0.012

References

[1] BOSE, P., MORIN, P., STOJMENOVIĆ, I. and URRUTIA, J. (1999) Routing with guaranteed delivery in *ad hoc wireless networks*. In *Workshop on Discrete Algorithms and Methods for Mobile Computing and Communications (DialM)*.

[2] BOSE, P., MORIN, P., STOJMENOVIC, I. and URRUTIA, J. (2001) Routing with guaranteed delivery in ad hoc wireless networks. *Wireless Networks* 7(6): 609–616.

[3] CHEN, D. and VARSHNEY, P. (2007) A survey of void handling techniques for geographic routing in wireless networks. *Communications Surveys & Tutorials, IEEE* 9(1): 50–67.

[4] CHEN, D. and VARSHNEY, P.K. (2007) On-demand geographic forwarding for data delivery in wireless sensor networks. *Elsevier Computer Communications* 30(14-15): 2954–2967.

[5] FAYED, M.M., CAIRNS, D.E. and MOUFTAH, H.T. (2010) *An Analysis of Planarity in Face-Routing*. Tech. Rep. TR-184, Computing Science & Math, University of Stirling.

[6] FREY, H. and STOJMENOVIC, I. (2006) On delivery guarantees of face and combined greedy-face routing in ad hoc and sensor networks. In *the 12th annual international conference on Mobile computing and networking (Mobicom)*: 390–401.

[7] FUBLER, H., WIDMER, J., KASEMANN, M., MAUVE, M. and HARTENSTEIN, H. (2003) Contention-Based Forwarding for Mobile Ad-Hoc Networks. *Elsevier's Ad Hoc Networks* 1(4): 351–369.

[8] GOVINDAN, Y.J.K.R., KARP, B. and SHENKER, S. (2006) Lazy cross-link removal for geographic routing. In *the 4th international conference on Embedded networked sensor systems (SenSys)*: 112–124.

[9] HEISSENBÄIJTTEL, M., BRAUN, T., BERNOULLI, T. and A, M.W. (2004) Blr: Beacon-less routing algorithm for mobile ad-hoc networks. *ElsevierâĂŹs Computer Communications Journal (Special Issue* 27: 1076–1086.

[10] HEISSENBÜTTEL, M., BRAUN, T., WÄLCHLI, M. and BERNOULLI, T. (2007) Evaluating the limitations of and alternatives in beaconing. *Elsevier Ad Hoc Networks* 5(5): 558–578.

[11] KALOSHA, H., NAYAK, A., RUHRUP, S. and STOJMENOVIC, I. (2008) Select-and-protest-based beaconless georouting with guaranteed delivery in wireless sensor networks. In *IEEE 27th Conference on Computer Communications (INFOCOM)*. (Pheonix, AZ, USA): 346–350.

[12] KARP, B. (2001) Challenges in geographic routing: Sparse networks, obstacles, and tra ffic provisioning, presented at DIMACS Workshop on Pervasive Networking.

[13] KARP, B. and KUNG, H. (2000) Gpsr: Greedy perimeter stateless routing for wireless networks. In *Proceedings of ACM MobiCom* (Boston, MA).

[14] KIM, Y.J., GOVINDAN, R., KARP, B. and SHENKER, S. (2005) Geographic routing made practical. In *Proceedings of the 2nd USENIX Symposium on Networked Systems Design and Implementation (NSDI)* (Boston, MA, USA).

[15] KIM, Y.J., GOVINDAN, R., KARP, B. and SHENKER, S. (2005) On the pitfalls of geographic face routing. In *Proceedings of the 2005 joint workshop on Foundations of mobile computing (DIALM-POMC*: 34–43.

[16] KUHN, F., WATTENHOFER, R., ZHANG, Y. and ZOLLINGER, A. (2003) Geometric Ad-Hoc Routing: Of Theory and Practice. In *22nd ACM Symposium on the Principles of Distributed Computing (PODC)* (Boston, Massachusetts, USA).

[17] KUHN, F., WATTENHOFER, R. and ZOLLINGER, A. (2002) Asymptotically Optimal Geometric Mobile Ad-Hoc Routing. In *6th International Workshop on Discrete*

Algorithms and Methods for Mobile Computing and Communications (DIALM) (Atlanta, Georgia).

[18] KUHN, F., WATTENHOFER, R. and ZOLLINGER, A. (2003) Worst-Case Optimal and Average-Case Efficient Geometric Ad-Hoc Routing. In *4th ACM International Symposium on Mobile Ad Hoc Networking and Computing (MOBI-HOC)* (Annapolis, Maryland, USA).

[19] LEONG, B., LISKOV, B. and MORRIS, R. (2006) Geographic routing without planarization. In *Proceedings of the 3rd USENIX Symposium on Networked Systems Design and Implementation (NSDI)* (San Jose, CA, USA).

[20] LEONG, B., LISKOV, B. and MORRIS, R. (2007) Greedy virtual coordinates for geographic routing. In *Proceedings of the IEEE International Conference on Network Protocols (ICNP)*.

[21] LOCHERT, C., MAUVE, M., FÃıJÃ§LER, H. and RTENSTEIN, H.H. (2005) Geographic Routing in City Scenarios. *ACM SIGMOBILE Mobile Computing and Communications Review (MC2R)* 9(1): 69–72.

[22] SANCHEZ, J., MARIN-PEREZ, R. and RUIZ, P. (2007) Boss: Beacon-less on demand strategy for geographic routing in wireless sensor networks. In *IEEE Internatonal Conference on Mobile Adhoc and Sensor Systems (MASS).*: 1–10.

[23] SCHINDELHAUER, C., VOLBERT, K. and ZIEGLER, M. (2007) Geometric spanners with applications in wireless networks. *Computational Geometry: Theory and Applications* 36(3): 197–214.

[24] SEADA, K., HELMY, A. and GOVINDAN, R. (2004) On the effect of localization errors on geographic face routing in sensor networks. In *Proceedings of the 3rd international symposium on Information processing in sensor networks (IPSN)*: 71–80.

[25] ZHANG, F., LI, H., JIANG, A., CHEN, J. and LUO, P. (2007) Face tracing based geographic routing in nonplanar wireless networks. In *IEEE 26th Conference on Computer Communications (INFOCOM).* (Anchorage, AK, USA).

[26] ZHAO, G., LIU, X., SUN, M.T. and MA, X. (2008) Energy-efficient geographic routing with virtual anchors based on projection distance. *Elsevier Computer Communications* 31(10).

11

Analytical modeling of address allocation protocols in wireless *ad hoc* networks[☆]

Ahmad Radaideh, John N. Daigle*

University of Mississippi, University, MS 38677, USA

Abstract

Detailed descriptions of Internet Protocol Address Assignment (IPAA) and Mobile Ad Hoc Network Configuration (MANETconf) are presented and state diagrams for their behavior are constructed. Formulae for the expected latency and communication overhead of the IPAA protocol are derived, with the results being given as functions of the number of nodes in the network with message loss rate, contention window size, coverage ratio, and the counter threshold as parameters. Simulation is used to validate the analytical results and also to compare performance of the two protocols. The results show that the latency and communication overhead for MANETconf are significantly higher than the measures of the IPAA protocol. Results of extensive sensitivity analyses for the IPAA protocol are also presented.

Keywords: address assignment protocols, mobile *ad hoc* networks, performance evaluation

1. Introduction

In this paper, detailed descriptions for two dynamic and distributed protocols proposed for address allocation in wireless *ad hoc* networks are presented, and analytical derivations for the expectations of performance measures—latency and communication overhead—for one of these protocols are carried out. Latency is the amount of time required for a newly joining node to obtain a network address, and communication overhead represents the number of messages sent during the allocation process. Expected values of the performance measures as a function of the number of network nodes at the time the new node joins the network are given. Analytical and simulation results are presented to show the effect of changes in message loss rate, coverage ratio, and the contention window size on the performance measures.

A Mobile Ad Hoc Network (MANET) has neither permanent infrastructure, nor centralized servers, nor connectivity to external networks. It consists of end systems, or nodes, that communicate with each other over a wireless medium. A node has a limited transmission range due to its limited power and it communicates directly with nodes within its transmission range. The MANET topology changes dynamically; nodes are free to move within the network, join, and leave at any time. Each node in the network runs a routing protocol so that a message from a source node can be transmitted to a destination node even if that node is outside the source's transmission range.

Address configuration can be performed using address mapping, static configuration, or dynamic configuration approaches. Address mapping uses a mapping function to derive a network address from a hardware interface identifier (MAC address) as in [1]. However, MANET nodes are not restricted to using 48-bit MAC addresses. Also the mapping is not guaranteed to be unique in IPv4 networks which use shorter 32-bit network addresses. The static configuration approach needs a user interaction and knowledge of the network's current configuration,

---

[☆]A preliminary version of this paper was presented at ICST ADOC-NETS 2010, Victoria, BC.

*Corresponding author. Email: wcdaigle@olemiss.edu

which is not practical for a dynamic topology network. In the dynamic approach, a dynamic configuration protocol is used to assign a network address. Configuration protocols such as the Dynamic Host Configuration Protocol (DHCP) [2], which run on centralized machines, cannot be extended to MANETs because of their distributed and dynamic nature. Hence, a distributed dynamic configuration protocol is required for a MANET.

Some characteristics that need to be taken into account when designing address configuration protocols include multi-hop communication, dynamic topology, and network merging and partitioning [3]. Some proposed solutions for the address allocation problem in MANETs use the duplicate address detection (DAD) mechanism to verify the uniqueness of a network address throughout the network. Others use the binary-split idea [4] to distribute the address block among network nodes so that each node has a disjoint subset of network addresses and therefore DAD is avoided. Most of these mechanisms are classified in [5] based on the following factors: network scenario (stand-alone or connected to Internet), routing protocols' dependency, address uniqueness (DAD or non-DAD), distributed or centralized address allocation, and MANET characteristics support.

Perkins [6] configures by first choosing a random address then performing a DAD procedure within the MANET. To perform the uniqueness check, the node sends an address request (AREQ) message including the randomly selected address. This message is broadcasted to all nodes in the network. The source address of the AREQ is another temporary IP address used only for sending this message. It has a different non-overlapping prefix than the prefix of the address selected for allocation and it is selected randomly so that the duplicate probability is very low. An address reply (AREP) message is sent only if the address of the receiving node matches the address in the AREQ message. The new node concludes that the selected address is unique if there is no AREP message received after sending the AREQ a finite number of retries denoted as REQUEST_RETRIES. Since the protocol performs DAD only when assigning an IP address to a new node, the proposed protocol lacks support for partitioning and merging in MANET.

Jeong [7] proposes a protocol with two address detection mechanisms. A strong DAD, based on the protocol proposed in [6], is performed in the initial phase to verify the uniqueness of the randomly selected address and a weak DAD, based on [8], which is always executed in order to prevent address conflicts with existing nodes. The weak DAD uses the concept of a virtual address, which is a combination of an address and a key. The key, which is assumed to be unique in the network, is appended to the address in the routing messages as well as the routing table. The weak DAD identifies duplicate addresses by monitoring routing information and reports the address duplication by sending out an error (AERR) message to

one of the conflicting nodes to change its address. This protocol monitors and changes the routing messages and therefore is considered routing protocol dependent.

A passive DAD approach [9] has been adopted in some proposed solutions for dynamic address configuration, such as in [10] and [11]. Passive DAD enables nodes to detect duplicate addresses in the network by analyzing received routing protocol messages. One way to detect address conflicts is based on the sequence number of a link-state routing message. The sequence number is always incremented and used to distinguish fresh from old routing information. Given that two messages with the same sequence number and source address are copies of the same message, a node may detect an address conflict if it receives a message with its own address as source address and a sequence number higher than its own counter. Other ways to detect address conflicts are based on locality and neighborhood and they are all based on analyzing routing information, which makes these solutions routing protocol dependent.

Zhou *et al.* [12] proposed a mechanism based on a stateful function, $f(n)$, to derive addresses with low probability of address duplication and therefore it avoids the use of DAD. The initial state of $f(n)$ is a seed that generates a sequence of unique numbers that can be used as network addresses. The function has to be designed carefully such that the interval between two occurrences of the same number in the generated sequence is extremely long and the probability of generating the same number in finite number of sequences initiated by different seeds is extremely low. A proposed solution based on the stateful function $f(n)$ works as follows: The first node in the network, say A, chooses a random number as its address and uses a random or default state value as a seed of its $f(n)$. When a new node, say B, joins the network and asks node A for an address, A generates an address using its state function and new state value and assigns them to B. Node A updates its state value accordingly. Node B uses the generated address as its address and the state value as a seed for its function so it can assign address to other nodes. The main concern about this protocol is designing an $f(n)$ that satisfies the properties mentioned above, which is considered a hard mathematical problem. Additional approaches based on genetic algorithms and a quadratic residue approach are presented in [13] and [14], respectively.

In MANETconf [15], an existing node is in charge of unique address allocation for a new joining node. Each configured node maintains state information of the currently assigned addresses so it can choose, based on its knowledge, an available address and verify its uniqueness throughout the network. When a new node joins the network, it asks one of its neighbors to perform the address allocation process on its behalf. The selected neighbor chooses an available address and performs the DAD procedure across the network to verify the uniqueness of the selected address. In case the new joining node is unable to find a neighbor, it concludes it is the first node in the network and performs address configuration for itself. The proposed protocol handles network partitioning and merging by performing address recovery and duplicate detection procedures.

Mohsin and Parakash [16] propose the IPAA protocol which is based on a dynamic configuration of addresses using the concept of binary split. The protocol is classified as a proactive approach because each node can independently assign a unique address to a new node without consulting any other node in the MANET. Each node in the network has a disjoint subset of the address space. When a new node joins the network, it tries to find a neighbor node that can perform an address configuration on its behalf. If a neighbor is found, the new node asks that neighbor for an address allocation. The neighbor splits its available address space into two halves and sends one half to the new node. The new node then assigns itself the first IP address in the received address space and keeps the rest with itself to configure other nodes in the future. If the new node could not find a neighbor, it concludes it is the first node in the network. So, it assigns itself the first address in the address space and keeps the rest of the address space for itself. The protocol handles network partitioning and merging as well as address recovery due to node departures. A very similar approach to IPAA, which is based on the binary-split idea, is proposed in Tayal and Patnaik [17].

The main contributions in this work are building state diagrams that represent the behavior of the two address allocation protocols proposed in [15] and [16], deriving analytical formulations for the expectations of latency and communication overhead for the protocol in [16], and presenting analytical and simulation results for the performance measures of that protocol for different numbers of existing nodes in the network. The protocol in [15] represents the DAD-based allocation schemes with an advantage of selecting a network address based on state information a node maintains to enhance the protocol performance[1]. The second protocol, which is proposed in [16], performs address allocation through local communications with neighbor nodes and that is achieved through distributing the address space among network nodes based on the binary-split idea.

The paper is organized as follows. In Section 2, detailed descriptions of the proposed address allocation protocols in [15] and [16] are presented. For each protocol, two state diagrams are derived based on the protocol specifications. The state diagrams give a complete picture of the protocol behavior, messages, timers, and handshakes. One diagram shows the state of the new node during the address allocation process while the other one shows the state of an existing node that performs the address allocation for the new node. In Section 3, analytical formulas for latency and communication overhead for the protocol in [16] are derived. The analytical formulas represent the expected values of the performance measures as a function of the number of nodes in the network and with message loss rate, contention window size, coverage ratio, and counter threshold as parameters. Section 4 presents the analytical as well as the simulation results for different values of the message loss rate, contention window size, and the coverage ratio. Section 5 concludes the paper and suggests future research.

2. Detailed descriptions of two address allocation protocols

This section presents detailed descriptions of MANETconf [15] and IPAA [16] in separate subsections. Both protocols are designed with the following network operating characteristics in mind. Nodes are free to move in the network, join, and leave at any time. Address allocation and maintenance are the responsibility of existing nodes in the network and have to be performed when the topology changes to maintain uniqueness of allocated addresses. The MANET is configured as a private IPv4 network in which the participating nodes are configured in advance to use a specific private address block. At any given time a group of connected nodes forms a network partition that has a universal unique identifier (UUID). As time evolves, a partition could either split or merge with another partition; the UUID is the key to managing the partitions. The nodes in MANET communicate with each other using IP datagrams. Communication between distant nodes in a partition is carried over intermediate nodes running an *ad hoc* routing protocol.

2.1. MANETconf protocol

In MANETconf protocol, previously configured nodes in the network manage address allocation for newly joining nodes. A newly joining node, a *requester*, chooses one of its configured neighbors, an *initiator*, to perform the address allocation on its behalf. The initiator first selects a candidate address and then broadcasts a message to all nodes in its partition to verify the uniqueness of the selected address. If verification is successful, the initiator allocates the address to the requester and informs all other nodes of the address assignment. Otherwise, the initiator repeats this address selection-verification mechanism for a finite number of times before giving up.

The protocol performs maintenance operations that clean up addresses of departed nodes due to node crashes or network partitioning, resolves address conflicts after network merging, maintains correct allocation information in each node using a soft state mechanism (i.e. timers), and resolves concurrent allocation conflicts using a priority

---

[1]The protocol proposed in [18] is compared to MANETconf via simulation with latency reported at about one half of that of MANETconf.

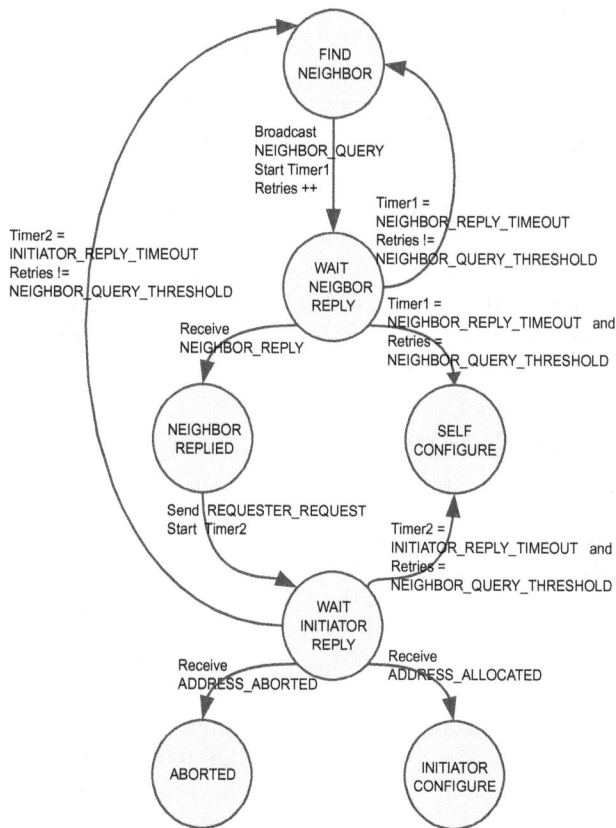

Figure 1. State diagram of a requester node during the address allocation process of MANETconf protocol.

mechanism based on the initiator address. New node address allocation and network partitioning and merging are described in separate subsections below.

New node address allocation. Figures 1 and 2 show the state diagrams of the address allocation process for a requester and an initiator nodes, respectively. The requester broadcasts a NEIGHBOR_QUERY message. If at least one NEIGHBOR_REPLY message to this query is received within a timeout value of NEIGHBOR_REPLY_TIMEOUT, the requester selects one of the responders as an initiator; otherwise, the requester retries until NEIGHBOR_QUERY_THRESHOLD is reached and then allocates an address and forms its own partition.

When the REQUESTER_REQUEST message is received, the initiator starts the address allocation process. As stated above, address allocation is a two-phase process that starts by selecting an address and verifying its uniqueness over the network partition and then confirms the allocation of the unique address to the requester. In order to minimize the conflict probability, each configured node maintains state information of address allocation in two data sets. An *Allocated* set which lists all allocated addresses, and an *Allocate_Pending* set which lists the addresses that are being allocated to newly joining nodes. Each entry in Allocate_Pending shows the address being allocated and the initiator node performing the allocation

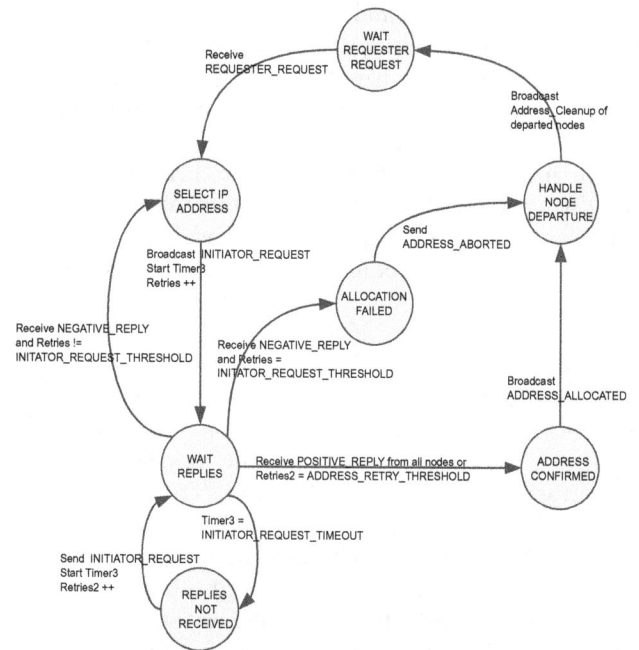

Figure 2. State diagram of an initiator node during the address allocation process of MANETconf protocol.

process to that address. A soft state maintenance to Allocate_Pending set is considered where each entry is automatically deleted after it times out.

The initiator selects an address that is neither in its Allocated nor Allocate_Pending sets. It inserts the allocation information {selected address, initiator address} to its Allocate_Pending set and broadcasts an INITIATOR_REQUEST message to all nodes in its partition to verify the address uniqueness. Address uniqueness is successfully verified when all nodes that are listed in the initiator Allocated set send a POSITIVE_REPLY message back to the initiator. This message verifies that the address is unique based on the state information of the replied node. If positive replies from all nodes have been received, the initiator sends the selected address in an ADDRESS_ALLOCATED message to all nodes in the partition including the requester. Each node inserts the allocated address in its Allocated set and deletes the address from its Allocate_Pending set.

Due to message losses, node departure, or mobility some replies may not be received. If all received replies are positive, the initiator verifies the selected address again by sending INITIATOR_REQUEST to nodes that did not reply to the previous request for ADDRESS_RETRY_THRESHOLD number of times. The address is considered unique if no negative reply has been received. The nodes that have not replied to the initiator request are considered departed nodes and their addresses are cleaned up.

The uniqueness verification of the selected address fails when a node finds the address in its Allocated set. It also fails if the address found in a node Allocate_Pending set is being allocated by a higher priority initiator. A node reports the failure by sending a NEGATIVE_REPLY message back to the initiator. If at least one negative reply has been received before INITIATOR_REQUEST_TIMEOUT, the initiator selects another address and repeats the allocation process again for INITIATOR_REQUEST_THRESHOLD times. If all trials have failed, the initiator sends an ADDRESS_ABORTED message to the requester indicating that all address allocation retries have failed. Receiving an ADDRESS_ABORTED message or no messages at all within the timeout period of INITIATOR_REPLY_TIMEOUT, the requester searches for another initiator for a maximum number of trails equal to NEIGHBOR_QUERY_THRESHOLD to perform the allocation process again.

Network partitioning and merging. The topology of a MANET changes dynamically due to node movement. The network may split into two or more partitions and partitions may also merge together to form a bigger partition. Address maintenance has to take place after these operations to solve address leak and duplicate problems.

In a MANET, each partition has a UUID, which is the lowest IP address of the partition. A newly joining node is provided with its partition UUID as well as its own IP address during its address allocation process, whereas a node that configured itself sets its partition's UUID to its own IP address.

When a group of nodes splits into two separate partitions due to nodes movement, nodes in both partitions will detect departure of other nodes during the address allocation process of a newly joining node. When a partition detects the partitioning event, address cleanup procedure is performed. The node that detects the partitioning event manages address cleanup of departed nodes, so it broadcasts an ADDRESS_CLEANUP message to all other nodes in its partition. A node that receives the address cleanup message deletes the addresses listed in the message from its Allocated set. Since the UUID is determined by the lowest address in the partition, one of the partitions has to be assigned a new UUID. In a partition that the lowest address has been deleted, each node selects the lowest remaining address allocated in that partition to be the new UUID.

Figure 3 shows the state diagram of the partitions merging process. When two nodes i and j from two different partitions get close to each other, they exchange their partition identifiers. If the received partitions identifier is different from a node partitions identifier then a node detects the merging of two partitions. Both i and j will detect merging of their partitions in that case. After detecting the merging event, both i and j exchange their Allocated sets. Each one of them broadcasts the Allocated set of the other to all nodes in its partition. Each node in both partitions takes the union of its Allocated set and the received Allocated set.

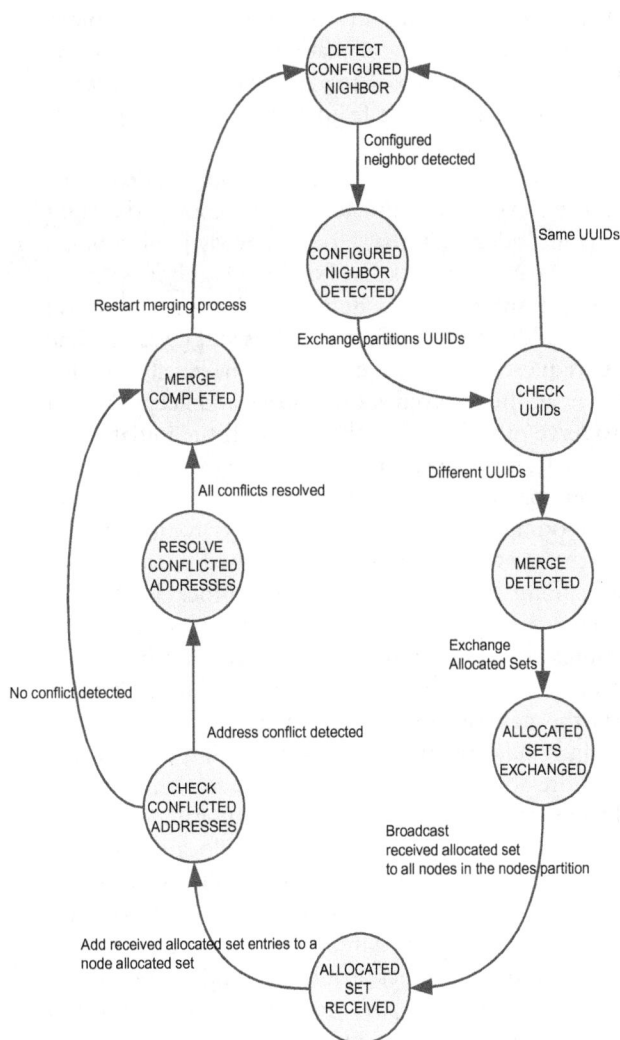

Figure 3. State diagram of the merging process of MANETconf protocol.

If the received Allocated set contains a node address, then there is a node in the other partition that has the same address. Those two nodes are called *conflicting* nodes. One of the conflicting nodes has to give up the conflicting address and ask existing neighbors for a unique address. The address allocation process has to be performed for each conflicting address and this time a global agreement on the selected address has to be granted from all nodes in both merged partitions. Nodes in one partition have to update their UUID to the lowest allocated address in both partitions. The merging of two partitions is completed when address conflict has been resolved.

2.2. IPAA protocol

The IPAA protocol presents a distributed dynamic address allocation protocol for a stand-alone MANET. The protocol avoids the DAD process by employing a proactive approach using the binary-split idea. The binary-split idea is that each node has a disjoint subset of the address block and it can independently allocate a unique address and hand half of its address space to a newly joining node without getting an agreement from every other node in its partition.

When a new partition is formed, the only node in that partition reserves the complete address block and assigns itself the first address in that block. A newly joining node, a requester (or client), asks an existing neighbor node, an initiator (or server), for a network address. The server divides its address space into two halves and gives one half to the requester. The requester assigns itself the first address from the received address space and keeps the rest of it to serve other nodes in the future. If the initiator has no space left, it borrows an address space from an existing node and forwards it to the requester. This procedure avoids flooding the entire partition to verify the uniqueness of the selected address as DAD-based protocols do.

Maintaining the complete Address_Block is the key issue in this protocol. The protocol performs maintenance procedures to avoid address leak and conflict problems. The address leak problem happens when a node abruptly departs the network or moves out its partition without returning its address space. Without cleaning up departed nodes addresses, these addresses will be considered allocated to existing nodes and cannot be allocated to newly joining nodes. Nodes keep track of allocated address blocks to resolve the address leak problem. Graceful departure is provided where nodes that want to leave the network return their address blocks and confirm their departure. Address conflict problems could happen when two partitions merge together since each partition reserves the entire address block for itself. Nodes with the same address in both partitions are allocated the same address space when configured. The conflicting node that has a bigger address space should give up its allocated space and ask other nodes for new allocation.

New node address allocation. Figures 4 and 5 show the state diagrams of a client and a server nodes during address allocation process, respectively. When a client joins the network it broadcasts a one-hop REQUEST message. A neighbor replies with a REPLY message to the client. The client selects one of its existing neighbors that replied to its request message to be its server. The client sends an acknowledgement (ACK) message to the selected server asking for a unique address. When receiving the ACK message, the server starts the allocation process for the requested client. Neighbors are expected to reply to the client request within REPLY_TIMEOUT amount of time. If the timer expires without receiving any reply, the client repeats searching for neighbors for NEIGHBOR_ REQUEST_THRESHOLD number of times. If all trials have failed, the client reserves the entire address block for itself, allocates itself the first address in that block, and sets a UUID for that partition.

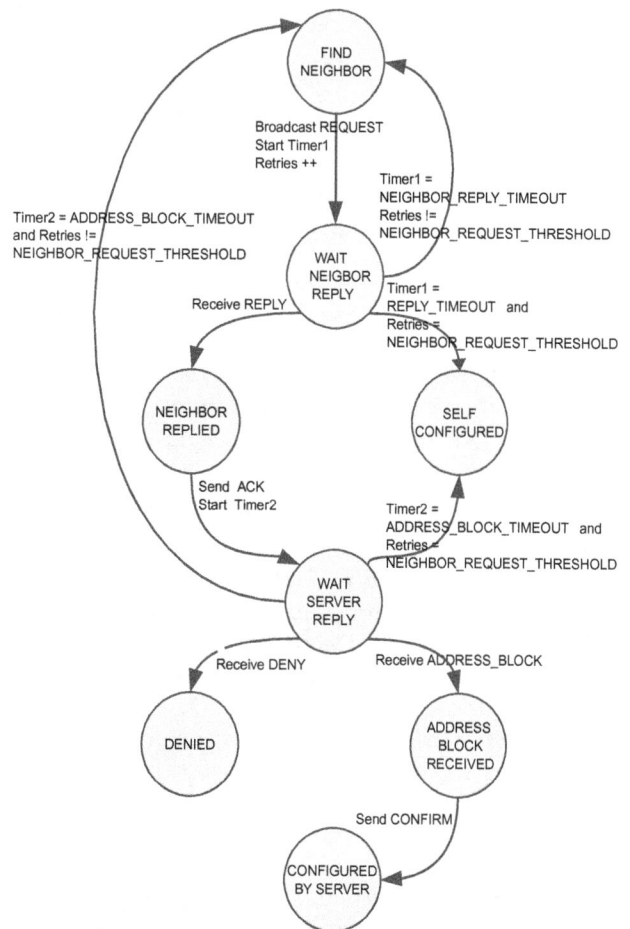

Figure 4. State diagram of a client node during the address allocation process of IPAA protocol.

If reply messages have been received, the client selects one of its neighbors to perform the address allocation process by sending an ACK message and starting a timer with a timeout value ADDRESS_BLOCK_TIMEOUT. When a node receives the ACK message, it divides its available address space into two disjoint subsets and sends one subset to the client in the ADDRESS_BLOCK message. The client receives the ADDRESS_BLOCK and the partition UUID, configures itself the first address in that block, and keeps the rest of its address space to configure newly joining nodes in the future. The client confirms a successful address allocation by sending CONFIRM message back to the server. If the client timer expires before receiving the ADDRESS_BLOCK message, the client considers that the server is no longer existing and searches for another server to perform the address allocation process again up to NEIGHBOR_REQUEST_THRESHOLD number of times. If the ADDRESS_BLOCK message has not been received in all trials, the client performs self-allocation to configure itself an address as described above.

In case the selected server has no available addresses to serve the client, the server searches for an existing node in

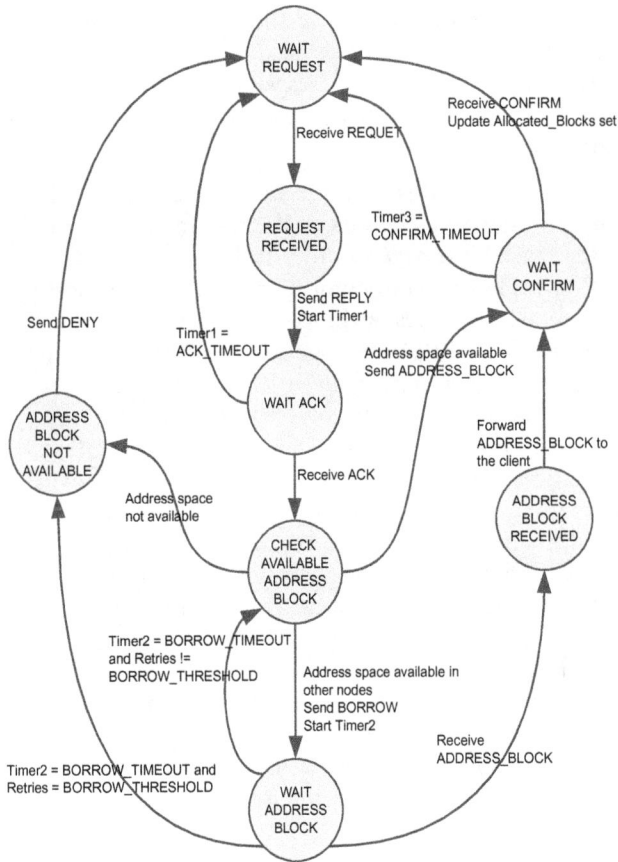

Figure 5. State diagram of a server node during the address allocation process of IPAA protocol.

the partition that has an available address. For this purpose, each node maintains state information about the allocated address blocks in Allocated_Blocks data set. The Allocated_Blocks set lists the configured nodes in the partition with their available Address_Blocks. The server selects the node with the largest available Address_ Block and sends it a BORROW message requesting half of its available space. Once the ADDRESS_BLOCK message is received from the requested node, the server forwards the message to the client. If all nodes in the server Allocated_Blocks set have no available address space, the server sends a deny (DENY) message to the client indicating that addresses are not available in this partition.

Nodes could depart the network or move out their partitions at any time. If a node does not respond to the BORROW message within a timeout period of BORROW_TIMEOUT, the server sends the borrow message to the node that has the second largest address space. The borrow process is repeated up to BORROW_THRESHOLD number of times. If all trials have failed or no address space is available in the rest of the partition nodes, the server sends DENY message to the client indicating that addresses are not available in this partition.

Network partitioning and merging. Nodes may depart the network or move out of their partitions at any time. Departure of a node leads to address leak problem where the nodes address block will not be used by other existing nodes. Each node is responsible for cleaning up the Address_Block of its missing buddy node. To achieve this, the Allocated_Blocks set in each node has to be updated regularly. Each node in the partition broadcasts its Allocated_Blocks set to every other node in the partition. A node updates its Allocated_Blocks set when receiving other nodes' sets to keep its information up to date.

Since the state information a node has is assumed up to date, each node looks up its Allocated_Blocks set from time to time to check the existence of its buddy node. If the buddy node is missing from the set, a node claims that its buddy has departed the network. Therefore, it merges its buddy node Address_Block with its block. Each partition has a UUID to be identified from other existing partitions. Partitioning is detected if the node with the lowest address is missing. When detecting the partitioning event, each node sets the partition UUID to the lowest address currently allocated to a node in their partition.

It is possible that a partition merges with another partition in the network. The merging process performed in this protocol is similar to the process described in Figure 3. When two configured nodes get close to each other, they exchange their partition UUIDs. If the nodes UUIDs are different, a merging event is detected. Those nodes that detect the merging event exchange their Allocated_Blocks sets. Each node broadcasts the other nodes' set to all nodes in its partition. When a node receives the Allocated_Blocks set, it searches the set to check if a node with the same address exists in the other partition. If an address conflict is detected, one of the two nodes with a larger address space gives up its address space and asks existing nodes for a new allocation. Merging of two partitions ends when all address conflicts are resolved. The partition UUID maintained by each node is updated to the lowest address allocated in the resulting partition.

3. Analytical modeling of address allocation protocols

In this section, we derive analytical expressions of the expected values of latency and communication overhead that are used to evaluate the performance of IPAA [16]. The main objective of such analytical derivations is to obtain mathematical formulations that can clarify the impact of network characteristics and the number of existing nodes in the network on the performance measures of the address allocation process under consideration. The network characteristics that have an impact on the performance measures are the network area, the node's coverage area, collisions, and message loss rate. The derived formulas also show the impact of the protocol parameters which are the timeout values and the counter values on the performance measures of the selected protocol.

The derivations of the expected latency and communication overhead are first carried out for small number of existing nodes in the network and then generalized for an arbitrary number of nodes. Section 3.1 presents the network model under consideration. It defines the network boundary, the node's coverage area, the message loss as well as collision for the incoming traffic at the new node. In Section 3.2, derivations for the expected latency and communication overhead are conducted.

3.1. The network model

Let A represent the total area of the network. Each node in the network is located at a random position and nodes are assumed to be uniformly distributed over the network area. Nodes are assumed to have a coverage area of A_x. Let $\tilde{n}$ denote the number of existing nodes in the network at the time a new node wishes to join. Let $\tilde{x}_i$ indicate the presence of node i, $i = 1, 2, \ldots, n$, within the transmission range of the new node; that is, $\tilde{x}_i = 1$ is a neighbor of the newly entering node and $\tilde{x}_i = 0$ otherwise.

Since the existing nodes are uniformly distributed over the network area independent of the placement of all other nodes, $\tilde{x}_i$, $i = 1, 2, \ldots, n$, is a set of identical, independent, Bernoulli trials with success probability $\frac{A_x}{A}$. Thus, the number of existing nodes that are within the transmission range of the new node, denoted as $\tilde{x}$, is a binomial random variable with parameters $\left(n, \frac{A_x}{A}\right)$ and

$$E[\tilde{x}|\tilde{n} = n] = n\,P\{\tilde{x}_i = 1|\tilde{n} = n\} = n\,\frac{A_x}{A}. \quad (1)$$

The new node communicates with a neighbor node over a wireless channel. A bad channel condition results in a low signal to interference plus noise ratio (SINR) for the received message and the message is considered lost if its SINR is below a given threshold. Assume that the wireless channel between the new node and a neighbor node has a message loss rate equal to ε. To avoid collision, each node in the network has a contention window of size W. A node that has data to send chooses a random slot number uniformly from $\{1, 2, \ldots, W\}$ and sends its data in that selected slot. Collision in a given slot could happen if two or more nodes transmit in the same slot despite the SINR value of their messages. A message is received successfully if it does not collide with other messages in the transmission slot and has a good SINR value.

3.2. Expectations of the performance measures

The amount of time for the new node to be configured with a network address is denoted as $\tilde{\ell}$ and the number of messages sent during the address allocation process is denoted as $\tilde{c}$. The derivations for the expected values of latency and communication overhead for a given number of nodes, $\tilde{\ell}|\tilde{n} = n$ and $\tilde{c}|\tilde{n} = n$, respectively, are first constructed for the case $\tilde{n} = 0$ and then generalized to an arbitrary number of nodes $\tilde{n} = n$.

The $\tilde{n} = 0$ case. In this case the new node is the only node in the network. The new node starts by broadcasting a REQUEST message in order to be allocated by an address through local communication with its neighbors. Since there are no nodes in the neighborhood, the timer for receiving the REPLY message will reach the timeout value T_{NR} without receiving any REPLY message. Since no REPLY messages have been received during the first timeout period, the new node sends another REQUEST message and again waits for timeout. The new node will repeat sending the REQUEST message and waiting for timeout for a maximum of K_{NR} trials. After the last timeout period, the new node considers itself the first node in the network and performs address configuration itself. It allocates the entire address space for itself and assigns itself the first address in that space.

The expected latency for a new node to be allocated a network address in this case, denoted as $E[\tilde{\ell}|\tilde{n} = 0]$, is the sum of the timeout value for all trials. The timeout value for each trial is T_{NR} and the maximum number of trials is K_{NR}. Therefore,

$$E[\tilde{\ell} \mid \tilde{n} = 0] = E[\tilde{\ell} \mid \tilde{x} = 0] = K_{NR}\,T_{NR}. \quad (2)$$

The expected communication overhead in this case, which is denoted as $E[\tilde{c}|\tilde{n} = 0]$, is the number of REQUEST messages sent by the new node. The new node sends one REQUEST message in each trial for a maximum number of K_{NR} without getting any replies. Therefore,

$$E[\tilde{c} \mid \tilde{n} = 0] = E[\tilde{c} \mid \tilde{x} = 0] = K_{NR}. \quad (3)$$

General formulas for $\tilde{n} = n$. In this section we derive general formulas for the expectations of latency and communication overhead for the address allocation protocol presented in [12]. The latency, as defined earlier, is the amount of time required for a new node to be allocated a network address, and the communication overhead is the number of messages sent during the address allocation process. Here, we derive expectations for the latency and communication overhead given that the number of nodes in the network is $\tilde{n} = n$.

Since the selected protocol performs address allocation through local communications with the new node neighbors, the expectations for its latency and communication overhead are derived from their conditional values on the number of nodes that are within the transmission range of the new node. The general formulas for the expected latency and communication overhead are given by

$$E[\tilde{\ell} \mid \tilde{n} = n] = \sum_{x=0}^{n} E[\tilde{\ell} \mid \tilde{x} = x] \, P\{\tilde{x} = x \mid \tilde{n} = n\} \quad (4)$$

and

$$E[\tilde{c} \mid \tilde{n} = n] = \sum_{x=0}^{n} E[\tilde{c} \mid \tilde{x} = x] \, P\{\tilde{x} = x \mid \tilde{n} = n\}. \quad (5)$$

Recall that $\tilde{x}$ is a binomial random variable with parameters $\left(n, \frac{A_x}{A}\right)$. Thus,

$$P\{\tilde{x} = x \mid \tilde{n} = n\} = \binom{n}{x} \left(\frac{A_x}{A}\right)^x \left(1 - \frac{A_x}{A}\right)^{n-x}. \quad (6)$$

For the special case when there is no node within the transmission range of the new node, that is ($\tilde{x} = 0$), the expected latency and communication overhead as derived in (2) and (3) are given by

$$E[\tilde{\ell} \mid \tilde{x} = 0] = K_{\text{NR}} \, T_{\text{NR}} \text{ and } E[\tilde{c} \mid \tilde{x} = 0] = K_{\text{NR}}. \quad (7)$$

For all other cases, where $\tilde{x} > 0$, the expectations for the latency and communication overhead for a given number of neighbors are calculated from their conditional values on the number of trials, denoted by $\tilde{\kappa}$, required to successfully allocate a network address to the new node. Conditioning on the value of $\tilde{\kappa}$ yields

$$
\begin{aligned}
E[\tilde{\ell} \mid \tilde{x} = x] = \sum_{\kappa=1}^{K_{\text{NR}}} E[\tilde{\ell} \mid \tilde{x} = x, \tilde{\kappa} = \kappa] \\
\times P\{\tilde{\kappa} = \kappa \mid \tilde{x} = x\} + E[\tilde{\ell} \mid \tilde{x} = x, \tilde{\kappa} > K_{\text{NR}}] \\
\times P\{\tilde{\kappa} > K_{\text{NR}} \mid \tilde{x} = x\},
\end{aligned}
$$

and

$$(8)$$

$$
\begin{aligned}
E[\tilde{c} \mid \tilde{x} = x] = \sum_{\kappa=1}^{K_{\text{NR}}} E[\tilde{c} \mid \tilde{x} = x, \tilde{\kappa} = \kappa] \\
\times P\{\tilde{\kappa} = \kappa \mid \tilde{x} = x\} + E[\tilde{c} \mid \tilde{x} = x, \tilde{\kappa} > K_{\text{NR}}] \\
\times P\{\tilde{\kappa} > K_{\text{NR}} \mid \tilde{x} = x\}.
\end{aligned}
$$

$$(9)$$

The kth trial is successful if an ADDRESS_BLOCK message is received from a neighbor node in response to the ACK message. The ADDRESS_BLOCK message contains a disjoint subset of the address space where the new node selects the first address from that space for itself and keeps the rest for serving other joining nodes in future. Assuming that successive trials are independent and identical, the probability that a successful address allocation occurs on the kth trial is given by

$$P\{\tilde{\kappa} = \kappa \mid \tilde{x} = x\} = (P\{\tilde{a} = 0 \mid \tilde{x} = x\})^{\kappa-1} P\{\tilde{a} = 1 \mid \tilde{x} = x\}, \quad (10)$$

where $\tilde{a}$ is the indicator random variable for the event of a successful allocation. As discussed earlier in previous sections, the trial fails if the ADDRESS_BLOCK message is

not received when at least one REPLY message is received or when no REPLY message is received. Define $\tilde{s}$ as the indicator random variable for the event of the reception of at least one reply message from a neighbor node and that message is transmitted with no other messages in a slot and has a good SINR value. Then, the probability of a failed trial in the presence of x neighbors is given by

$$
\begin{aligned}
P\{\tilde{a} = 0 \mid \tilde{x} = x\} &= P\{\tilde{a} = 0 \mid \tilde{x} = x, \tilde{s} = 0\} \, P\{\tilde{s} = 0 \mid \tilde{x} = x\} \\
&\quad + P\{\tilde{a} = 0 \mid \tilde{x} = x, \tilde{s} = 1\} \, P\{\tilde{s} = 1 \mid \tilde{x} = x\} \\
&= P\{\tilde{s} = 0 \mid \tilde{x} = x\} + (\varepsilon + (1 - \varepsilon)\varepsilon) \\
&\quad \times P\{\tilde{s} = 1 \mid \tilde{x} = x\}.
\end{aligned}
$$

$$(11)$$

The failure to receive a REPLY message is due either to loss of the REQUEST message in transmission channels between the new node and all neighbor nodes or due to loss of the responders' REPLY messages due to either collisions or external noise. Each neighbor node that received a REQUEST message will send its reply in a randomly chosen slot from the range $(1, 2, \ldots, W)$, where W denotes the contention window size. A REPLY message is received successfully if it is transmitted with no other messages in the same slot and has a good SINR value. Define $\tilde{r}$ to be the number of responders that received a REQUEST message and $\tilde{r}_i$ to be the number of responders that responded in slot number i. Then the probability of receiving at least one REPLY message successfully given that there are x neighbors is given by

$$P\{\tilde{s} = 1 \mid \tilde{x} = x\} = \sum_{r=1}^{x} P\left\{\bigcup_{i=1}^{W} \{\tilde{r}_i = 1\} \mid \tilde{r} = r\right\} P\{\tilde{r} = r \mid \tilde{x} = x\}. \quad (12)$$

The first term of the equation represents the probability that at least one transmission slot has exactly one REPLY message and that message has a good SINR value. The second term represents the probability that $\tilde{r}$ neighbors received the REQUEST message out of the total number of neighbors. Since message transmissions are assumed to have identical failure probabilities of ε and messages are transmitted independently, $\tilde{r}$ is a binomial random variable with parameters $(x, (1 - \varepsilon))$. The probability that r nodes will receive the message out of the total number of x neighbors is given by

$$P\{\tilde{r} = r \mid \tilde{x} = x\} = \binom{x}{r}(1 - \varepsilon)^r \varepsilon^{x-r}. \quad (13)$$

The probability that at least one slot has exactly one reply message given that there are r responders is equal to[2]

---

[2]This is simply the probability of the union of arbitrary events as given in any probability book [19].

$$P\left\{\bigcup_{i=1}^{W}\{\tilde{r}_i=1\}\mid \tilde{r}=r\right\}$$

$$= \sum_i P\{\tilde{r}_i=1\mid\tilde{r}=r\} - \sum_{i<j} P\{\tilde{r}_i=1,\tilde{r}_j=1\mid\tilde{r}=r\}$$

$$+ \sum_{i<j<k} P\{\tilde{r}_i=1,\tilde{r}_j=1,\tilde{r}_k=1\mid\tilde{r}=r\} - \cdots$$

$$+ (-1)^{W+1} P\{\tilde{r}_i=1,\tilde{r}_j=1,\tilde{r}_k=1,\cdots,\tilde{r}_W=1\mid\tilde{r}=r\}. \tag{14}$$

The first term of (14) represents the probability that slot i has exactly one message, in other words, exactly one responder transmitted the message in slot number i and that message when received had a good SINR value. That is

$$P\{\tilde{r}_i=1\mid\tilde{r}=r\} = \binom{r}{1}\frac{1}{W}\left(1-\frac{1}{W}\right)^{r-1}(1-\varepsilon). \tag{15}$$

The other terms of (14) represent joint probabilities of having more than one slot with exactly one message. Note that the sum of the slots that have exactly one reply message, $\tilde{r}_i=1$, is equal to the total number of replies r. That is the probability in the second term of the equation evaluates to zero if the number of replies is less than two. In general, the joint probability of having exactly one reply in each of the slots i, j, ..., z where the $\tilde{r}_i+\tilde{r}_j+\cdots+\tilde{r}_z=h$ is equal to zero for $h>r$ and for the $h\le r$ case it is calculated as

$$P\{\tilde{r}_{i_1}=1,\cdots,\tilde{r}_{z_h}=1\mid\tilde{r}=r\}$$

$$= \binom{r}{h} h! \frac{(W-h)^{r-h}}{(W)^r}(1-\varepsilon)^h. \tag{16}$$

Substituting this result in (14) and performing some algebra yield

$$P\left\{\bigcup_{i=1}^{W}\{\tilde{r}_i=1\}\mid\tilde{r}=r\right\}$$

$$= \sum_{h=1}^{\min\{r,W\}} (-1)^{h+1}\binom{W}{h}\binom{r}{h} h! \frac{(W-h)^{r-h}}{W^r}(1-\varepsilon)^h. \tag{17}$$

Now, we consider the conditional expectations of the latency and communication overhead on the number of trials required for the new node to be allocated a network address. Recall that $\tilde{\kappa}$ represents the number of trials that has been performed, the expected latency when the address allocation succeeded on the kth trial, where $\kappa\le K_{NR}$ is the sum of the timeouts for the first $(k-1)$

failed trials plus the timeout of receiving a REPLY message and the timeout of receiving the ADDRESS_BLOCK message in the last successful trial. The latency for a failed trial is the sum of the timeout to receive the REPLY message plus the timeout to receive the ADDRESS_BLOCK message if at least one reply is successfully received. That is

$$E[\tilde{\ell}\mid\tilde{x}=x,\tilde{\kappa}=\kappa] = (\kappa-1)(T_{NR}+P\{\tilde{s}=1\mid\tilde{x}=x\}\,T_{AR})$$
$$+ T_{AR}+T_{AR} = \kappa\,T_{NR}$$
$$+ T_{AR}(1+(\kappa-1)\,P\{\tilde{s}=1\mid\tilde{x}=x\}), \tag{18}$$

where $P\{\tilde{s}=1\mid\tilde{x}=x\}$ is presented in (12).

The expected communication overhead for a successful address allocation on the kth trial is the sum of messages sent during the first $(k-1)$ trials and they are one REQUEST message, the expected number of REPLY messages, and one ACK message that is sent if at least one reply is received correctly. In the last trial, the new node sent a REQUEST and received at least one REPLY so it sent out an ACK message to one of the responders and received an ADDRESS_BLOCK message that contains a portion of the address space. Therefore, the expected communication overhead is equal to

$$E[\tilde{c}\mid\tilde{x}=x,\tilde{\kappa}=\kappa] = (\kappa-1)\,(1+E[\tilde{r}\mid\tilde{x}=x]$$
$$+ P\{\tilde{s}=1\mid\tilde{x}=x\})$$
$$+ (1+E[\tilde{r}\mid\tilde{x}=x]+2) \tag{19}$$
$$= \kappa(1+E[\tilde{r}\mid\tilde{x}=x])$$
$$+ (\kappa-1)P\{\tilde{s}=1\mid\tilde{x}=x\}+2.$$

In the case where all trials have failed, that is $\kappa>K_{NR}$, the expected latency and communication overhead are given by

$$E[\tilde{\ell}\mid\tilde{x}=x,\tilde{\kappa}>K_{NR}]$$
$$= K_{NR}\,(T_{NR}+P\{\tilde{s}=1\mid\tilde{x}=x\}\,T_{AR}) \tag{20}$$

and

$$E[\tilde{c}\mid\tilde{x}=x,\tilde{\kappa}>K_{NR}] = K_{NR}\,(1+E[\tilde{r}\mid\tilde{x}$$
$$= x]+P\{\tilde{s}=1\mid\tilde{x}=x\}). \tag{21}$$

Substituting (10), (18), (19), (20), and (21) into (8) and (9) gives

$$E[\tilde{\ell}\mid\tilde{x}=x] = \sum_{\kappa=1}^{K_{NR}}\kappa\,T_{NR}+T_{AR}\,(1+(\kappa-1)$$
$$\times P\{\tilde{s}=1\mid\tilde{x}=x\})\cdot(P\{\tilde{a}=0\mid\tilde{x}=x\})^{\kappa-1}$$
$$\times P\{\tilde{a}=1\mid\tilde{x}=x\}+K_{NR}$$
$$\times (T_{NR}+P\{\tilde{s}=1\mid\tilde{x}=x\}\,T_{AR})$$
$$\times (P\{\tilde{a}=0\mid\tilde{x}=x\})^{K_{NR}}, \tag{22}$$

$$E[\tilde{c} \,|\, \tilde{x} = x] = \sum_{\kappa=1}^{K_{NR}} (\kappa \,(1 + E[\tilde{r}|\tilde{x} = x]) + (\kappa - 1)$$

$$\times P\{\tilde{s} = 1|\tilde{x} = x\} + 2) \cdot (P\{\tilde{a} = 0 \,|\, \tilde{x} = x\})^{\kappa-1}$$

$$\times P\{\tilde{a} = 1 \,|\, \tilde{x} = x\} + K_{NR} \,(1 + E[\tilde{r}|\tilde{x} = x]$$

$$\times +P\{\tilde{s} = 1|\tilde{x} = x\}) \,(P\{\tilde{a} = 0 \,|\, \tilde{x} = x\})^{K_{NR}},$$

$$(23)$$

where $P\{\tilde{a} = 0 \,|\, \tilde{x} = x\}$ is from (11) and $P\{\tilde{s} = 1|\tilde{x} = x\}$ is from (12).

4. Numerical results

In this section, we present numerical results for latency and communication overhead based on analysis and simulation, and in addition, we consider the sensitivity of these measures to message loss rate, contention window size, and coverage ratio. In Section 4.1, a description of the simulation carried out for the IPAA and MANETconf protocols is presented. Then, Section 4.2 presents a comparison of the IPAA and MANETconf protocols. Section 4.3 presents the analytical and simulation results, which show good agreement, for the IPAA protocol as functions of network size with different parameter values. A summary of results is deferred to Section 5.

4.1. Simulation description

The network dimensions are set to 100 m × 100 m. Each node is placed at a uniformly distributed location within the network area. Results are collected for node populations from 0 to 50. For a given number of nodes in the network, the number of nodes within the transmission range of the new node depends on its coverage area. Results are collected for coverage ratios of 10%, 15%, and 20%. Each message sent from one node to another is subject to loss with rate ε; loss rates examined were 0, 0.1, and 0.2. Contention window sizes, W, considered were 10, 20, and 30. For the IPAA protocol, the timeout value to find a neighbor T_{NR} was set to 0.2 s and the timeout to receive the Address_Block message T_{AR} was set to 0.02 s, and K_{NR} was varied from 3 to 5 to show the effect of the counter threshold on the performance measures of the protocol. For MANETconf protocol T_{NR} was set to 0.2 s and T_{AR} was set to 2 s, and K_{NR} was set to 3 and K_{AR} was set to 5.

The simulation results presented are the average values of 10000 simulation runs. For large number of simulation runs, the sample mean of a performance measure for any network size follows the normal distribution $N(\mu, \frac{s}{\sqrt{n}})$, where μ is the true population mean, s is the sample standard deviation, and n is the sample size. Throughout simulation, we have found that the maximum value of the sample standard deviation is 0.002 for latency samples and it is 0.1 for communication overhead samples for all values of network size. The 95% confidence interval, that

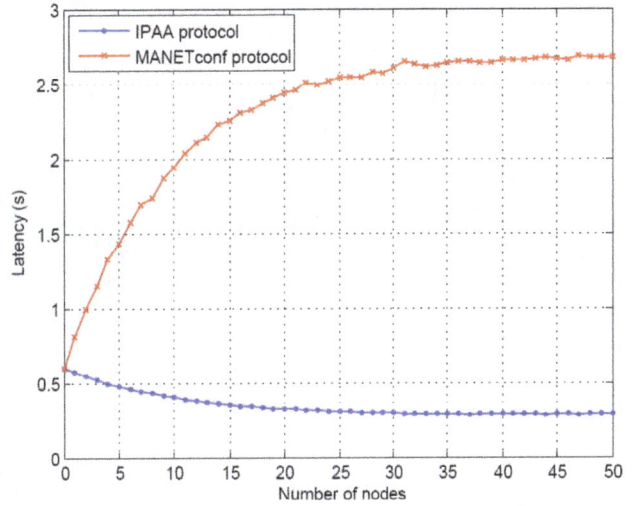

Figure 6. Latency of the IPAA and MANETconf protocols as a function of network size.

is likely to include the true population mean of a performance measure, is equal to $(\bar{a} \pm 1.96 \frac{s}{\sqrt{n}})$, where $\bar{a}$ represents the sample mean for the performance measure. Therefore, the 95% confidence interval for latency samples is $(\bar{x} - 3.92 \times 10^{-5}, \bar{x} + 3.92 \times 10^{-5})$ and for communication overhead it is $(\bar{y} - 0.002, \bar{y} + 0.002)$.

4.2. IPAA versus MANETconf protocol

Figure 6 compares the latency of IPAA protocol to the latency of the MANETconf protocol. In MANETconf, the initiator node selects an address and performs DAD process throughout the network. Therefore, the latency of the allocation process increases as the number of nodes in the network increases. In IPAA, the address allocation process is carried out by an existing neighbor or it may be carried out by the new node itself in case no neighbor exists. For large number of nodes, the probability of finding a node within the transmission range of the new node is high so that address allocation can be carried out by a neighbor node and therefore it decreases as the number of nodes increases.

Figure 7 presents the communication overhead for IPAA and MANETconf protocols. Since MANETconf performs the DAD process throughout the network, the number of messages sent increases as the number of nodes increases. The IPAA protocol performs address allocation through local communications with neighbor nodes only. Therefore, the number of messages sent is less than that for the MANETconf protocol.

4.3. Analytical and simulation results for the performance measures of the IPAA protocol

Figures 8 and 9 represent the latency and communication overhead at loss rates $\varepsilon = \{0, 0.1, 0.2\}$. The contention

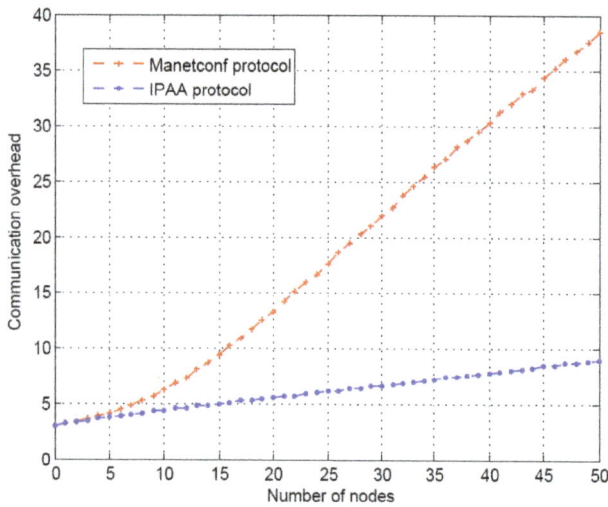

Figure 7. Communication overhead of the IPAA and MANET-conf protocols as a function of network size.

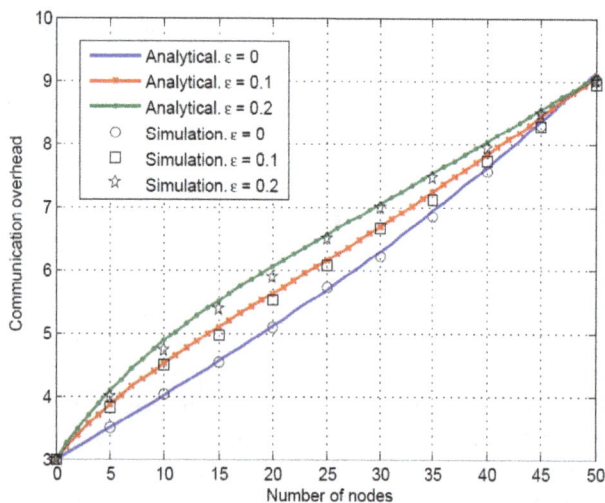

Figure 8. Latency of the IPAA protocol as a function of network size with message loss rate as a parameter.

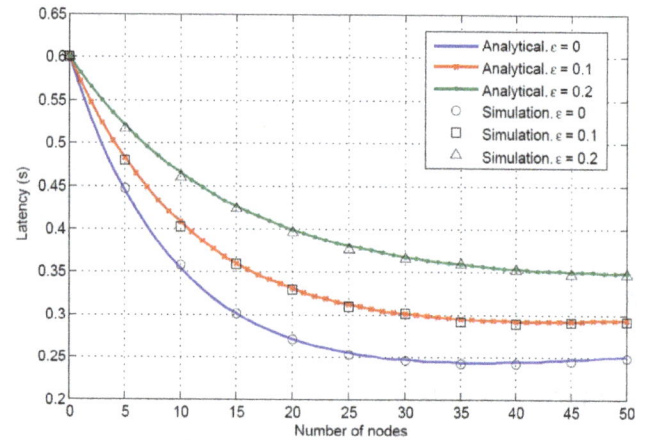

Figure 9. Communication overhead of the IPAA protocol as a function of network size with message loss rate as a parameter.

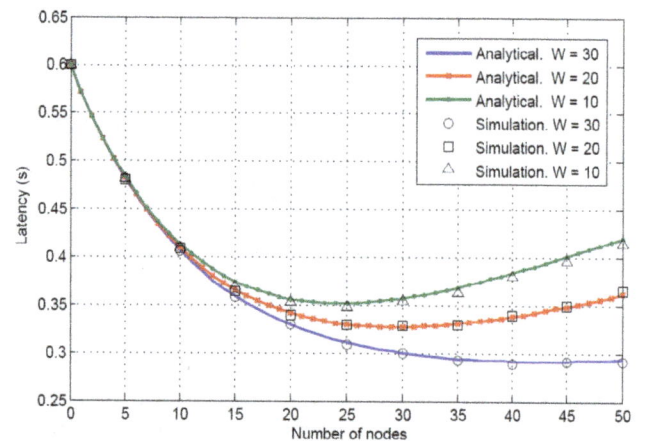

Figure 10. Latency of the IPAA protocol as a function of network size with contention window size as a parameter.

window size is set to $W = 30$, K_{NR} is set to 3, and the coverage ratio is set to $\frac{A_x}{A} = 10\%$. The results show that the latency increases as the loss rate increases since a loss of the protocol messages may result in a failed trial and force the new node to start the process again. For a relatively small number of nodes in the network, the communication overhead increases as the loss rate increases since there exist a small number of neighbor nodes within the transmission range of the new node so the probability of loss of the protocol message is high. For a large number of neighbors, the probability of loss for all messages is lower and the number of messages becomes closer to the case where no loss is assumed.

Figures 10 and 11 show the effect of the contention window size W on the latency and the communication

overhead, respectively, when $\varepsilon = 0.1$, $\frac{A_x}{A} = 10\%$, and $K_{NR} = 3$. The window size has a greater effect on latency and communication overhead for a large number of nodes since the collision probability increases as the number of nodes increases. For a given large number of nodes, decreasing the window size results in more collisions for the incoming reply messages at the new node which may result in a failed allocation trial. More failed trials result in more latency and more communication overhead as shown in the figures.

Numerous additional analytical and simulation results were carried out and are given in [20]. Also analyzed were the effect of the coverage ratio $\frac{A_x}{A}$ on the latency and communication overhead, effect of the protocol parameter K_{NR} on the latency and communication overhead, and effect of the loss ratio on the number of neighbor allocations. In all cases analytical and simulation results showed close agreement.

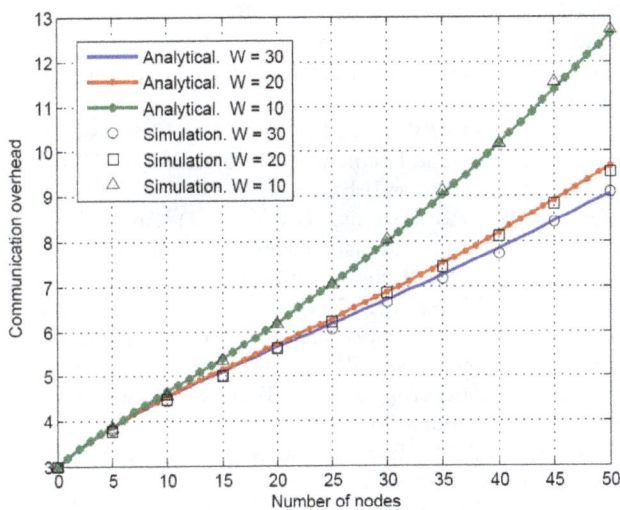

Figure 11. Communication overhead of the IPAA protocol as a function of network size with contention window size as a parameter.

5. Conclusions

Detailed descriptions for two address allocation protocols in wireless *ad hoc* network were presented. One of the protocols, the MANETconf protocol of [15], represents the DAD-based address allocation scheme and the other, the IPAA protocol of [16], represents the neighbor-based scheme. State machines that show the protocols' behavior are constructed for the two protocols mentioned above. The state machine gives a complete picture of the states, handshakes, timers, and types of messages for a protocol. For IPAA, analytical formulas for the expectation of latency and communication overhead are carried out as a function of the number of nodes in the network with message loss rate, contention window size, coverage ratio, and the counter threshold as parameters. Latency is the amount of time required for a new node to be allocated a network address, and communications overhead is the number of messages sent during the address allocation process.

Extensive numerical results were collected. The results show that the latency and communication overhead for MANETconf are higher than those measures for the IPAA protocol and that is due to performing the DAD process in MANETconf throughout the network. In IPAA, address allocation is performed through local communication between the new node and neighbor nodes and therefore its latency and communication overhead are low. It has been shown that the latency and communication overhead of the allocation process for the IPAA protocol increase as the message loss rate increases. Loss of REPLY or ADDRESS_BLOCK messages may result in a failed trial and requires the new node to repeat the allocation process again. The contention window size has an effect on the performance measures for relatively large number of nodes in the network. As the number of nodes in the network increases, the number of neighbor nodes increases and

therefore the probability of collision increases which may result in a failed trial and forces the new node to perform the allocation process again. Additional numerical results given in [20] show that increasing the coverage ratio has basically the same effect as increasing the number of nodes within the transmission range of the new node which leads to high collision probability as mentioned above. Numerical results presented in [20] also show that for loss rate equal to 0.1, contention window size of 30, and coverage ratio of 10%, address allocation is carried out by neighbor nodes for almost 95% of the time when the number of nodes in the network is more than 30.

The address allocation problem for *ad hoc* networks is still unsolved. The work done in this paper provides understanding of the nature of the problem and the way to evaluate the performance of an allocation protocol. The objective of future research will be to contribute to the solution of the address allocation problem in *ad hoc* networks. What is needed is a protocol that handles all types of exceptions such as message losses, node departures, network partitioning, and merging. An address allocation protocol for an *ad hoc* network has to be a dynamic and distributed protocol that guarantees address uniqueness for each node in the network and scalable to large networks. Scalability means that the protocol should perform well as the network size increases. The protocol should perform the address allocation for a node with minimum communication overhead and in a timely fashion. Besides the verbal description of the protocol, state machines that show the protocol states, handshakes, types of messages, and timers should be constructed. For completeness, the performance of the protocol should be analyzed and compared to the performance of existing protocols.

References

[1] THOMSON, S. and NARTEN, T. (1998) Ipv6 stateless address autoconfiguration, RFC 2462, IETF Zeroconf Working Group.

[2] DROMS, R. (1997) Dynamic host configuration protocol, RFC 2131, IETF Network Working Group.

[3] BACCELLI, E. (2007, September) Address autoconfiguration for MANET: terminology and problem statement, IETF draft-ietfautoconf-statement-02.

[4] KNOWLTON, K.C. (1965) A fast storage allocator. *Commun. ACM* **8**(10): 623–624.

[5] BERNARDOS, C., CALDERON, C. and MOUSTAFA, H. (2008, April) Survey of IP address autoconfiguration mechanisms for MANETS, IETF draft-bernardos-manet-autoconf-survey-03.

[6] PERKINS, C. (2001, November) IP address autoconfiguration for ad hoc networks, IETF draft-perkins-manet-autoconf-01.

[7] JEONG, J. (2006, January) Ad hoc IP address autoconfiguration, IETF draft-jeong-adhoc-ip-addr-autoconf-06.

[8] VAIDYA, N. (2002) Weak duplicate address detection in mobile ad hoc networks. In *Proceedings of ACMMOBIHOC* (Lausanne).

[9] WENIGER, K. (2003, March) Passive duplicate address detection in mobile ad hoc networks. In *Proceedings of IEEE WCNC, IEEE* (New York: IEEE Press).

[10] MASE, K. and ADJIH, C. (2006, April) No overhead autoconfiguration OLSR, IETF draft-mase-manet-auto-conf-noaolsr-01.

[11] WENIGER, K. (2005) PACMAN: passive autoconfiguration for mobile ad hoc networks. *IEEE J. Sel. Areas Commun.* **23**(3): 507–519.

[12] ZHOU, H., NI, L. and MUTKA, M. (2003) Prophet address allocation for large scale MANETs. In *Proceedings of IEEE INFOCOM* (New York: IEEE INFOCOM).

[13] YONG, L., PING, Z. and JIAXIONG, L. (2009) Dynamic address allocation protocols for mobile ad hoc networks based on genetic algorithm. In *Proceedings of 5th International Conference on Wireless Communications, Networking and Mobile Computing (WiCOM)* (Piscataway, NJ: IEEE Press).

[14] CHU, X., SUN, Y., XU, K., SAKANDER, Z. and LIU, J. (2008) Quadratic residue based address allocation for mobile ad hoc networks. In *Proceedings of IEEE ICC* (Beijing, China), 2343–2347.

[15] NESARGI, S. and PARAKASH, R. (2002) MANETconf: configuration of hosts in a mobile ad hoc network. In *Proceedings of IEEE INFOCOM, IEEE* (New York: IEEE Press).

[16] MOHSIN, M. and PARAKASH, R. (2002) IP address assignment in a mobile ad hoc network. In *Proceedings of IEEE MILCOM* (New York: IEEE Press).

[17] TAYAL, A.P. and PATNAIK, L.M. (2004) An address assignment for the automatic configuration of mobile ad hoc networks. *Pers. Ubiquitous Comput.* **8**(1): 47–54.

[18] XU, T. and WU, J. (2007) Quorum based IP address autoconfiguration in mobile ad hoc networks. In *ICDCSW '07: Proceedings of the 27th International Conference on Distributed Computing Systems Workshops* (Washington, DC: IEEE Computer Society), 1.

[19] ROSS, S.M. (2008) Introduction to probability models (Academic Press), 9th ed.

[20] RADAIDEH, A.M. (2008, December) Analytical modeling of address allocation protocols in wireless ad hoc networks. Master's thesis, The University of Mississippi.

12

Cross–layer cross–domain adaptation of mobile video services

Jose Oscar Fajardo*, Ianire Taboada, Fidel Liberal

NQaS Research Group, Department of Electronics and Telecommunications, University of the Basque Country (UPV/EHU), ETSI Bilbao, 48013 Bilbao, Bizkaia, Spain

Abstract

This paper deals with the analysis of user perceived visual quality for mobile multimedia services. H.264/AVC is selected for low resolution video encoding, and a mobile UMTS network is considered for the access to services. First, from subjective tests results, the combined impact of different service– and radio network–level parameters is inferred. As a result, different cross–layer adaptation alternatives are proposed to maximize the perceived quality level under different service conditions. Afterwards, the backhaul segment is considered taking into account possible congestion effects after the aggregation of multi–user media stream. The proposed dynamic adaptation process determines the best combination of media bitrates in order to optimise the overall quality, by making use of information related to network performance both at the radio and backhaul segments.

Keywords: adaptation, cross–layer, cross-domain, H.264/AVC, mobile multimedia services, UMTS radio, UMTS backhaul

1. Introduction

Multimedia applications have become increasingly popular over Internet. Furthermore, the wide–spreading of enhanced mobile Internet access and the continuous evolution of multimedia encoding techniques allow provisioning video services over mobile data connections at acceptable quality levels. For instance, the mobile version of the YouTube video sharing platform currently offers H.264/AVC–encoded video clips, which can be accessed from a mobile handset via RTSP with a standard video client. Typically, two versions of the video clips are available. On one hand, the normal version is based on the low spatial resolution (SR) of QCIF (176x144) and low encoding bitrate (about 80 kbps). On the other one, the High Quality (HQ) versions offer a higher SR at QVGA –a.k.a. square pixel SIF– (320x240) at higher encoding bitrates (about 250 kbps).

The choice of the most suitable version depends on different parameters. The screen resolution of the mobile device imposes a first requirement. Nowadays,

typical screen resolutions are QCIF, QVGA and VGA. As well, the performance variability associated to mobile data services must be considered. Taking into account a typical 3G Universal Mobile Telecommunications System (UMTS) service, the 384kbps bearer supports the requirements of HQ versions under perfect reception conditions. However, the variable radio conditions may introduce degradations in the transmission, and the normal version could be preferred.

As widely studied, an adaptive approach could provide the best quality level all over the service time by introducing real–time modifications in the service provision configuration. Particularly, combined service– and network–level actions are of paramount interest in order to dynamically perform cross–layer adaptations in the service provision.

In [1] we addressed this topic, focusing on the provision of H.264/AVC–based video services over 3G UMTS data connections. Specifically, this work presents a consolidated performance study by taking into account the combined effects of the specific characteristics of the UMTS Terrestrial Radio Access Network (UTRAN) on one hand, and the specific characteristics of the H.264 encoding on the other

---

*Corresponding author. Email: joseoscar.fajardo@ehu.es

one. In order to perform the most suitable adaptation actions, the proposed decision making process is driven by the expected visual quality level as perceived by end users. Thus, instead of using classical objective visual quality assessment metrics such as PSNR, VQM and SSIM [2], subjective video quality perception approach is considered in this work, which is essential for network providers in order to achieve an acceptable user satisfaction.

As a step forward, in this paper we take into consideration the possible effects of the backhaul segment in the mobile infrastructure. In this case, the decision making process is not only driven by the service configuration and the experienced performance in the radio link, but it also takes into account the most suitable adaptation actions in order to overcome possible congestion situations in the mobile backhaul. This new functionality is stated in this paper as a cross–domain feature, assuming that the radio and wired segments of a mobile network are considered as two different domains.

1.1. Background

The analysis of user perceived visual quality has been subject of many studies in recent years. With regard to H.264/AVC at mobile resolutions, several studies present their subjective tests results for service contexts similar to the considered in this paper.

From the subjective tests results presented in [3, 4], the H.264/AVC codec can be assumed as the best performing codec for QCIF mobile video services. As well, the most suitable audio–video (A/V) bitrate ratio is inferred as a function of the content type (CT). In [5] authors study the combined effects of the source bitrate (SBR) and frame rate (FR) for mobile resolutions. The main outcome is a regression–based expression which relates the Mean Opinion Score (MOS) to the SBR and FR values for different CT. The FR is considered in the range of 5 fps to 15 fps, while the target SBR varies from 24 kbps to 105 kbps. In this case, all the subjective test sequences are presented at QCIF resolution screen, which justifies the low values considered for SBR. This range of SBR values is considered insufficient for the aims of this paper, where higher SR values are considered. As well, in these studies only the encoding parameters are considered, assuming ideal transmission conditions.

Concerning higher SR, e.g. results in [6] present quality evaluations of CIF (352x288) video sequences in terms of blocking, blurring and flickering artifacts. The SBR is considered from 100 kbps to 300 kbps for this SR at 25 fps. However, only the encoding effects are considered for the quality analysis as well.

Both H.264/AVC CIF video sequences and transmission effects are considered in the analyses of perceived

video quality presented in [7, 8]. In both cases, a 2–state Markov model is used to implement the bursty packet losses. However, in both cases the loss model is implemented at IP level. In our case, the bursty error pattern is implemented at UMTS Radio Link Control (RLC) level following the results shown in [9], where the error pattern is obtained from live UMTS network traces. This feature entails a better emulation of the combined effects of the service–level settings and the experienced UMTS performance conditions.

As a result, although all the reviewed studies are close to the case study, none of them covers all the objectives proposed in this paper and hence new subjective tests have been performed.

1.2. Scope and objectives

The main objective of the study is to in–depth study the combined impact of different service– and network–level parameters into the experienced visual quality from a consolidated standpoint, in order to analyze the most suitable cross–layer adaptations that maximize the Quality of Experience (QoE). In this sense, the main contributions of this paper are twofold:

(i) A thorough analysis of the visual quality which could be expected in a mobile multimedia service, as currently being provided in real–world services.

(ii) A dynamic cross–layer cross–domain adaptation mechanism, aimed at maximizing the QoE under variable conditions in the mobile network, including both the radio and the backhaul segments.

One of the main novelties of the paper is the mobile multimedia service awareness. All the subjective tests have been performed resembling actual service conditions, including mobile–oriented media encoding and presentation to end users. As well, the used UMTS reception patterns related to the radio interface are based on real–world measurements under different mobility scenarios. Finally, we include in the analysis possible congestion situations in the mobile backhaul, taking into account the bitrate ranges related to mobile video streams.

The remainder of this paper is structured as follows. Section 2 presents the methodology followed for performing the subjective tests. Section 3 focuses on the analysis of perceived visual quality, in function of the encoding settings. From subjective tests results, the evolution of expected MOS with the SBR is inferred for each CT and SR. This way, the most suitable SR can be identified for the achievable SBR and per CT. Likewise, Section 4 focuses on the analysis of perceived visual quality in function of the UMTS performance. From subjective tests results, the evolution of expected

MOS with the experienced UMTS conditions is inferred for each CT. This way, the most suitable SBR can be identified for the experienced Block Error Rate (BLER) and per CT. Section 5 illustrates the considered cross–layer adaptation capabilities, which provides improvement in terms of QoE based on the knowledge of the service and radio network states. Finally, Section 6 introduces the effects of the mobile backhaul in the decision making process. Section 7 gathers the main conclusions to this paper.

2. Subjective testing methodology

The experimental design for the subjective video tests is mainly based on ITU recommendations and tutorials [10–12]. Different aspects are taken into account for planning the subjective tests for this kind of multimedia applications.

Concerning the viewing conditions, video sequences are presented to end users resembling a mobile UMTS service in order to enhance the accuracy of results [13]. All video sequences are displayed in a mobile handset and users are asked to hold it in their hands. The device used in the tests is a Nokia N95-8Gb, which provides a 320x240 screen resolution. The tests are carried out with RealPlayer for s60, which uses image re–scaling for presenting QCIF sequences at full screen.

The video sequences are a priori generated including both encoding and transmission effects, and they are stored in the mobile device for its presentation to subjects. So, an appropriate device and displaying format is used to achieve the proposed objectives in a fixed environment conditions for all the subjects. In order to capture the combined effects of the specific encoding and transmission techniques, long duration video sequences (about 2 minutes) have been used instead of the typical short (about 10s) reference video sequences.

Taking as reference the proposed test structure in [10–12], before starting with test sessions, written instructions were shown to subjects and a training phase was done in which some videos are presented and evaluated, without taking into account these results. Then this is followed by several test sessions. In each test session different types of test scenes are shown. These are presented in random order and some implicit replications are included to check coherence [10]. Due to fatigue issues, break periods between sessions are introduced [12]. For carrying out the different experiments proposed in this study, each subject participates in the experiments for three different days.

In order to reproduce viewing conditions that are as close as possible to real–world contexts, Single Stimulus (SS) tests are used and the audio track is also included in the multimedia stream. The evaluations are based on Absolute Category Rating (ACR). Thus, after each test

Table 1. Complexity–motion metrics and snapshot of videos used.

CT	Complexity–motion metrics	Snapshot
LM	C=0.46, M=0.28	
MM	C=0.49, M=0.42	
HM	C=0.62, M=0.96	

sequence presentation the subjects are asked to evaluate the quality of the presented sequence in the MOS scale of 1 to 5.

3. Service–level parameters

All the video sequences have been encoded with the H.264/AVC Joint Model Reference Software. Three different CT have been considered: low–motion (LM), medium–motion (MM), and high–motion (HM) video sequences. The considered content is described as follows:

(i) LM sequence is made up of news video clip with two people, including change of planes and a commercial in middle of the sequence.

(ii) MM sequence is a typical TV series scene, featuring different people and including change of planes.

(iii) HM sequence is taken from a basketball top 10 best plays video sequence, with changes of planes from field to face planes.

In order to evaluate spatio-temporal activity level of video sequences the parameters proposed in [14] are used. This approach consists on associating an scene complexity (C) and level of motion (M) value to each video sequence based on the average bits per frame and the average quantization parameter (QP) for I and P frames respectively. Table 1 shows the associated complexity-motion metrics for each CT as well as a snapshot of the video signals used.

Table 2. Encoding settings for subjective tests.

CT	SBR (kbps)	SR	FR (fps)
LM	{80, 130, 200}	320x240	10
	{48, 88, 128}	176x144	10
MM	{80, 130, 200}	320x240	12.5
	{48, 88, 128}	176x144	12.5
HM	{80, 200, 256}	320x240	15
	{48, 88, 128}	176x144	15

Table 3. Subjective tests results for QVGA.

SBR (kbps)	LM	MM	HM
80	3.49	2.90	1.98
130	4.26	3.92	–
200	4.53	4.49	4.26
256	–	–	4.53

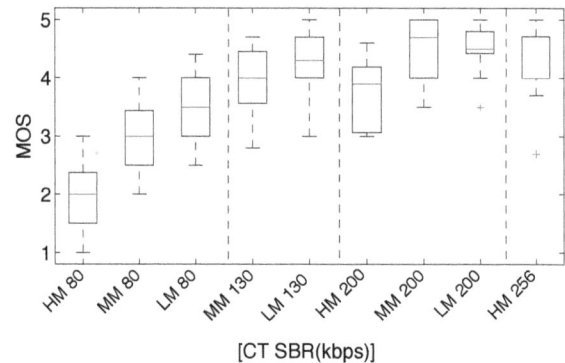

Figure 1. Boxplot of subjective testing results for encoding of QVGA sequences.

Table 4. Subjective tests results for QCIF.

SBR (kbps)	LM	MM	HM
48	3.49	2.75	1.83
88	3.52	3.46	2.65
128	3.66	3.60	2.91

Table 5. Experimental coefficients for the fitting function.

	LM QVGA	MM QVGA	HM QVGA	LM QCIF	MM QCIF	HM QCIF
a	1.141	1.739	2.274	0.159	0.608	1.124
b	-1.442	-4.665	-7.954	2.858	0.556	-2.48

Instead of analyzing the visual quality as a function of the combined setting of FR and SBR, in this study only the SBR is modified. The FR values are set up in function of the content dynamics: 10, 12.5 and 15 fps are established for LM, MM and HM respectively. Table 2 summarizes the considered values.

The H.264/AVC encoder is set up to its Baseline Profile at level 1.2. The frame structure is IPPP, with a Group of Pictures (GOP) size of 10 seconds. Thus, each 10 seconds the video sequence is refreshed with a new I frame.

3.1. Encoding quality

First, subjective tests were performed for evaluating the impact of SBR into the encoding quality of QVGA sequences. A total of 20 people evaluated each sequence, for a total of 180 tests. The obtained results are presented in Figure 1, which gathers the box plots of the subjective testing results for the whole set of considered CT–SBR conditions. For each combination of CT and SBR considered, the median, Q1 and Q3 percentiles and minimum/maximum values are illustrated. As can be observed, there is no too much variability in the quality scores provided among users. As well, those values which are considered outliers in the data sample are individually plotted. Outlier values are determined by the Chauvenet's Criterion and they are not taken into account in the computation of the MOS. Table 3 shows the average quality values derived from the subjetive tests for QVGA resolutions.

Similarly, new tests were performed for the QCIF versions of the same videos. In order to resemble actual service conditions, QCIF video sequences are presented at full screen in the mobile handset with RealPlayer, so image scaling is required. A total of 20 people evaluated

each sequence, for a total of 180 tests. In this case only the averaged MOS values are provided in Table 4.

The QCIF versions, being half the size than the QVGA sequences, require less SBR to achieve an acceptable visual quality. However, the increase of SBR does not entail a proportional increase in the perceived quality for different SR values. As described in [8], the video encoding process may reach the visual quality threshold and above this threshold a higher SBR is not captured by the human visual system.

3.2. Impact on service–level adaptation

In order to evaluate the evolution of both alternatives, we applied fitting techniques to the subjective results and obtained an approximation of the relationship between MOS and SBR values for both SR. The fitting function is given in Equation (1) as proposed in [8].

$$MOS = a \cdot log(SBR) + b \qquad (1)$$

Parameters a and b are related to the activity level of the content and spatial resolution, and the obtained values are illustrated in Table 5.

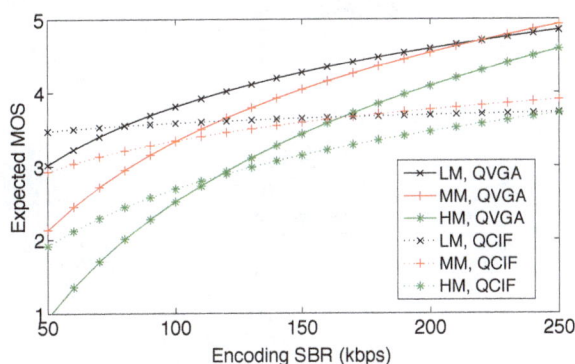

Figure 2. Evolution of expected MOS vs. SBR per CT and SR.

By using these results, Figure 2 presents the expected evolution of the subjective visual quality in terms of MOS for each pair of considered CT–SR values. For each CT, the two resulting curves cross at a specific point (SBR_{th}) and two differentiated regions are identified. For SBR values lower than SBR_{th}, it is preferable to switch to a lower SR in order to control the impact of the limited SBR. The SBR_{th} value is higher for more dynamic sequences. For the three CT considered in this study, we find that this threshold is around 80, 100 and 115 kbps for LM, MM and HM respectively.

4. UMTS radio access

The scope of this paper regards to the provision of mobile video streaming services over a typical wide–area 3G UMTS data service, as defined in 3GPP TR 25.993 [15] for the *Interactive or Background/UL : 64DL : 384kbps/PS RAB*. This kind of bearer service provides a maximum downlink bitrate (DLBR) of 384 kbps. A detailed description of the considered service provision and the impact of the UTRAN is given in [16].

From the total bitrate, the final amount available for the video encoding process is reduced by several factors. First of all, the RTP/UDP/IP packetization introduces an overhead in the data transmission. Second, we must consider the effect of the audio stream within the multimedia transmission. The analysis of the audio quality and the integral quality (as shown e.g. in [17, 18]) is out of the scope of this paper. Yet, the impact of its transmission is simulated by adding a 64 kbps stream to the considered video stream. Finally, part of the DLBR is used by the RLC AM functions for the local recovery of lost MAC PDUs. In this sense, the performance of the video streaming service highly depends on the radio error pattern.

As cited in Sect. 1.1, one of the novelties of this paper is that the UMTS error model is implemented at RLC level from real–world measurement results, instead of using a typical 2–state Markov model for simulating the IP–level loss events. For the simulation of different

network conditions, the implemented error model is a 2–state Markov model with variable Block Error Rate (BLER) values.

Two characteristics are adopted from the results presented in [9]:

(i) For mobile users, the radio errors can be grouped at Transmission Time Interval (TTI) level.

(ii) The Mean Burst Length (MBL) of erroneous TTIs can be approximated to 1.75.

The error model, as well as the simulation methodology, is further detailed in [19]. The RLC–level error model, in combination with the application–level settings, determines the performance of the service. The RLC errors may derive to additional delays if the RLC is able to recover the lost PDU, or to video frame losses otherwise.

Taking into account the relevance of the different frame types, a content–aware scheduling is implemented in a similar way to the concepts proposed in [20] for 3G UMTS, in [21] for HSDPA and in [22] for LTE. In this case, the priority of different RLC retransmissions is modified in order to implement an enhanced protection for I frames. This way, we prevent severe degradations in the initial picture of each GOP and its propagation all over the 10 seconds period.

4.1. 3G UMTS transmission quality

Considering the mentioned characteristics for 3G UMTS mobile multimedia services, different combinations of service and network conditions are simulated. All the simulations are run with OPNET Modeler, where both the specific UMTS error pattern and a H.264/AVC RTP trace injector have been implemented. QVGA video traces corresponding to the three CT have been used in the simulations with several SBR values. 80, 130 and 200 kbps have been considered for LM and MM video sequences, while an additional 256 kbps version has been used for HM traces. All the traces have been transmitted several times from a video server to the mobile endpoint, traversing the UMTS network segment. The downlink BLER value in the radio part is set up from 1% to 30% at 5% steps. At 30% of BLER, all the sequences experience high degradations except for the LM 80 kbps versions.

From the whole set of results obtained, a mapping between different BLER values and experienced IP Loss Ratio (IPLR) patterns is established. For those points with negligible IPLR values (under 0.1%), the service performance is considered accurate. Similarly, high IPLR values (over 5%) indicate unaffordable service conditions.

The rest of intermediate points are considered for subjective evaluation of the visual quality perceived by

Table 6. SBR and BLER settings for subjective tests.

CT	SBR (kbps)	BLER(%)
LM	130	15, 20, 25
	200	5, 10
MM	130	20, 25
	200	5, 10
HM	256	5

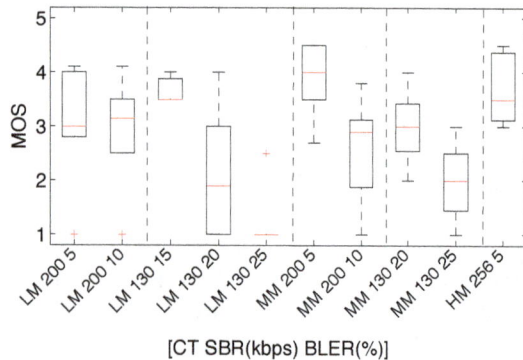

Figure 3. Boxplot of subjective testing results for UMTS transmission of QVGA sequences.

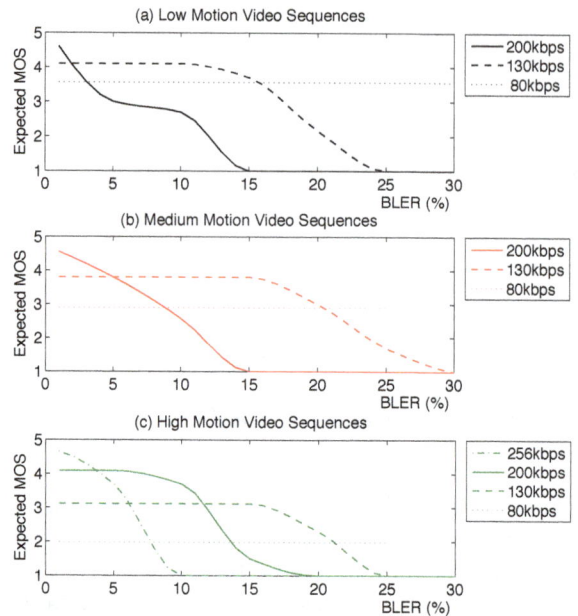

Figure 4. Evolution of expected MOS vs. BLER per CT and SBR for QVGA sequences: (a) low motion, (b) medium motion, (c) high motion.

users. Table 6 presents all the service– and network–level conditions that have been included in this set of subjective tests. The application–level performance (in terms of IPLR) is not only determined by the experienced BLER values, but also by the different traffic patterns associated to the different CT. Thus, different conditions require subjective evaluation in function of the CT. For the aims of this study, a total of 114 subjective tests were performed by 20 people.

Figure 3 illustrates the results obtained from subjective tests for the considered encoding/transmission conditions. In this plot, the same blox plot parameters as in Figure 1 are depicted. As can be observed, especially two conditions (namely LM–130–20 and MM–200–10) show a high variability in the quality scores provided by users. In those cases, the associated MOS are 2.15 and 2.58 respectively.

4.2. Impact on cross–layer service adaptation

The extensive results obtained both from the analysis of simulation results and from subjective tests allow us depicting a mapping between the application and transmission conditions to the expected video quality in terms of MOS, as shown in Figure 4 for each considered CT.

From the behavior illustrated in Figure 4, each combination of CT, SBR and BLER determines the expected visual quality value. Thus, for a specific BLER

value, the video streaming session can be set up to the new SBR value that maximizes the expected MOS. For the aims of this paper, this adaptation is considered as a standalone decision making process.

The BLER value is not modified (e.g. by modifying the power control functions of the link layer) so the impact of the adaptation of a video stream on the performance experienced by other users in the same cell is limited in this case.

If power or rate control mechanisms are considered in multi–user environments, where several users are contending for the access to limited cell resources at the same time, the optimization problem can be studied as shown in [23, 24].

Another alternative in the standalone management of mobile services is the capability of modifying the Radio Bearer settings in function of the experienced conditions. If for the experienced BLER condition none of the highest considered SBR values is suitable, the UMTS Radio Bearer can be switched to a DLBR value of 128 kbps, which exhibits a better resilience to noise and interference at the same transmission power levels. Thus, low BLER values can be expected in order to guarantee no further impairments than the encoding process itself.

At the same time, the multimedia streams are switched to a lower SBR in order to cope with the new DLBR requirements. As a result, this approach entails a combined encoding/network service adaptation. As in the previous case, the impact on other users is limited and each user can be adapted by itself.

5. Evaluation of cross–layer adaptations

Based on the results obtained from the subjective tests, different adaptation approaches can be evaluated. Four alternative approaches are considered, each option including a new adaptation capability from zero to three configurable parameters.

(i) *No Adaptation (NA)*. In this case, no adaptations actions are considered. Thus, the achieved MOS values are those corresponding to the initial service conditions. In order to offer accurate quality levels, for LM and MM sequences the initial SBR is set up to 200 kbps, while HM sequences are configured to 256 kbps.

(ii) *Network–Aware Bitrate Adaptation (NABA)*. This alternative considers the adaptation of SBR to the value that maximizes the expected MOS for the specific CT and experienced BLER, driven by the estimation curves shown in Figure 4. For the purposes of this paper, the allowed SBR values are 256, 200, 130 and 80 kbps.

(iii) *Network–Aware Bitrate and Spatial Adaptation (NABSA)*. This approach takes into account the two configurable service–level parameters considered in this paper: SBR and SR. Thus, based on the resulting SBR adaptation, the SR is also switched from QVGA to QCIF if the optimal SBR is lower than the SBR_{th} for the specific CT.

(iv) *Network–Aware Cross–Layer Adaptation (NACLA)*. In this case, a cross–layer adaptation is adopted by taking into account the three parameters considered. Besides the application–level adaptation (combined SBR/SR), the DLBR can be decreased to overcome severe degradation conditions in the UMTS data connection.

Figure 5 shows the obtained results for each dynamic adaptation approach, taking into account the different CT considered.

For LM sequences, only two curves are differentiated. If no dynamic adaptation is applied, the mobile video service gets completely degraded around the 15% of BLER. However, the service can be kept at suitable quality levels (MOS=3.5) just with bitrate adaptation under the analysed UMTS conditions. The LM sequences exhibit no degradations at BLER=30%, and both SR versions provide similar quality levels at SBR = 80 kbps. Hence, the three dynamic adaptation approaches offer analogous behaviours at the whole range of the BLER conditions studied. Thus, just SBR adaptation provides the maximum achievable quality under different conditions in this case.

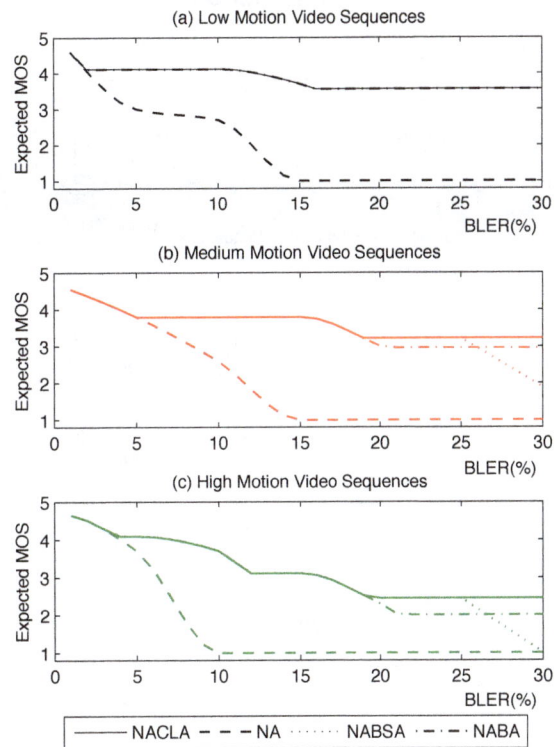

Figure 5. Quality achieved by the different adaptation approaches for (a) low motion, (b) medium motion, (c) high motion sequences.

As can be observed, this is not case for MM and HM sequences. On one hand, the SBR_{th} is located above 80 kbps in both cases, and thus the SR should be switched to QCIF when SBR is set up to 80 kbps. On the other hand, the QCIF versions exhibit frame losses when the experienced BLER is above 25%. As a result, the three dynamic adaptation approaches provide different quality levels under severe UMTS degradations and different type of adaptations are required in order to maximize the QoE.

From the analysis of results, a dynamic network–aware cross–layer adaptation mechanism can be proposed, as illustrated in Figure 6. Following the depicted logic, a mobile endpoint could be capable of launching the required adaptation procedures to keep the mobile video service in the maximum achievable quality level along the service time.

6. Flow aggregation at the mobile backhaul

In the previous sections, we have analysed the behaviour of each mobile user in a standalone fashion. Based on the performance experienced in the radio access network, the system is capable of adapting the service– and network–level parameters regardless the rest of the users.

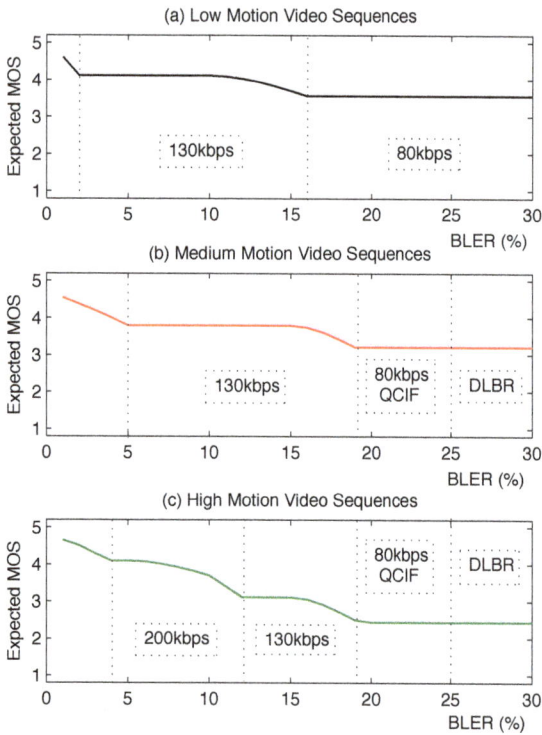

Figure 6. Dynamic configuration and maximum achievable quality for (a) low motion, (b) medium motion, (c) high motion sequences.

In this section we introduce a new variable by taking into account the backhaul link, where the traffic of all users is aggregated. In such context, the analysis of the backhaul capacity is critical for the study of the QoE. In the simplest formulation, traffic losses will appear whenever the aggregated traffic results in a higher bitrate than the actual link capacity. Thus, we introduce a new relevant system parameter: the traffic loss ratio (TLR):

$$TLR = \frac{\sum_{n=1}^{N} SBR_n}{R} \qquad (2)$$

being N the total number of mobile multimedia users in the system and R the capacity of the aggregation link.

To overcome this problem, the system could decide to implement some kind of admission control based on users' required bitrate. In a more advanced strategy, the system could propose service–level adaptations in order to decrease the bitrates of the different sessions to suitable levels. The key issue is to identify which users are more suitable for adaptation and how much bitrate can be decreased, in order to maximise the general offered QoE.

Therefore, the decision making process becomes a QoE-driven multi-user resource allocation problem,

where the QoE is also subject to the service and radio network conditions specific to each user.

Instead of the typical MOS averaging between all users, the proposed system is aimed at maximising the overall MOS while keeping the worst–case users at the maximum possible quality level. Thus, the proposed objective function is expressed as:

$$F(\tilde{x}) = \frac{1}{N} \sum_{n=1}^{N} \left(\frac{1}{MOS_n(\tilde{x})} \right)^2 \qquad (3)$$

where the parameter tuple $\tilde{x} \in \tilde{\chi}$, being $\tilde{\chi}$ the set of all accepted value tuples in accordance to the definitions of the parameter set {CT, SBR, SR, BLER, DLBR, TLR}.

The decision making process can be expressed as:

$$\tilde{x}_{opt} = \arg\min_{\tilde{x} \in \tilde{\chi}} F(\tilde{x}) \qquad (4)$$

This is, $\tilde{x}_{opt}$ is the parameter tuple which maximizes the proposed objective function and therefore the system performance.

In the case study here presented, the parameters CT and BLER are non–variable. From results in Section 5, the tupla CT–BLER determines the maximum value for SBR, and consequently the optimal values for SR and DLBR parameters. As a result, the parameters to study are the individual SBR values and the experienced TLR, being TLR a function of the sum of individual SBR values. In the context of this paper, the decision making process is aimed at keeping TLR=0. As a result, the objective parameter becomes the vector of individual SBR values.

We can state the problem as:
Minimise

$$\frac{1}{N} \sum_{n=1}^{N} \left(\frac{1}{MOS(SBR_n)} \right)^2 \qquad (5)$$

subject to

$$\sum_{n=1}^{N} (SBR_n) \leq R \qquad (6)$$

6.1. Adaptation decision making algorithm

The optimization problem formulated in Equations (5) and (6) is solved by means of a genetic algorithm [25]. This way, in order to efficiently solve the proposed adaptation problem in real–time, this approach allows us to obtain optimal or suboptimal solutions in lower execution time than exact optimization methods.

Figure 7 summarizes the implemented genetic algorithm based adaptation logic.

Each individual represents a possible system state that can be reached by performing SBR adaptation over the existing multimedia sessions. The initial population is generated at the moment of the invocation of the adaptation.

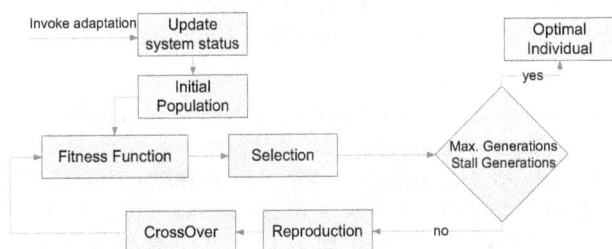

Figure 7. Genetic algorithm based adaptation logic.

Table 7. Initial service and radio access conditions.

Session ID	CT	SBR (kbps)	BLER(%)
1,2,3	LM	200	0, 10, 25
4,5,6	LM	130	0, 10, 25
7,8,9	MM	200	0, 10, 25
10,11,12	MM	130	0, 10, 25
13,14	HM	256	0, 10
15,16,17	HM	200	0, 10, 25
18,19,20	HM	130	0, 10, 25

The fitness function provides an assessment of how accurate the evaluated solution is. The implemented fitness function receives two parameters: the SBR for each media flow and all the information required for the computation of the expected MOS of each flow. The output of the fitness function is determined by all the MOS values associated to each multimedia session, and the fitness score is computed based on the average and minimum MOS values obtain for the whole set of multimedia sessions.

In the selection module two outcomes are possible. If the algorithm reaches the Maximum number of established generations (5) or when the best score has not changed over the last Stall generations (2), the best scoring individual from the current generation is returned. Otherwise, the algorithm goes on the reproduction phase which determines how the next generations of the individuals are generated. On this case study, the three best scoring individuals survive in each generation, and the rest of individuals are generated by the cross–over function between the selected parents in a per flow basis.

6.2. Evaluation of cross–domain adaptations

In order to analyze the performance of the proposed cross–domain management system, we propose an scenario where 20 mobile users experience different service and network conditions, as shown in Table 7.

As well, we evaluate the performance of the system in different conditions of the backhaul segment by introducing several background traffic loads. Taking

Table 8. Background load conditions and backhaul traffic loss.

Background load	Total load	Traffic loss
0 Mbps	3.482 Mbps	0%
0.5 Mbps	3.982 Mbps	3.97%
1 Mbps	4.482 Mbps	17.01%
1.5 Mbps	4.982 Mbps	30.07%

into account SBR values in Table 7, the traffic load due to media flows becomes 3.482 Mbps. We assume that the backhaul capacity is overdimensioned 10% over the aggregation of initial service bitrates (thus, $R = 3.83 Mbps$). Following Equation (2), we compute the TLR associated to different background traffic loads as shown in Table 8.

Under these conditions and the adaptation procedure stated in Section 6.1, we compare the performance of three alternative approaches:

(i) *No adaptation*, thus the additional backhaul load makes the TLR increase until the overall quality is totally degraded for the multimedia sessions.

(ii) *NACLA*, the network–aware cross–layer adaptation approach as described in Section 5.

(iii) The new *Cross–Layer Cross-Domain Adaptation (CLCDA)*, which implements the multi–user adaptation actions as presented in Section 6 based on the performance of both radio and backhaul segments.

Figure 8 gathers the MOS values experienced by every multimedia session with the different approaches, each subplot representing a different backhaul load condition. As well, Figure 9 shows the resulting average and minimum MOS statistics.

As can be observed, the two adaptation approaches proposed highly outperform the no adaptation case. In the first case, there are no backhaul losses and thus both adaptation approaches exhibit similar performance. As the additional backhaul load increases, it is proven that cross–domain adaptation entails a better performance. While the average values are quite similar for both optimisation processes, the cross–domain approach always provides a better performance with regard to the minimum value, which means that the system will offer enhanced protection to severe degradations.

7. Conclusions

This paper deals with possible dynamic adaptations for H.264/AVC based video services over 3G UMTS mobile connections. In order to get an optimal configuration, different cross–layer adaptations are proposed and evaluated. Thus, both service– and

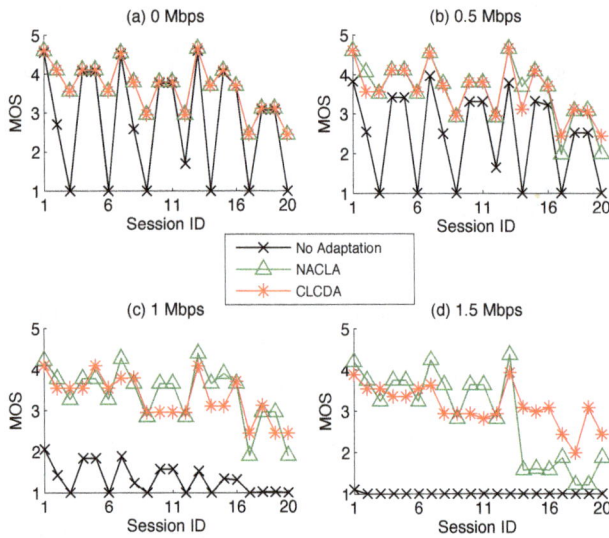

Figure 8. Evolution of MOS scores for different backhaul background traffic loads: (a) 0 Mbps, (b) 0.5 Mbps, (c) 1 Mbps, (d) 1.5 Mbps.

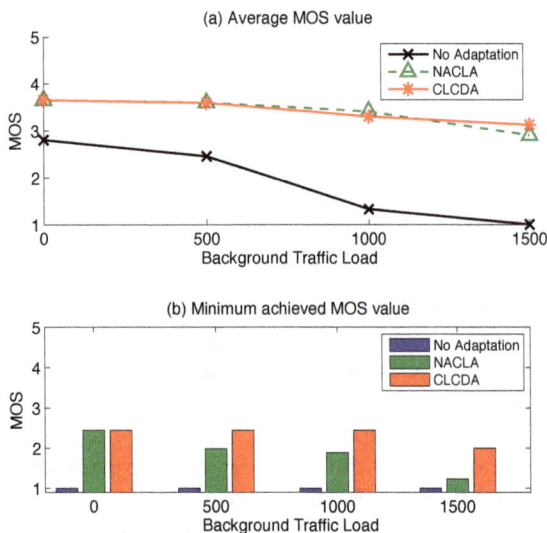

Figure 9. QoE statistics for each adaptation scheme: (a) average MOS, (b) minimum MOS.

network–level adaptations are considered to cooperate in a coordinated way in order to maximize the QoE.

Since the proposed adaptations are driven by the expected QoE, specific subjective tests have been performed to cope with the specific service context. Thus, both specific H.264/AVC parameters and specific error models for commercial UMTS networks have been used. The outcomes of the subjective testing phase are later considered as inputs for the dynamic service adaptation logic.

Different adaptation approaches have been evaluated, considering different number of variable parameters:

(i) On the one hand, we can state the enhancements of the three–parameter adaptation approach, based on the combination of encoding bitrate, spatial resolution and UMTS bearer bitrate. As described in Section 5, the considered adaptations do not have an impact on other users, so it can be implemented in a per–user basis.

(ii) On the other hand, once the service provision is optimised in terms of service and radio network, the multi–user problem is analysed. The aggregation of traffic at the mobile backhaul may introduce further degradations in the media streams. Thus, we propose an enhanced automated decision process which takes into account the performance of the different segments of the network (i.e., radio and backhaul segments) and maximises the overall QoE of the whole set of users. For this aim, the decision making process, driven by a genetic algorithm, considers the effect of bitrate reduction for each considered user, in terms of QoE, and determines the most suitable adaptation actions.

In order to obtain optimal results, the overall decision making process must be aware of different service– and network–level parameters. Thus, it could be deployed as a central management element which dynamically drives the provision of mobile media services and reacts to variations in the network performance. Regarding just the radio-aware adaptation, the adaptation logic could be implemented at the mobile endpoint if the mobile device requires access to low–level parameters. Currently, several Android–based commercial mobile devices are capable of providing BLER statistics in real-time. As a work in progress, we have developed the software to capture these statistics from the chipset in order to make them available for the applications.

Acknowledgement. The research leading to these results has received funding from the European Union Seventh Framework Programme (FP7/2007-2013) under grant agreement 284863 (FP7 SEC GERYON) and from the Basque Country under BASAJAUN project.

References

[1] Fajardo, J.O., Taboada, I. and Liberal, F. (2011) Cross-Layer Adaptation of H.264/AVC over 3G UMTS Mobile Video Services. In *Proceedings of the 3rd International ICST Conference on Mobile Lightweight Wireless Systems, MOBILIGHT 2011* (ICST). doi:10.1007/978-3-642-29479-2_8.

[2] Wang Y.(2006) Survey of objective video quality measurements. *EMC Corporation Hopkinton, MA* **1748**.

[3] Jumisko–Pyykkö, S. and Häkkinen, J. (2005) Evaluation of subjective video quality of mobile devices. In *Proceedings of the 13th Annual ACM International Conference on Multimedia, MULTIMEDIA '05* (New York: ACM), 535–538. doi:10.1145/1101149.1101270.

[4] Khan, A., Sun, L. and Ifeachor, E. (2011) QoE Prediction Model and its Application in Video Quality Adaptation over UMTS Networks . *Multimedia IEEE Transactions on*, **99**: 431–442.

[5] Ries, M., Nemethova, O. and Rupp, M. (2008) Video Quality Estimation for Mobile H.264/AVC Video Streaming. *Journal of Communications* 3(1): 41–50 doi:10.4304/jcm.3.1.41-50.

[6] Ho, H.H., Wolff, T., Salatino, M., Foley, J.M., Mitra, S.K., Yamada, T. and Harasaki, H. (2007) An investigation on the subjective quality of H. 264 compressed/decompressed videos. In *Proceedings of the Third International Workshop on Video Processing and Quality Metrics for Consumer Electronics, VPQM 2007*.

[7] De Simone, F., Naccari, M., Tagliasacchi, M., Dufaux, F.D. Tubaro, S. and Ebrahimi, T. (2009) Subjective assessment of H.264/AVC video sequences transmitted over a noisy channel. In *Proceedings IEEE International Workshop on Quality of Multimedia Experience, QoMEX 2009* (IEEE), 204–209. doi:10.1109/QOMEX.2009.5246952.

[8] Koumaras, H., Lin, C.H., Shieh, C.K. and Kourtis, A. (2010) A Framework for End–to–End Video Quality Prediction of MPEG Video. *J. Visual Communication and Image Representation* **21**(2): 139–154. doi:10.1016/j.jvcir.2009.07.005.

[9] Karner, W., Nemethova, O. Svoboda, P. and Rupp, M. (2007) Link Error Analysis and Modeling for Video Streaming Cross–Layer Design in Mobile Communication Networks. *ETRI Journal* 29(5): 569 – 595. doi:10.4218/etrij.07.0107.0102.

[10] ITU-T (2003) Rec. J.148: Requirements for an objective perceptual multimedia quality model.

[11] ITU-T (2001) P.862 Perceptual evaluation of speech quality (PESQ), an objective method for end–to–end speech quality assessment of narrowband telephone networks and speech codecs.

[12] Beerends, J.G. and Stemerdink, J.A.: A perceptual audio quality measure based on a psychoacoustic sound representation (1992) *Audio Engineering Society* **40**: 963–974.

[13] Jumisko–Pyykkö, S. and Hannuksela, M.M (2008) Does Context Matter in Quality Evaluation of Mobile Television? In: *Proceedings of the 10th international conference on Human computer interaction with mobile devices and services, MobileHCI '08* (New York: ACM), 63–72. doi:10.1145/1409240.1409248.

[14] Hu, J.and Wildfeuer, H.(2009) Use of content complexity factors in video over IP quality monitoring. In: *International Workshop on Quality of Multimedia Experience – QoMEX* IEEE: 216–221. doi: 10.1109/QOMEX.2009.5246950.

[15] 3GPP (2008) TR 25.993: Typical examples of Radio Access Bearers (RABs) and Radio Bearers (RBs) supported by Universal Terrestrial Radio Access (UTRA).

[16] Fajardo, J.O., Liberal, F. and Bilbao, N. (2009) Impact of the video slice size on the visual quality for H.264 over 3G UMTS services. In: *Proceedings of the Sixth International Conference on Broadband Communications, Networks, and Systems, BROADNETS 2009*, 1–8. doi:10.4108/ICST.BROADNETS2009.7022.

[17] Winkler, S. and Faller, C. (2006) Perceived audiovisual quality of low-bitrate multimedia content. *IEEE Trans. Multimedia* 8(5): 973–980. doi:10.1109/TMM.2006.879871.

[18] Fajardo, J.O., Taboada, I and Liberal, F. (2011) QoE-driven and network-aware adaptation capabilities in mobile multimedia applications. *Multimedia Tools and Applications* , 1–22. doi:10.1007/s11042-011-0825-y.

[19] Khan, A., Sun, L., Ifeachor, E., Fajardo, J.O. and Liberal, F. (2010) Video Quality Prediction Model for H.264 Video over UMTS Networks and their Application in Mobile Video Streaming. In: *Proceedings of the 2010 IEEE International Conference on Communications, ICC* (IEEE), 1–5. doi:10.1109/ICC.2010.5502195.

[20] Benayoune, S., Achir, N., Boussetta, K., and Chen, K. (2008) Content–aware ARQ for H.264 Streaming in UTRAN. In: *Proceedings of the IEEE Wireless Communication and Networking Conference, WCNC 2008*, 1956–1961. doi:10.1109/WCNC.2008.348.

[21] Superiori, L., Wrulich, M., Svoboda, P. and Rupp, M. (2009) Cross–Layer Optimization of Video Services over HSDPA Networks. In: *First International ICST Conference on Mobile Lightweight Wireless Systems, MOBILIGHT 2009* (ICST), 135-146. doi:10.1007/978-3-642-03819-8-14.

[22] Li, L. and Song, J. (2012) Research of Scheduling for Video Communication in OFDM Systems. In: *Video Engineering*. doi:CNKI:SUN:DSSS.0.2012-05-020.

[23] Khan, S., Thakolsri, S., Steinbach, E. and Kellerer, W. (2008) QoE–based Cross-layer Optimization for Multiuser Wireless Systems. In: *Proceedings of the 18th ITC Specialist Seminar on Quality of Experience*, 63–72.

[24] Xie, L., Hu, C., Wu, W. and Shi, Z. (2011) QoE-aware Power Allocation Algorithm in Multiuser OFDM Systems. In: *Mobile Ad-hoc and Sensor Networks (MSN), 2011 Seventh International Conference on IEEE*, 418–422.

[25] Affenzeller, M.and Winkler, S.(2009) Genetic algorithms and genetic programming: modern concepts and practical applications. *Chapman & Hall/CRC* **6**.

13

On Cooperative Relay in Cognitive Radio Networks

Donglin Hu   and   Shiwen Mao*

Department of Electrical and Computer Engineering, Auburn University, Auburn, AL 36849-5201, USA

Abstract

Cognitive radios (CR) and cooperative communications represent new paradigms that both can effectively improve the spectrum efficiency of future wireless networks. In this paper, we investigate the problem of cooperative relay in CR networks for further enhanced network performance. We investigate how to effectively integrate these two advanced wireless communications technologies. In particular, we focus on the two representative cooperative relay strategies, *decode-and-forward* (DF) and *amplify-and-forward* (AF), and develop optimal spectrum sensing and *p*-Persistent CSMA for spectrum access. We develop an analysis for the comparison of these two relay strategies in the context of CR networks, and derive closed-form expressions for network-wide throughput achieved by DF, AF and direct link transmissions. Our analysis is validated by simulations. We find each of the strategies performs better in a certain parameter range; there is no case of dominance for the two strategies. The significant gaps between the cooperative relay results and the direct link results exemplify the diversity gain achieved by cooperative relays in CR networks.

Keywords: Cognitive radio networks, cooperative communications, relay, cooperative diversity, spectrum sensing, performance modeling.

1. Introduction

According to Cisco's recent study, wireless data traffic is expected to increase by a factor of 66 times by 2013. Much of this future wireless data traffic will be video based services driven by the need for ubiquitous access to wireless multimedia content. Such drastic increase in traffic demand will significantly stress the capacity of future wireless networks.

Cognitive radios (CR) provide an effective solution to meeting this critical demand by exploiting co-deployed networks and aggregating underutilized spectrum for future wireless networks [2–6]. CR was motivated by the spectrum measurements by the FCC, where a significant amount of the assigned spectrum is found to remain underutilized. CR represents a paradigm change in spectrum regulation and access, from exclusive use by primary users to shared spectrum for secondary users, which can enhance spectrum utilization and achieve high throughput capacity.

Cooperative wireless communications represents another new paradigm for wireless communications [7–9]. It allows wireless nodes to assist each other in data delivery, with the objective of achieving greater reliability and efficiency than each of them could attain individually (i.e., to achieve the so-called *cooperative diversity*). Cooperation among wireless nodes enables opportunistic use of energy and bandwidth resources in wireless networks, and can deliver many salient advantages over conventional point-to-point wireless communications.

Recently, there has been some interesting work on cooperative relay in CR networks [10, 11]. In [10], the authors considered the case of two single-user links, one primary and one secondary. The secondary transmitter is allowed to act as a "transparent" relay for the primary link, motivated by the rationale that helping primary users will lead to more transmission opportunities for CR nodes. In [11], the authors presented an excellent overview of several cooperative relay scenarios and various related issues. A new MAC protocol was proposed and implemented in a testbed to select a spectrum-rich CR node as relay for a CR transmitter/receiver pair.

---

*Part of this work was presented at IEEE GLOBECOM 2010, Miami, FL, USA, December 2010 [1].

*Corresponding author.   URL: http://www.eng.auburn.edu/~szm0001/, Email: smao@ieee.org.

In this paper, we investigate the problem of cooperative relay in CR networks. We assume a primary network with multiple licensed bands and a CR network consisting of multiple cooperative relay links. Each cooperative relay link consists of a CR transmitter, a CR relay, and a CR receiver. The objective is to develop effective mechanisms to integrate these two wireless communication technologies, and to provide an analysis for the comparison of two representative cooperative relay strategies, i.e., *decode-and-forward* (DF) and *amplify-and-forward* (AF), in the context of CR networks. We first consider cooperative spectrum sensing by the CR nodes. We model both types of sensing errors, i.e., miss detection and false alarm, and derive the optimal value for the sensing threshold. Next, we incorporate DF and AF into the *p*-Persistent Carrier Sense Multiple Access (CSMA) protocol for channel access for the CR nodes. We develop closed-form expressions for the network-wide capacities achieved by DF and AF, respectively, as well as that for the case of direct link transmission for comparison purpose.

Through analytical and simulation evaluations of DF and AF-based cooperative relay strategies, we find the analysis provides upper bounds for the simulated results, which are reasonably tight. We also find cross-point with the AF and DF curves when some system parameter is varied, indicating that each of them performs better in a certain parameter range. There is no case that one completely dominates the other for the two strategies. The considerable gaps between the cooperative relay results and the direct link results exemplify the diversity gain achieved by cooperative relays in CR networks.

The remainder of this paper is organized as follows. The system model is described in section 2. We analyze the two CR cooperative relay strategies in Section 3. Our simulation evaluations are presented in Section 4. Related work is discussed in Section 5 and Section 6 concludes the paper. The notation used in the paper is summarized in Table 1.

2. System Model

We assume a primary network and a spectrum band that is divided into M licensed channels, each modeled as a time slotted, block-fading channel. The state of each channel evolves independently following a discrete time Markov process [3]. The status of channel m in time slot t is denoted as $S_m(t)$, for $m = 1, 2, \cdots, M$. When $S_m(t) = 0$, the channel is in the idle state; when $S_m(t) = 1$, the channel is in the busy state (i.e., being used by primary users). Let λ_m and μ_m be the transition probability to remain in state 0 and the transition probability from state 1 to 0 for channel m, respectively. The channel model is illustrated in Fig. 1. The utilization of channel m with respect to primary

Table 1. Notation

Symbol	Definition
M	number of licensed channels
$S_m(t)$	status of channel m at time slot t
λ_m	transition probability of channel m from idle to idle
μ_m	transition probability of channel m from busy to idle
η_m	utilization of channel m
γ_m	maximum allowable collision probability on channel m
N	number of CR cooperative relay links
N_m	number of CR nodes sensing channel m
H_0^m	hypothesis that channel m is idle
H_1^m	hypothesis that channel m is busy
ϵ_m	false alarm probability on channel m
δ_m	miss detection probability on channel m
Θ_i^m	sensing result of channel m at CR node i
D_m	decision variable for channel m
τ_m	sensing threshold for channel m
$a_m(\vec{\Theta}_m)$	probability that channel m is idle
$a_m^{(j)}$	the jth largest value of $a_m(\vec{\Theta}_m)$
$\vec{\theta}_m^{(j)}$	argument of $a_m(\vec{\Theta}_m)$ achieving the jth largest value $a_m^{(j)}$
P_s	source transmit power
P_r	relay transmit power
G_0^k	path gain of kth relay link from transmitter to receiver
G_1^k	path gain of kth relay link from transmitter to relay
G_2^k	path gain of kth relay link from relay to receiver
$\sigma_{r,k}^2$	noise at the kth relay
$\sigma_{d,k}^2$	noise at the kth receiver
κ	decoding threshold for received SNR
$\bar{F}_{G_0^k}(x)$	CCDF of G_0^k
$\bar{F}_{G_1^k}(x)$	CCDF of G_1^k
$\bar{F}_{G_2^k}(x)$	CCDF of G_2^k
L	packet length
P_{DF}^k	decoding probability of DF
P_{AF}^k	decoding probability of AF
P_{DL}^k	decoding probability of DL
N_{DF}	no. received frames in two consecutive time slots with DF
N_{AF}	no. received frames in two consecutive time slots with AF
N_{DL}	no. received frames in two consecutive time slots with DL
C_{DF}	network-wide capacity with DF
C_{AF}	network-wide capacity with AF
C_{DL}	network-wide capacity with DL

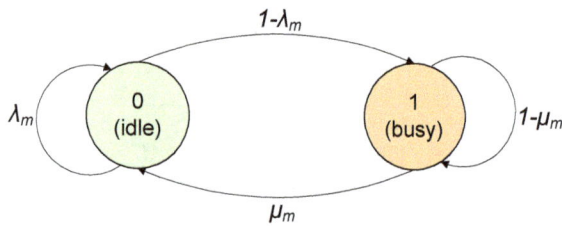

Figure 1. The discrete-time two-state Markov model for channel m, $m = 0, 1, \ldots, M - 1$.

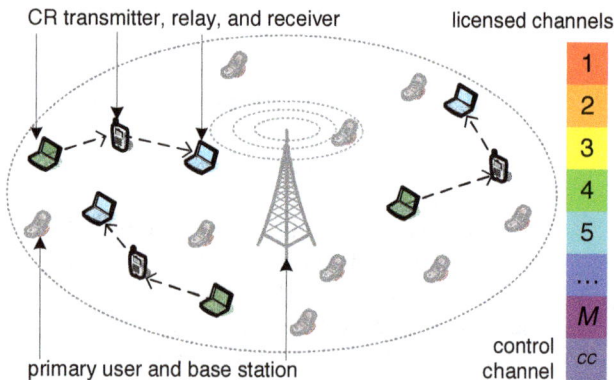

Figure 2. Illustration of colocated primary and CR networks. The CR network consists of a number cooperative relay links, each consisting of a CR transmitter, a CR relay and a CR receiver.

user transmissions, denoted by $\eta_m = \Pr\{S_m(t) = 1\}$, can be written as:

$$
\begin{aligned}
\eta_m &= \lim_{T \to \infty} \frac{1}{T} \sum_{t=1}^{T} S_m(t) \\
&= \frac{1 - \lambda_m}{1 - \lambda_m + \mu_m}.
\end{aligned} \tag{1}
$$

As illustrated in Fig. 2, there is a CR network colocated with the primary network. The CR network consists of N sets of cooperative relay links, each including a CR transmitter, a CR relay, and a CR receiver. Each CR node (or, secondary user) is equipped with two transceivers, each incorporating a software defined radio (SDR) that is able to tune to any of the M licensed channels and a control channel and operate from there.

We assume CR nodes access the licensed channels following the same time slot structure [3]. Each time slot is divided into two phases, the *sensing phase* and the *transmission phase*, as shown in Fig. 3. In the sensing phase, a CR node chooses one of the M channels to sense using one of its transceivers, and then exchanges sensed channel information with other CR nodes using the other transceiver over the control channel. During the transmission phase, the CR transmitter and/or relay transmit data frames on licensed channels that are believed to be idle based on sensing results, using one or

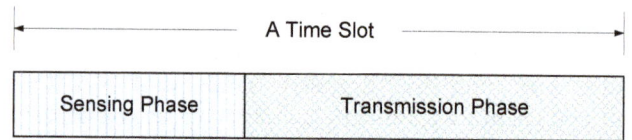

Figure 3. Time slot structure: a time slot consists of a sensing phase and a transmission phase.

both of the transceivers. We consider cooperative relay strategies AF and DF, and compare their performance in the following sections.

3. Cooperative Relay in CR Networks

In this section, we investigate how to effectively integrate the two advance wireless communication technologies, and present an analysis of the cooperative relay strategies in CR networks. We first examine cooperative spectrum sensing and derive the optimal sensing threshold. We then consider cooperative relay and spectrum access, and derive the network-wide throughput performance achievable when these two technologies are integrated.

3.1. Spectrum Sensing

Although precise and timely channel state information is desirable for spectrum access and primary user protection, continuous, full-spectrum sensing is both energy inefficient and hardware demanding [3]. Recall that each CR node is equipped with two transceivers and one has to be used for exchanging spectrum sensing results over the control channel. Thus each CR node can sense only one of the licensed channels at a time, using the remaining transceiver. Without loss of generality, we assume that each CR node senses a fixed channel throughout the time slots and receives from the control channel sensing results on other channels, i.e., distributed by other CR nodes.

During the sensing process, two kinds of detection errors may occur. A *false alarm* refers to the case when an idle channel is considered busy. Consequently, the CR nodes will not attempt to access that channel and a spectrum opportunity will be wasted. A *miss detection* refers to the case when a busy channel is considered idle. Since CR nodes will attempt to access this channel in the transmission phase, collisions with primary user transmissions will occur subsequently. Such spectrum sensing errors have been considered in the design of MAC protocols for CR networks [12, 13].

In this paper, we adopt hypothesis test to detect the availability of channel m. The null hypothesis H_0^m is "channel m is idle." The alternative hypothesis H_1^m is "channel m is busy." Let ϵ_i^m and δ_i^m be the probabilities of false alarm and miss detection, respectively, when CR node i senses channel m. We have

$$\epsilon_i^m = \Pr\left\{\Theta_i^m = 1 | H_0^m\right\} \quad \text{and} \tag{2}$$

$$\delta_i^m = \Pr\left\{\Theta_i^m = 0 | H_1^m\right\}, \tag{3}$$

where $\Theta_i^m \in \{0,1\}$ is the channel m sensing result of channel m at node i.

Assume there are N_m CR nodes sensing channel m. After the sensing phase, each CR node obtains a *sensing result vector* $\vec{\Theta}_m = [\Theta_1^m, \Theta_2^m, \cdots, \Theta_{N_m}^m]$ for channel m. The conditional probability $a_m(\vec{\Theta}_m)$ on channel m availability is

$$
\begin{aligned}
&a_m\left(\Theta_1^m, \Theta_2^m, \cdots, \Theta_{N_m}^m\right) \\
\cong\ & \Pr\left\{H_0^m | \Theta_1^m, \Theta_2^m, \cdots, \Theta_{N_m}^m\right\} \\
=\ & \frac{\Pr\left\{\Theta_1^m, \Theta_2^m, \cdots, \Theta_{N_m}^m | H_0^m\right\} \Pr\left\{H_0^m\right\}}{\sum_{j\in\{0,1\}} \Pr\left\{\Theta_1^m, \Theta_2^m, \cdots, \Theta_{N_m}^m | H_j^m\right\} \Pr\left\{H_j^m\right\}} \\
=\ & \frac{\prod_{i=1}^{N_m} \Pr\left\{\Theta_i^m | H_0^m\right\} \Pr\left\{H_0^m\right\}}{\sum_{j\in\{0,1\}} \prod_{i=1}^{N_m} \Pr\left\{\Theta_i^m | H_j^m\right\} \Pr\left\{H_j^m\right\}} \\
=\ & \left[1 + \frac{\Pr\left\{H_1^m\right\}}{\Pr\left\{H_0^m\right\}} \prod_{i=1}^{N_m} \frac{\Pr\left\{\Theta_i^m | H_1^m\right\}}{\Pr\left\{\Theta_i^m | H_0^m\right\}}\right]^{-1} \\
=\ & \left[1 + \frac{\eta_m}{1-\eta_m} \prod_{i=1}^{N_m} \frac{(\delta_i^m)^{1-\Theta_i^m}(1-\delta_i^m)^{\Theta_i^m}}{(\epsilon_i^m)^{\Theta_i^m}(1-\epsilon_i^m)^{1-\Theta_i^m}}\right]^{-1}. \tag{4}
\end{aligned}
$$

If $a_m(\vec{\Theta}_m)$ is greater than a *sensing threshold* τ_m, channel m is believed to be idle; otherwise, channel m is believed to be busy. The decision variable D_m is defined as follows.

$$D_m = \begin{cases} 0, & \text{if } a_m(\vec{\Theta}_m) > \tau_m \\ 1, & \text{if } a_m(\vec{\Theta}_m) \le \tau_m. \end{cases} \tag{5}$$

CR nodes only attempt to access channel m where D_m is 0. Since function $a_m(\vec{\Theta}_m)$ in (4) has N_m binary variables, there can be 2^{N_m} different combinations corresponding to 2^{N_m} values for $a_m(\vec{\Theta}_m)$. We sort the 2^{N_m} combinations according to their $a_m(\vec{\Theta}_m)$ values in the non-increasing order. Let $a_m^{(j)}$ be the jth largest function value and $\vec{\theta}_m^{(j)}$ the argument that achieves the jth largest function value $a_m^{(j)}$, where

$$\vec{\theta}_m^{(j)} = [\theta_1^m(j), \theta_2^m(j), \cdots, \theta_{N_m}^m(j)].$$

In the design of CR networks, we consider two objectives:

1. how to avoid harmful interference to primary users, and

2. how to fully exploit spectrum opportunities for the CR nodes.

For primary user protection, we limit the collision probability with primary user with a threshold. Let γ_m be the *tolerance threshold*, i.e., the maximum allowable interference probability with primary users on channel m. The probability of collision with primary users on channel m is given as $\Pr\{D_m = 0 | H_1^m\}$; the probability of detecting an available transmission opportunity is $\Pr\{D_m = 0 | H_0^m\}$. Our objective is to maximize the probability of detecting available channels, while keeping the collision probability below γ_m. Therefore, the optimal spectrum sensing problem can be formulated as follows.

$$\max_{\tau_m} \quad \Pr\left\{D_m = 0 | H_0^m\right\} \tag{6}$$

$$\text{subect to:} \quad \Pr\left\{D_m = 0 | H_1^m\right\} \le \gamma_m. \tag{7}$$

From their definitions, both $\Pr\{D_m = 0 | H_1^m\}$ and $\Pr\{D_m = 0 | H_0^m\}$ are decreasing functions of τ_m. As $\Pr\{D_m = 0 | H_1^m\}$ approaches its maximum allowed value γ_m, $\Pr\{D_m = 0 | H_0^m\}$ also approaches its maximum. Therefore, solving the optimization problem (6) $\sim$ (7) is equivalent to solving

$$\Pr\left\{D_m = 0 | H_1^m\right\} = \gamma_m.$$

If $\tau_m = a_m^{(j)}$, we have

$$
\begin{aligned}
&\Pr\left\{D_m = 0 | H_1^m\right\}\left(a_m^{(j)}\right) \\
=\ & \Pr\left\{a_m(\vec{\Theta}_m) > a_m^{(j)} | H_1^m\right\} \\
=\ & \sum_{l=1}^{j-1} \Pr\left\{a_m(\vec{\Theta}_m) = a_m^{(l)} | H_1^m\right\} \\
=\ & \sum_{l=1}^{j-1} (\delta_i^m)^{1-\theta_i^m(l)}(1-\delta_i^m)^{\theta_i^m(l)}. \tag{8}
\end{aligned}
$$

Obviously, $\Pr\{D_m = 0 | H_1^m\}(a_m^{(j)})$ is an increasing function of j. The optimal sensing threshold τ_m^* can be set to $a_m^{(j)}$, such that

$$\Pr\left\{D_m = 0 | H_1^m\right\}\left(a_m^{(j)}\right) \le \gamma_m$$

and

$$\Pr\left\{D_m = 0 | H_1^m\right\}\left(a_m^{(j+1)}\right) > \gamma_m.$$

The algorithm for computing the optimal sensing threshold τ_m^* is presented in Algorithm 1.

Once the optimal sensing threshold τ_m^* is determined, $\Pr\{D_m = 0 | H_1^m\}$ can be computed as given in (8) and

---

Algorithm 1: Algorithm for Computing the Optimal Sensing Threshold

1 Compute $a_m^{(j)}$ and the corresponding $\vec{\theta}_m^{(j)}$, for all j ;

2 Initialize $p_c = \Pr\left\{a_m(\vec{\Theta}_m) = a_m^{(1)}|H_1^m\right\}$ and $\tau_m = a_m^{(1)}$;

3 Set $j = 1$;

4 **while** $p_c \leq \gamma_m$ **do**

5 | $j = j + 1$;

6 | $\tau_m = a_m^{(j)}$;

7 | $p_c = p_c + \Pr\left\{a_m(\vec{\Theta}_m) = a_m^{(j)}|H_1^m\right\}$;

8 **end**

	Odd time slot	Even time slot

DF:

	Odd time slot	Even time slot
Channel 1	Busy	$R_1 \rightarrow D_1$
Channel 2	$S_1 \rightarrow R_1$	Busy
Channel 3	$S_2 \rightarrow R_2$	$R_2 \rightarrow D_2$
Channel 4	Busy	Idle

AF:

	Odd time slot	Even time slot
Channel 1	Busy	$S_2 \rightarrow R_2$
Channel 2	$S_1 \rightarrow R_1$	Busy
Channel 3	$R_1 \rightarrow D_1$	$R_2 \rightarrow D_2$
Channel 4	Busy	Idle

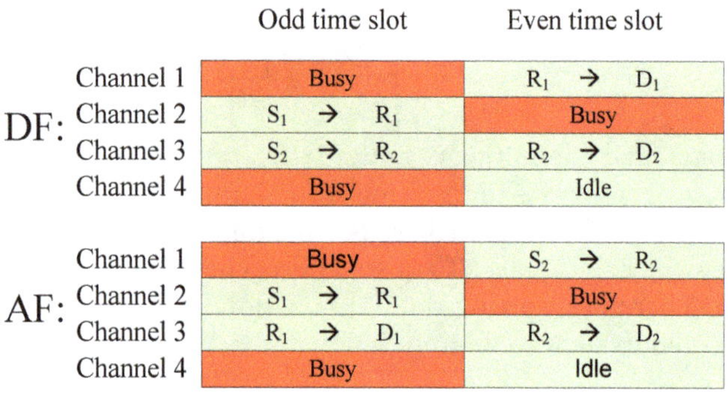

Figure 4. Illustration of the protocol operation of AF and DF, where $S_i \Rightarrow R_i$ represents the transmission from source to relay and $R_i \Rightarrow D_i$ represents the transmission from relay to destination, for the ith cooperative relay link.

$\Pr\{D_m = 0 \,|\, H_0^m\}$ can be computed as:

$$
\begin{aligned}
&\Pr\left\{D_m = 0|H_0^m\right\} \\
=\ & \Pr\left\{a_m(\vec{\Theta}_m) > \tau_m^*|H_0^m\right\} \\
=\ & \sum_{l=1}^{j-1} \Pr\left\{a_m(\vec{\Theta}_m) = a_m^{(l)}|H_0^m\right\} \\
=\ & \sum_{l=1}^{j-1} (\epsilon_i^m)^{\theta_i^m(l)}(1 - \epsilon_i^m)^{1-\theta_i^m(l)}. \quad (9)
\end{aligned}
$$

3.2. Cooperative Relay Strategies

During the transmission phase, CR transmitters and relays attempt to send data through the channels that are believed to be idle. We assume fixed length for all the data frames. Let G_1^k and G_2^k denote the path gains from the transmitter to relay and from the relay to receiver, respectively, and let $\sigma_{r,k}^2$ and $\sigma_{d,k}^2$ denote the noise powers at the relay and receiver, respectively, for the kth cooperative relay link. We examine the two cooperation relay strategies DF and AF in the following. For comparison purpose, we also consider direct link transmission below.

Decode-and-Forward (DF). With DF, the CR transmitter and relay transmit separately on consecutive odd and event time slots: the CR transmitter sends data to the corresponding relay in an *odd* time slot; the relay node then decodes the data and forwards it to the receiver in the following *even* time slot, as shown in Fig. 4.

Without loss of generality, we assume a data frame can be successfully decoded if the received signal-to-noise ratio (SNR) is no less than a *decoding threshold* κ. That is, outage probability of the cooperative channel is used to approximate packet loss probability. We assume gains on different links are independent to each other. The receiver can successfully decode the frame if it is not lost or corrupted on both links. The *decoding rate* of DF at the kth receiver, denoted by P_{DF}^k, can be computed as,

$$
\begin{aligned}
P_{DF}^k &= \Pr\left\{\left(\frac{P_s G_1^k}{\sigma_{r,k}^2} \geq \kappa\right) \text{ and } \left(\frac{P_r G_2^k}{\sigma_{d,k}^2} \geq \kappa\right)\right\} \\
&= \bar{F}_{G_1^k}\left(\frac{\sigma_{r,k}^2 \kappa}{P_s}\right)\bar{F}_{G_2^k}\left(\frac{\sigma_{d,k}^2 \kappa}{P_r}\right), \quad (10)
\end{aligned}
$$

where P_s and P_r are the transmit powers at the transmitter and relay, respectively, $\bar{F}_{G_1^k}(x)$ and $\bar{F}_{G_2^k}(x)$ are the complementary cumulative distribution functions (CCDF) of path gains G_1^k and G_2^k, respectively.

Amplify-and-Forward (AF). With AF, the CR transmitter and relay transmit simultaneously in the same time slot on different channels. A pipeline is formed connecting the CR transmitter to the relay and then to the receiver; the relay amplifies the received signal and immediately forwards it to the receiver in the same time slot, as shown in Fig. 4. Recall that the CR relay has two transceivers. The relay receives data from the transmitter using one transceiver operating on one or more idle channels; it forwards the data simultaneously to the receiver using the other transceiver operating on one or more *different* idle channels.

With this cooperative relay strategy, a data frame can be successfully decoded if the SNR at the receiver is no less than the decoding threshold κ. Then the decoding rate of AF at the kth receiver, denoted as P_{AF}^k, can be computed as,

$$
\begin{aligned}
P_{AF}^k &= \Pr\left\{\frac{P_r}{G_1^k P_s + \sigma_{r,k}^2}\frac{P_s G_1^k G_2^k}{\sigma_{d,k}^2} \geq \kappa\right\} \\
&= \int_0^{+\infty} \bar{F}_{G_2^k}\left(\frac{(P_s x + \sigma_{r,k}^2)\sigma_{d,k}^2 \kappa}{P_s P_r x}\right)dF_{G_1^k}(x). \quad (11)
\end{aligned}
$$

Direct Link Transmission. For comparison purpose, we also consider the case of direct link transmission (DL). That is, the CR transmitter transmits to the receiver via the direct link; the CR relay is not used in this case. Let the path gain be G_0^k with CCDF $\bar{F}_{G_0^k}(x)$, and recall that the noise power is $\sigma_{d,k}^2$ at the receiver, for the kth direct link transmission.

Following similar analysis, the decoding rate of DL at the kth receiver, denoted as P_{DL}^k, can be computed as

$$
\begin{aligned}
P_{DL}^k &= \Pr\left\{\frac{P_s G_0^k}{\sigma_{d,k}^2} \geq \kappa\right\} \\
&= \bar{F}_{G_0^k}\left(\frac{\sigma_{d,k}^2 \kappa}{P_s}\right).
\end{aligned}
\tag{12}
$$

3.3. Opportunistic Channel Access

We assume greedy transmitters that always have data to send. The CR nodes use p-Persistent CSMA for channel access. At the beginning of the transmission phase of an odd time slot, CR transmitters send Request-to-Send (RTS) with probability p over the control channel. Since there are N CR transmitters, the transmission probability p is set to $1/N$ to maximize the throughput (i.e., to maximize P_1 in (13) given below).

The following three cases may occur:

- *Case 1*: none of the CR transmitters sends RTS for channel access. The idle licensed channels will be wasted.

- *Case 2*: only one CR transmitter sends RTS, and it successfully receives Clear-to-Send (CTS) from the receiver over the control channel. It then accesses some of or all the licensed channels that are believed to be idle for data transmission in the transmission phase.

- *Case 3*: more than one CR transmitters send RTS and collision occurs on the control channel. No CR node can access the licensed channels, and the idle licensed channels will be wasted.

Let P_0, P_1 and P_2 denote the probability corresponding to the three cases enumerated above, respectively. We then have

$$
P_0 = (1-p)^N = \left(1 - \frac{1}{N}\right)^N
\tag{13}
$$

$$
P_1 = Np(1-p)^{N-1} = \left(1 - \frac{1}{N}\right)^{N-1}
\tag{14}
$$

$$
P_2 = 1 - P_0 - P_1.
\tag{15}
$$

The CR cooperative relay link that wins the channels in the odd time slot will continue to use the channels in the following even time slot. A new round of channel competition will start in the next odd time slot following these two time slots.

Since a licensed channel is accessed with probability P_1 in the odd time slot, we modify the tolerance threshold γ_m as $\gamma_m' = \gamma_m/P_1$, such that the maximum allowable collision requirement can still be satisfied.

In the even time slot, the channels will continue to be used by the winning cooperative relay link, i.e., to be accessed with probability 1. Therefore, the tolerance threshold is still γ_m for the even time slots.

3.4. Capacity Analysis

Once the CR transmitter wins the competition, as indicated by a received CTS, it begins to send data over the licensed channels that are inferred to be idle (i.e., $D_m = 0$) in the transmission phase. We assume the *channel bonding and aggregation* technique is used, such that multiple channels can be used collectively by a CR node for data transmission [14, 15].

With DF, the winning CR transmitter uses all the available channels to transmit to the relay in the odd time slot. In the following even time slot, the CR transmitter stops transmission, while the relay uses the available channels in the even time slot to forward data to the receiver. If the number of available channels in the even time slot is equal to or greater than that in the odd time slot, the relay uses the same number of channels to forward all the received data. Otherwise, the relay uses all the available channels to forward part of the received data; the excess data will be dropped due to limited channel resource in the even time slot. The dropped data will be retransmitted in some future odd time slot by the transmitter.

With AF, no matter it is an odd or even time slot, the CR transmitter always uses half of the available licensed channels to transmit to the relay. The relay uses one of its transceivers to receive from the chosen half of the available channels. Simultaneously, it uses the other transceiver to forward the received data to the receiver using the remaining half of the available channels.

Let D_m^{od} and D_m^{ev} be the decision variables of channel m in the odd and even time slot, respectively (see (5)). Let S_m^{od} and S_m^{ev} be the status of channel m in the odd and even time slot, respectively. We have,

$$
\Pr\left\{D_m^{od} = i, S_m^{od} = j, D_m^{ev} = k, S_m^{ev} = l\right\}
\tag{16}
$$
$$
= \Pr\left\{D_m^{ev} = k | S_m^{ev} = l\right\} \Pr\left\{D_m^{od} = i | S_m^{od} = j\right\} \times
$$
$$
\Pr\left\{S_m^{ev} = l | S_m^{od} = j\right\} \Pr\left\{S_m^{od} = j\right\}, \text{ for } i, j, k, l \in \{0, 1\}.
$$

where $\Pr\{S_m^{od} = j\}$ are the probabilities that channel m is busy or idle, $\Pr\{S_m^{ev} = l | S_m^{od} = j\}$ are the channel m transition probabilities. $\Pr\{D_m^{ev} = k | S_m^{ev} = l\}$ and $\Pr\{D_m^{od} = i | S_m^{od} = j\}$ can be computed as in (8) and (9).

Let N_{DF}, N_{AF} and N_{DL} be the number of frames successfully delivered to the receiver in the two consecutive time slots using DF, AF and DL, respectively. Define $\bar{S}_m^{od} = 1 - S_m^{od}$, $\bar{S}_m^{ev} = 1 - S_m^{ev}$, $\bar{D}_m^{od} = 1 - D_m^{od}$ and

$\bar{D}_m^{ev} = 1 - D_m^{ev}$. We have

$$N_{DF} = \left(\sum_{m=1}^{M} \bar{S}_m^{od} \bar{D}_m^{od}\right) \wedge \left(\sum_{m=1}^{M} \bar{S}_m^{ev} \bar{D}_m^{ev}\right) \quad (17)$$

$$N_{AF} = \left\lfloor \frac{1}{2} \sum_{m=1}^{M} \bar{S}_m^{od} \bar{D}_m^{od} \right\rfloor + \left\lfloor \frac{1}{2} \sum_{m=1}^{M} \bar{S}_m^{ev} \bar{D}_m^{ev} \right\rfloor \quad (18)$$

$$N_{DL} = \left(\sum_{m=1}^{M} \bar{S}_m^{od} \bar{D}_m^{od}\right) + \left(\sum_{m=1}^{M} \bar{S}_m^{ev} \bar{D}_m^{ev}\right), \quad (19)$$

where $x \wedge y$ represents the minimum of x and y, and $\lfloor x \rfloor$ means the maximum integer that is not larger than x.

As discussed, the probability that a frame can be successfully delivered is P_{DF}^k, P_{AF}^k, or P_{DL}^k for the three schemes, respectively. Recall that spectrum resources are allocated distributedly for every pair of two consecutive time slots. We derive the capacity for the three cooperative relay strategies as

$$C_{DF} = \mathrm{E}\left[N_{DF}\right] \times \sum_{k=1}^{N} \frac{P_{DF}^k P_1 L}{2N T_s} \quad (20)$$

$$C_{AF} = \mathrm{E}\left[N_{AF}\right] \times \sum_{k=1}^{N} \frac{P_{AF}^k P_1 L}{2N T_s} \quad (21)$$

$$C_{DL} = \mathrm{E}\left[N_{DL}\right] \times \sum_{k=1}^{N} \frac{P_{DL}^k P_1 L}{2N T_s}, \quad (22)$$

where L is the packet length and T_s is the duration of a time slot. The expectations are computed using the results derived in (16) ∼ (19).

4. Performance Evaluation

We evaluate the performance of the cooperative relay strategies with analysis and simulations. The analytical capacities of the schemes are obtained with the analysis presented in Section 3. The actual throughput is obtained using MATLAB simulations. The simulation parameters and their values are listed in Table 2, unless specified otherwise. We consider $M = 5$ licensed channels and a CR network with seven cooperative relay links. The channels have identical parameters for the Markov chain models. Each point in the simulation curves is the average of 10 simulation runs with different random seeds. We plot 95% confidence intervals for the simulation results, which are negligible in all the cases.

We first examine the impact of the number of licensed channels. To illustrate the effect of spectrum sensing, we let the decoding rate P_{AF}^k be equal to P_{DF}^k. In Fig. 5, we plot the throughput of AF, DF, and DL under increased number of licensed channels. The analytical curves are upper bounds for the simulation curves in all the cases, and the gap between the two is reasonably

Table 2. Simulation Parameters and Values

Symbol	Value	Definition
M	5	number of licensed channels
λ	0.7	channel transition probability from idle to idle
μ	0.2	channel transition probability from busy to idle
η	0.6	channel utilization
γ	0.08	maximum allowable collision probability
N	7	number of CR cooperative relay links
P_s	10 dBm	transmit power of the CR transmitters
P_r	10 dBm	transmit power of CR relays
L	1 kb	packet length
T_s	1 ms	duration of a time slot

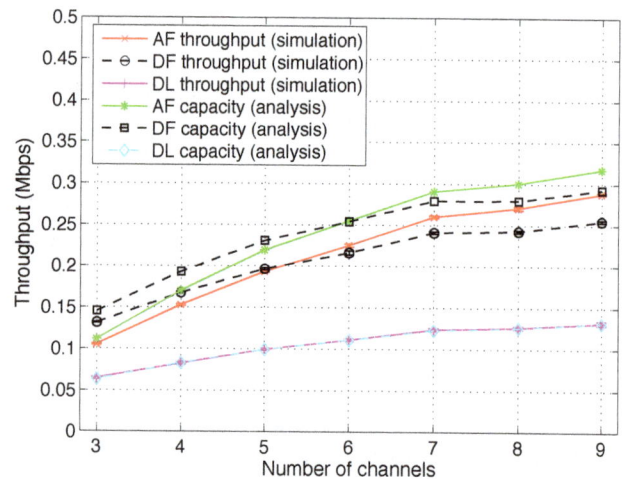

Figure 5. Throughput performance versus number of licensed channels.

small. Furthermore, as the number of license channels is increased, the throughput of both AF and DF are increased. The slope of the AF curves is larger than that of the DF curves. There is a cross point between five and six, as predicted by both simulation and analysis curves. This indicates that AF outperforms DF when the number of channels is large. This is because AF is more flexible than DF in exploiting the idle channels in the two consecutive time slots. The DL analysis and simulation curves also increases with the number of channels, but with the lowest slope and the lowest throughput values.

In Fig. 6, we demonstrate the impact of channel utilization on the throughput of the schemes. The channel utilization η is increased from 0.3 to 0.9, when primary users get more active. As η is increased, the

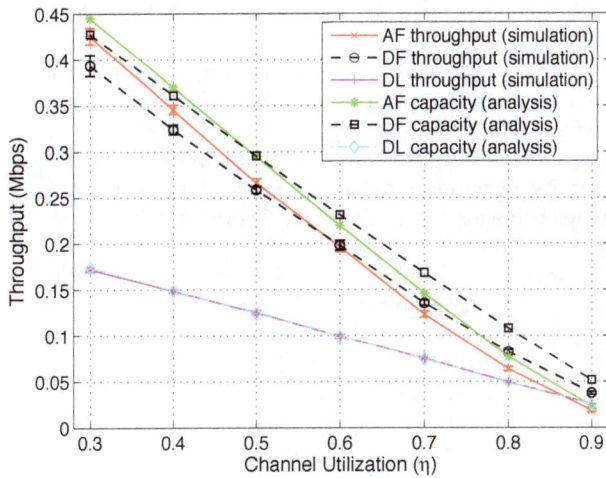

Figure 6. Throughput performance versus primary user channel utilization.

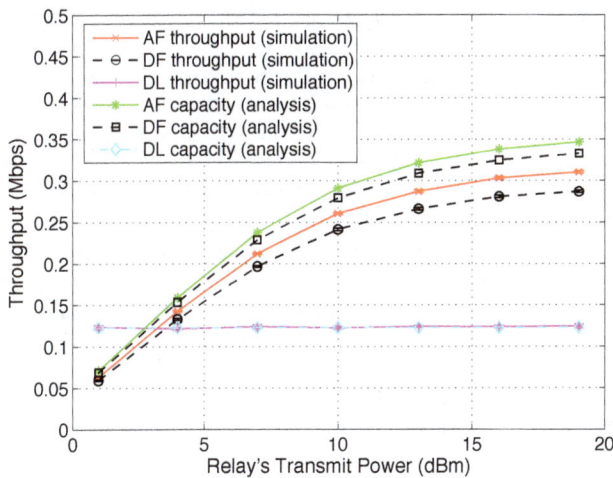

Figure 7. Throughput performance versus transmit power of relay nodes.

transmission opportunities for CR nodes are reduced and all the throughputs are degraded. We find the throughputs of AF and DF are close to each other when the channel utilization is high. AF outperforms DF in the low channel utilization region, but is inferior to DF in the high channel utilization region. There is a cross point between the AF and DF curves between $\eta = 0.5$ and $\eta = 0.6$. When the channel utilization is low, there is a big gap between the cooperative relay curves and the DL curves.

In Fig. 7, we examine the channel fading factor. We consider Rayleigh block fading channels, where the received power is exponentially distributed with a distance-dependent mean. We fix the transmitter power at 10 dBm, and increase the relay power from one dBm to 18 dBm. As the relay power is increased, the throughput is also increased since the SNR at the

receiver is improved. We can see the increasing speed of AF is larger than that of DF, indicating that AF has superior performance than DF when the relay transmit power is large. The capacity analysis also demonstrate the same trend. The throughput of DL does not depend on the relay node. Its throughput is better than that of AF and DF when the relay transmit power is low, since both AF and DF are limited by the relay-to-receiver link in this low power region. However, the throughputs of AF and DF quickly exceed that of DL and grow fast as the relay-to-receiver link is improved with the increased relay transmit power. The considerable gaps between the cooperative relay link curves and the DL curves in Figs. 5, 6 and 7 exemplify the diversity gain achieved by cooperative relays in CR networks.

5. Related Work

The theoretical foundation of relay channels was laid by the seminal work [16]. The capacities of the Gaussian relay channel and certain discrete relay channels are evaluated, and the achievable lower bound to the capacity of the general relay channel is established in this work. In [7, 8], the authors described the concept of cooperative diversity, where diversity gains are achieved via the cooperation of mobile users. In [9], the authors developed and analyzed low-complexity cooperative diversity protocols. Several cooperative strategies, including AF and DF, were described and their performance characterizations were derived in terms of outage probabilities.

In practice, there is a restriction that each node cannot transmit and receive simultaneously in the same frequency band. The "cheap" relay channel concept was introduced in [17], where the authors derived the capacity of the Gaussian degraded "cheap" relay channel. Multiple relay nodes for a transmitter-receiver pair are investigated in [18] and [19]. The authors showed that, when compared with complex protocols that involve all relays, the simplified protocol with no more than one relay chosen can achieve the same performance. This is the reason why we consider single relay in this paper.

In [20], Ng and Yu proposed a utility maximization framework for joint optimization of node, relay strategy selection, and power, bandwidth and rate allocation in a cellular network. Cai et al. [21] presented a semi-distributed algorithm for AF relay networks. A heuristic was adopted to select relay and allocate power. Both AF and DF were considered in [22], where a polynomial time algorithm for optimal relay selection was developed and proved to be optimal. In [23], a protocol is proposed for joint routing, relay selection, and dynamic spectrum allocation for multi-hop CR networks, and its performance is evaluated through simulations.

In [24], we investigate cooperative geographical routing in the context of wireless sensor networks. In a recent work [25], we investigate the problem of combing cooperative relay with CR for multiuser downlink video streaming, where interference alignment is incorporated to facilitate concurrent transmissions of multiple video packets.

6. Conclusion

In this paper, we studied the problem of cooperative relay in CR networks. We modeled the two cooperative relay strategies, i.e., DF and AF, which are integrated with p-Persistent CSMA. We analyzed their throughput performance and compared them under various parameter ranges. Cross-point with the AF and DF curves are found when some parameter is varied, indicating that each of them performs better in a certain parameter range; there is no case of dominance for the two strategies. Considerable gains were observed over conventional DL transmissions, as achieved by exploiting cooperative diversity with the cooperative relays in CR networks.

Acknowledgements. This work is supported in part by the US National Science Foundation (NSF) under Grants CNS-0953513 and DUE-1044021, and through the NSF Broadband Wireless Access & Applications Center (BWAC) at Auburn University. Any opinions, findings, and conclusions or recommendations expressed in this material are those of the author(s) and do not necessarily reflect the views of the Foundation.

References

[1] D. Hu and S. Mao, "Cooperative relay in cognitive radio networks: Decode-and-forward or amplify-and-forward?" in *IEEE GLOBECOM'10*, Miami, FL, Dec. 2010, pp. 1–5.

[2] I. Akyildiz, W. Lee, M. Vuran, and S. Mohanty, "NeXt Generation/dynamic spectrum access/cognitive radio wireless networks: A survey," *Elsevier Computer Networks*, vol. 50, no. 13, pp. 2127–2159, Sept. 2006.

[3] Q. Zhao and B. Sadler, "A survey of dynamic spectrum access," *IEEE Signal Process. Mag.*, vol. 24, no. 3, pp. 79–89, May 2007.

[4] Y. Zhao, S. Mao, J. Neel, and J. H. Reed, "Performance evaluation of cognitive radios: Metrics, utility functions, and methodologies," *Proc. IEEE*, vol. 97, no. 4, pp. 642–659, Apr. 2009.

[5] D. Hu, S. Mao, Y. T. Hou, and J. H. Reed, "Scalable video multicast in cognitive radio networks," *IEEE Journal on Selected Areas in Communications*, vol. 29, no. 3, pp. 334–344, Apr. 2010.

[6] D. Hu and S. Mao, "Streaming scalable videos over multi-hop cognitive radio networks," *IEEE Transactions on Wireless Communications*, vol. 9, no. 11, pp. 3501–3511, Nov. 2010.

[7] A. Sendonaris, E. Erkip, and B. Aazhang, "User cooperation diversity. part I. system description," *IEEE Trans. Commun.*, vol. 51, no. 11, pp. 1927–1938, Nov. 2003.

[8] ——, "User cooperation diversity. part II. implementation aspects and performance analysis," *IEEE Trans. Commun.*, vol. 51, no. 11, pp. 1939–1948, Nov. 2003.

[9] N. Laneman, D. Tse, and G. Wornell, "Cooperative diversity in wireless networks: Efficient protocols and outage behavior," *IEEE Trans. Inf. Theory*, vol. 50, no. 11, pp. 3062–3080, Nov. 2004.

[10] O. Simeone, Y. Bar-Ness, and U. Spagnolini, "Stable throughput of cognitive radios with and without relaying capability," *IEEE Trans. Commun.*, vol. 55, no. 12, pp. 2351–2360, Dec. 2007.

[11] Q. Zhang, J. ia, and J. Zhang, "Cooperative relay to improve diversity in cognitive radio networks," *IEEE Commun. Mag.*, vol. 47, no. 2, pp. 111–117, Feb. 2009.

[12] D. Hu and S. Mao, "Design and analysis of a sensing error-aware MAC protocol for cognitive radio networks," in *Proc. IEEE GLOBECOM'09*, Honolulu, HI, Nov./Dec. 2009.

[13] ——, "A sensing error aware mac protocol for cognitive radio networks," *ICST Trans. Mobile Commun. Appl.*, vol. 1, no. 2, 2012, in press.

[14] C. Corderio, K. Challapali, D. Birru, and S. Shankar, "IEEE 802.22: An introduction to the first wireless standard based on cognitive radios," *J. Commun.*, vol. 1, no. 1, pp. 38–47, Apr. 2006.

[15] H. Su and X. Zhang, "Cross-layer based opportunistic MAC protocols for QoS provisionings over cognitive radio mobile wireless networks," *IEEE J. Sel. Areas Commun.*, vol. 26, no. 1, pp. 118–129, Jan. 2008.

[16] T. Cover and A. Gamal, "Capacity theorems for the relay channel," *IEEE Trans. Inf. Theory*, vol. 25, no. 5, pp. 572–584, Sept. 1979.

[17] M. Khojastepour, A. Sabharwal, and B. Aazhang, "On capacity of Gaussian 'cheap' relay channel," in *Proc. IEEE GLOBECOM'03*, San Francisco, CA, Dec. 2003, pp. 1776–1780.

[18] Y. Zhao, R. Adve, and T. Lim, "Improving amplify-and-forward relay networks: Optimal power allocation versus selection," *IEEE Trans. Wireless Commun.*, vol. 6, no. 8, pp. 3114–3123, Aug. 2007.

[19] A. Bletsas, A. Khisti, D. Reed, and A. Lippman, "A simple cooperative diversity method based on network path selection," *IEEE J. Sel. Areas Commun.*, vol. 24, no. 3, pp. 659–672, Mar. 2006.

[20] T. C.-Y. Ng and W. Yu, "Joint optimization of relay strategies and resource allocations in cooperative cellular networks," *IEEE J. Sel. Areas Commun.*, vol. 25, no. 2, pp. 328–339, Feb. 2007.

[21] J. Cai, X. Shen, J. Mark, and A. Alfa, "Semi-distributed user relaying algorithm for amplify-and-forward wireless relay networks," *IEEE Trans. Wireless Commun.*, vol. 7, no. 4, pp. 1348–1357, Apr. 2008.

[22] Y. Shi, S. Sharma, Y. Hou, and S. Kompella, "Optimal relay assignment for cooperative communications," in *Proc. ACM MobiHoc'08*, Hong Kong, P. R. China, May 2008, pp. 3–12.

[23] L. Ding, T. Melodia, S. Batalama, and J. Matyjas, "Distributed routing, relay selection, and spectrum allocation in cognitive and cooperative ad hoc networks," in

IEEE SECON'10, Boston, MA, June 2010, pp. 1–9.

[24] M. Chen, V. C. Leung, and S. Mao, "Cooperative geographical routing in wireless sensor networks," in *Handbook on Sensor Networks*, Y. Xiao, H. Chen, and F. Li, Eds.   Hackensack, NJ: World Scientific Publishing Company, 2010, ch. 7, pp. 141–166.

[25] D. Hu and S. Mao, "Cooperative relay with interference alignment for video over cognitive radio networks," in *Proc. IEEE INFOCOM'12*, Orlando, FL, Mar. 2012.

14

A decentralized scheduling algorithm for time synchronized channel hopping

Andrew Tinka[1,*], Thomas Watteyne[2,3], Kristofer S. J. Pister[2], Alexandre M. Bayen[4]

[1]Electrical Engineering and Computer Sciences, University of California, Berkeley, CA, USA; [2]Berkeley Sensor & Actuator Center, University of California, Berkeley, CA, USA; [3]Currently with Dust Networks, Hayward, CA, USA; [4]Systems Engineering, Department of Civil and Environmental Engineering, University of California, Berkeley, CA, USA

Abstract

Time Synchronized Channel Hopping (TSCH) is an existing Medium Access Control scheme which enables robust communication through channel hopping and high data rates through synchronization. It is based on a time-slotted architecture, and its correct functioning depends on a schedule which is typically computed by a central node. This paper presents, to our knowledge, the first scheduling algorithm for TSCH networks which both is distributed and which copes with mobile nodes. Two variations on scheduling algorithms are presented. Aloha-based scheduling allocates one channel for broadcasting advertisements for new neighbors. Reservation-based scheduling augments Aloha-based scheduling with a dedicated timeslot for targeted advertisements based on gossip information. A mobile *ad hoc* motorized sensor network with frequent connectivity changes is studied, and the performance of the two proposed algorithms is assessed. This performance analysis uses both simulation results and the results of a field deployment of floating wireless sensors in an estuarial canal environment. Reservation-based scheduling performs significantly better than Aloha-based scheduling, suggesting that the improved network reactivity is worth the increased algorithmic complexity and resource consumption.

Keywords: decentralized scheduling, mobile *ad hoc* networks, simulation, time synchronized channel hopping

1. Introduction

The *Floating Sensor Network* (FSN) project built by UC Berkeley [1] includes autonomous, motorized floating sensor packages for deployments in rivers and estuaries (see Figure 1). The floating sensors (or 'drifters', in the terminology of the hydrodynamic community) are untethered; once deployed in the river, they are carried by the current and can modify their trajectory using limited actuation (differential drive propellers) to control their positioning. The Berkeley FSN drifters carry two communication systems: a GSM module for transmissions to a central server, and a low-power, low-range IEEE802.15.4-2006 [2] radio for communication between nodes.

The GSM communication channel is both expensive (both monetarily and in terms of energy consumption) and unreliable (due to variable GSM coverage, particu-

larly on the water). One strategy for delivering data from individual nodes to a remote server is to have one or several nodes with good GSM connections act as *ad hoc* sink nodes. Nodes connected by IEEE802.15.4-2006 links that do not have their own GSM connections available can send their data to one of the sinks, which retransmits the data *via* GSM to the server. Since it is not known *a priori* which nodes have GSM connectivity, the design objective for the IEEE802.15.4-2006 network must be to maximize point-to-point connectivity.

We define the *physical connectivity graph* to be the ensemble of wireless links 'good enough' to be used for communication at a given instant in time. We define the *logical connectivity graph* to be the set of links scheduled to be used at the same instant.

Due to the water currents, the mobility of the nodes means that the physical connectivity between nodes changes significantly over time. Global connectivity is not guaranteed. Therefore, centralized schemes for determining a

Figure 1. Prototype of a motorized drifter (*left*). Five passive drifters in a river (*right*).

communication schedule are poor fits for the problem. Our goal is to develop an algorithm which schedules intermittent bi-directional links between neighboring nodes as these links become available. We assess candidate schemes by evaluating how close the logical connectivity gets to the physical connectivity; that is, how many of the possible links are actually scheduled by the protocol.

Time Synchronized Channel Hopping (TSCH) is a *Medium Access Control* (MAC) scheme which enables robust communication through channel hopping and high data rates through synchronization. It is based on a time-slotted architecture, where a schedule indicates to the nodes on which timeslot and on which channel to transmit/receive data to/from which neighbor. Time Synchronized Channel Hopping is being standardized by the IEEE802.15.4e Working Group [3] and is expected to be included in the next revision of the IEEE802.15.4 standard. In this paper, the terms 'timeslot' and 'slot' are used interchangeably.

Time Synchronized Channel Hopping only defines the mechanism and makes no recommendation on how the schedule should be built. Typically, nodes report their communication needs (expressed in terms of throughput, reliability, and latency) to a central scheduler, which computes a schedule and injects this into the network. This technique has proven perfectly adequate for static networks such as industrial control *Wireless Sensor Networks* (WSNs). A distributed solution seems more appropriate for mobile networks. In those types of networks, each topological change would have to be reported to the central scheduler, which would have to re-compute a schedule and inform the nodes about the change. This is sometimes infeasible since this central scheduler may be disconnected from parts of the network.

This article presents two related distributed scheduling algorithms to be used on top of a TSCH MAC protocol. These algorithms are designed for the scheduling needs of IEEE802.15.4-2006 radios in applications with high mobility. In particular, these algorithms are purely decentralized. The first algorithm, 'Aloha-based scheduling', uses advertisements on a specific channel to discover neighbors and initiate schedule negotiations. The second algorithm, 'Reservation-based scheduling', augments the Aloha-based algorithm with a gossip mechanism that distributes the scheduling information to more nodes, speeding up the negotiation of a common schedule. In order to assess the performance of the scheduling algorithms, we present two metrics: 'relative connectivity', a static metric which evaluates how many feasible neighbors from the physical connectivity graph have been added to the schedule; and 'link duration', a dynamic metric that evaluates the lifetime of a link in the logical connectivity graph compared to its lifetime in the physical connectivity graph. We have evaluated the two algorithms in both a simulated environment and with a field experiment. Our field experiment features an interleaved implementation of the two algorithms, which allows us to compare their performance directly, without having to replicate the physical connectivity in separate experiments. By comparing the performance of the algorithms under different network density conditions, we can infer the importance of the different features of the two approaches, which gives insight into the design of future protocols.

The remainder of this article is organized as follows. Section 2 provides a comprehensive overview of MAC protocol approaches and standardization activities, and highlights the need for a distributed scheduling algorithm for TSCH. Section 3 then details the two scheduling algorithms proposed in this article, called 'Aloha-based scheduling' and 'reservation-based scheduling'. A simulation environment is described in Section 4, and an implementation and field experiment described in Section 5. Performance of the two algorithms in simulated and real environments is explored in Section 6. Finally, Section 7 concludes this article and presents directions for future work.

2. Time synchronized channel hopping

There are two main approaches for regulating access to a shared wireless medium: contention-based and reservation-based approaches. Any derived MAC protocol is based on one of those two approaches or a combination thereof.

Contention-based protocols are fairly simple, mainly because neither global synchronization nor topological knowledge is required. In a contention-based approach, nodes compete for the use of the wireless medium and only the winner of this competition is allowed to access the channel and transmit. Aloha and *Carrier Sense Multiple Access* (CSMA) are canonical representative schemes of contention-based approaches. They do not rely on a central entity and are robust to node mobility, which makes them intuitively a good candidate for dynamic mobile networks.

Preamble sampling is a low-power version of contention-based medium access, widely popular in WSNs. All nodes in the network periodically sample the channel for a short amount of time (at most a few milliseconds) to check whether a transmission is ongoing. Nodes do not need to be synchronized, but all use the same check interval. To ensure all neighbors are listening, a sender prepends a preamble which is at least as long as the check interval. Upon hearing the preamble, nodes keep listening for the data that follow. The optimal check interval, which minimizes the total energy expenditure, is a function of the average network degree and the load of the network. A check interval of 100 ms is typical. Numerous efforts have proposed ways to optimize the sampling [4], reducing the preamble length by packetization [5] or by synchronizing the nodes [6].

Despite their success, contention-based protocols suffer from degraded performance in terms of throughput when the traffic load increases. In addition, the uncoordinated nature of their resource allocation prevents them from achieving the same efficiency as ideal reservation-based protocols. Finally, frequency agility is hard to achieve by such protocols, as nodes are not synchronized.

Reservation-based protocols require the knowledge of the network topology to establish a schedule that allows each node to access the channel and communicate with other nodes. The schedule may have various goals such as ensuring fairness among nodes or reducing collisions by preventing nodes from transmitting at the same time. *Time Division Multiple Access* (TDMA) is a representative example for such a reservation-based approach.

In TDMA, time is divided into slots which are grouped into superframes which repeat over time. A schedule is used to indicate to each node when it has to transmit or receive, to/from which neighbor. Provided the schedule is correctly built, transmissions do not suffer from collisions, which guarantees finite and predictable scheduling delays and also increases the overall throughput in highly loaded networks.

Many approaches to MAC for WSNs combine some elements of contention-based protocols, especially for neighbor discovery or other startup tasks, with reservation-based scheduling for improved performance once neighbors are known. For example, in the PEDAMACS protocol [7], nodes transmit randomly using CSMA in order to discover the network topology and collect the topology information at a central node, which then computes all schedule information for all nodes in the network and distributes it. After this centralized schedule has been distributed, communication is governed by the schedule. In the TRAMA protocol [8], each TDMA superframe contains 'random-access' frames, where neighbors are discovered and local topological information is shared, and 'scheduled-access' frames, where nodes determine which of their two-hop neighbors has priority using a hash of frame number and node ID. In the Dozer protocol [9], new nodes use CSMA-like arbitration to respond to the beacon packets transmitted by nodes that have already joined the network; authority for setting the schedule is based on the tree hierarchy that emerges as 'child' nodes associate with the older 'parent' nodes. The SMACS protocol [10] uses a contention-based exchange of 'invitation' and 'response' packets to establish links between neighbors and to negotiate a transmit/receive schedule for that link for future communications. These four examples show the variety of approaches to scheduling that have been explored, from centralized (PEDAMACS) to purely decentralized (SMACS and TRAMA). The two algorithms presented in this article belong to the family of purely decentralized scheduling algorithms, and are designed specifically for the scheduling requirements of TSCH networks, and in particular the challenges of scheduling on a mobile network with connectivity that changes frequently.

The reliability of a wireless link is mainly challenged by external interference and multipath fading. Previous works [11, 12] show how channel hopping combats both of these, respectively. If a transmission fails, the sender retransmits the packet on a different frequency channel. Because this frequency change causes the wireless environment to be different, the retransmission has a higher probability of being successful than if it were retransmitted on the same channel.

Channel hopping was first applied to WSNs in a proprietary protocol called *Time Synchronized Mesh Protocol* (TSMP) [13]. In TSMP, nodes in the network are synchronized on a slotted time base. An individual timeslot is long enough for a sender to send a data frame, and for a receiver to acknowledge correct reception (a timeslot of 10 ms is common). L consecutive timeslots form a superframe, which repeats over time. A schedule of length L timeslots indicates, for each timeslot, whether the node is supposed to transmit or receive, to/from which neighbor and on which channel. TSMP runs on IEEE802.15.4-2006 [2] compliant radios, which offer 16 frequency channels in the 2.4GHz ISM band. A central scheduler is used to compute a schedule, which is then injected and used in the network.

TSMP makes a subtle difference between channel and frequency. The former is used in the schedule: node A schedules a link to node B on a given timeslot, and a given channel. This means that every superframe, node A will have the opportunity to use that link. The latter is the frequency nodes A and B communicate on. Nodes use the *Absolute Slot Number* (ASN) to keep track of which timeslot they are in. It is an ever-increasing number which is incremented at each timeslot, and which is shared by all nodes in the network as part of the synchronization procedure. TSMP uses the following function to obtain the frequency used for transmission from the channel in the schedule and the ASN. Per cent (%) is the modulo operator; 16 indicates that there are 16 available channels.

$$\text{frequency} = (\text{channel} + \text{ASN})\,\%16.$$

As a consequence, even when a link always appears at the same channel in the schedule, the operation described above ensures that communication happens in a channel hopping manner, thereby increasing the reliability of the link.

TSMP, which combines time synchronization and frequency agility, has been shown to achieve end-to-end reliabilities larger than 99.999% [14]. Its core idea has been standardized for industrial applications in WirelessHART [15–17] and ISA100.11a [18]. In 2009, it has been introduced in the draft standard IEEE802.15.4e under the name Time Synchronized Channel Hopping. This draft standard will replace the current IEEE802.15.4-2006 standard in its next revision.

All of the above standards rely on a central controller to compute a schedule for the network to use. The goal of this paper is to propose a distributed alternative, targeted at mobile nodes.

3. Distributed scheduling algorithms

3.1. Goal and metrics

The goal of the proposed schedule is full connectivity, which is achieved when each node in the network has established a bidirectional link to each of its physical neighbors. A bidirectional link is established between nodes A and B when, in the superframe, there is at least

Table 1. Variables used in this article.

Variable	Description
c	A channel
i, j, k, n	Slot numbers
L	Number of slots in a superframe
$S = \{S_0, S_1, \ldots S_{L-1}\}$	State for each slot
$C = \{C_0, C_1, \ldots C_{L-1}\}$	Data channel for each slot
$N = \{N_0, N_1, \ldots N_{L-1}\}$	Neighbor for each slot (can be NULL)
$P = \{(r, c)_1, (r, c)_2 \ldots\}$	List of potential neighbors (id and channel)
$D = \{(r, c)_1, (r, c)_2 \ldots\}$	List of neighbors self is connected to

one slot scheduled from A to B, and one from B to A. The unreliability of the wireless link and the movement of the nodes are challenges the scheduling algorithm needs to cope with.

If a link is present in the physical graph, it is *feasible*; if a link is present in the physical but not in the logical graph, it is said to be *unscheduled*; a link which still appears in the logical graph after it has disappeared from the physical graph is called *stale*. We use the ratio between the scheduled and feasible links as a metric for the static goodness of the scheduling algorithm.

Node mobility causes links to come and go. A link therefore has a finite lifetime, or *link duration*. To take advantage of a link, the scheduling algorithm needs to establish a logical link as soon as the physical link appears, and unscheduled it as soon as it disappears from the physical graph. We quantify the dynamic goodness of the scheduling algorithm by comparing the link duration between the physical and logical graphs.

Results presented in Section 6 are normalized against the optimal case, that is, the physical connectivity graph. The variables to be used in this article are listed in Table 1.

To be able to communicate, two nodes need to schedule a slot to one another. They hence need to communicate to agree which slot in the superframe to use, and which channel. We present two variants of the proposed scheduling mechanism. Aloha-based scheduling (Section 3.2) is a simple, canonical algorithm, in which neighbor nodes opportunistically discover each other and establish links. Reservation-based scheduling (Section 3.3) builds upon that. By adding an explicit reservation channel, nodes discover each other faster, which is desirable in the presence of mobile nodes.

3.2. Aloha-based scheduling

For each of the L slots in the superframe, the algorithm maintains a state S_i, a channel C_i, and a neighbor N_i. There are five states: '*Aloha*', '*Transmit Connection Request*', '*Receive Connection Request*', '*Transmit Data*', and '*Receive Data*'. A slot is assigned a channel C_i and a neighbor N_i only in the latter four states.

The *Aloha* state is the default. When establishing a uni-directional link from *A* to *B*, the scheduling algorithm causes a slot in *A*'s schedule to transition from *Aloha* to *Transmit Connection Request*, to *Transmit Data*. Simi-larly, the same slot in *B*'s schedule transitions from *Aloha* to *Receive Connection Request*, to *Receive Data*. When both *A* and *B*'s slots are in the *Transmit Data* and *Receive Data* state, respectively, data packets can be transmitted from *A* to *B*, once per superframe if exactly one slot is scheduled in the superframe. While communicating, *A* monitors whether its data packets are acknowledged; *B* monitors whether it receives data at all. If for five con-secutive superframes no data are successfully transmitted, the slot returns to the *Aloha* state; the connection is then lost. To ensure these statistics are up to date, if a sender has no data to send on a given slot, it sends an empty 'keep-alive' message.

Three types of packets move through the network:

(i) *Advertisement* packets contain a list of *Receive Con-nection Request* slots of the sender node. This can be used by neighbors to know where it can be reached to establish a link. Each entry is a tuple *(s,c)* of slot and associated channel. Advertisements are broadcast and always exchanged on channel 0.

(ii) *Connection Request* packets are sent in response to Advertisements; they are unicast on one of the slots announced in the Advertisement (at the announced channel, always different from channel 0).

(iii) *Data* packets flow over the slot when a link is established. Their content is determined by the application, but their successful transmission is monitored by the scheduling algorithm to detect stale links. An empty data packet is used as a keep-alive. Data packets are always sent on a chan-nel different from channel 0.

Note that there are L slots in a superframe, each of which can be used for an independent link. That is, an independent state machine is running for each slot. IEEE802.15.4-2006 compliant radios can transmit on 16 independent frequency channels. We dedicate channel 0 exclusively to Advertisements, and channels 1–15 exclu-sively to Connection Requests and Data packets.

Pseudocode listings for the two proposed algorithms are given below. The Aloha-based algorithm is described in Algorithm 1. The reservation-based algorithm has dif-ferent behaviors during timeslot 0 and other slots; its slot 0 behavior is given in Algorithm 2, while the behavior at other times is given in Algorithm 3.

Algorithm 1 in the Appendix presents Aloha-based scheduling in pseudocode. It is executed by every node in the network. Upon startup (lines 1–5), all the slots are set to the *Aloha* state. The main loop (lines 6–51) iter-ates at each slot; different actions are taken according to

the state of the slot. When in an *Aloha* slot, a node listens for Advertisements 90% of the time (on channel 0, lines 17–26), while 10% of the time it transmits an Advertise-ment (lines 10–15).

Sending advertisement packets. The idea of sending an Advertisement is for a node to announce different rendez-vous slot/channel tuples so that interested nodes can establish a link to it. When sending an Advertisement, a node converts all of its *Aloha* slots to the *Receive Connec-tion Request* state and assigns each of those a random channel other than channel 0 (lines 10–14). It puts that list in an Advertisement which it sends on channel 0. It then waits to be contacted on one of the *Receive Connec-tion Request* slots it just announced.

Receiving connection request packets. When reaching a slot in the *Receive Connection Request* state (lines 29–36), a node listens to the channel it has previously randomly picked and announced in its Advertisement. If it does not receive anything (line 35), it converts that slot back to *Aloha* state. If it does receive a Connection Request (lines 31–33), it converts that slot to *Receive Data* state and records the identifier of the requester.

Receiving advertisement packets. When receiving an Advertisement (lines 19–25), a node learns about the presence of a neighbor and is given the opportunity to contact it. If it has no slot scheduled to that neighbor, it picks one of the slots announced in the Advertisement where itself is in the *Aloha* state, that is, it picks a rendez-vous slot and channel. In case there are multiple slots which satisfy these requirements, it picks one of them ran-domly. It changes the state of that slot in its schedule to *Transmit Connection Request* (line 22), records the chan-nel announced in the Advertisement (line 23), and the sender of that packet (line 24).

Transmitting connection request packets. When reaching a slot *i* in the *Transmit Connection Request* state (lines 38–44) a node sends a Connection Request to the neigh-bor recorded in N_i, at the channel recorded in C_i (line 38). If it receives an acknowledgment, it puts that slot in the *Transmit Data* state, and the logical link is estab-lished. If the Connection Request is not established (e.g. due to a collision or nodes moving apart), the slot is reset to the *Aloha* state.

3.3. Reservation-based scheduling

The reservation-based scheduling protocol behaves like the Aloha-based protocol, with the following additions:

(i) Slot 0 is a permanent rendezvous slot, that is, only Advertisements can be exchanged. Unlike other slots, Advertisements can be exchanged on any of the 16 available channels, in slot 0. Each node picks a channel on which it listens for Advertise-ments. Using slot 0 as a reservation slot gives nodes more opportunities to establish links to one another.

(ii) In their Advertisements, nodes also include the list of neighbors they are connected to, and the channel those neighbors are listening on in slot 0. This means that nodes learn about their two-hop neighbors.

(iii) Each node maintains a list P of potential neighbors and the channel they are listening on in slot 0. This information is obtained by listening to Advertisements. Each node also maintains a list D of neighbors it is currently connected to. The scheduling algorithm tries to get as many nodes as possible from P to D.

(iv) A node only announces the *even* slots in its Advertisement. When the state of even slot i becomes *Transmit Data* (resp. *Receive Data*), the state of odd slot $i + 1$ is implicitly changed to *Receive Data* (resp. *Transmit Data*). This means that links are scheduled in pairs, one in each direction, establishing only bidirectional links.

Algorithms 2 and 3 in the Appendix present reservation-based scheduling in pseudocode. Algorithm 2 contains initialization, the main loop, and the behavior for slot 0, while Algorithm 3 contains the behavior for all other slots.

4. Simulation environment

We use a Python-based simulator[1] to model the mobility and RF propagation characteristics for a fleet of 25 mobile nodes. The superframe size was chosen to be 17 slots. The size must be co-prime with 16 in order to gain the benefits of the channel offset scheme; a relatively small superframe size was chosen to ensure that the scheduling constraints would be significant.

4.1. Propagation model

The design objective for the RF propagation model is to create a deterministic model which captures the variance of the distance-to-received-power relationship observed in empirical studies of static spatial configurations [4], while also providing plausible spatial correlation of link strength. Approximately 30% of the simulated environment is covered with obstacles. The radiated power from a transmitting antenna is attenuated by an inverse square law as it moves through 'obstacle-free' space, but is attenuated by an inverse fourth power law as it moves through 'obstacle' space. This 'higher power attenuation' scheme is inspired by empirical models of the effect of foliage

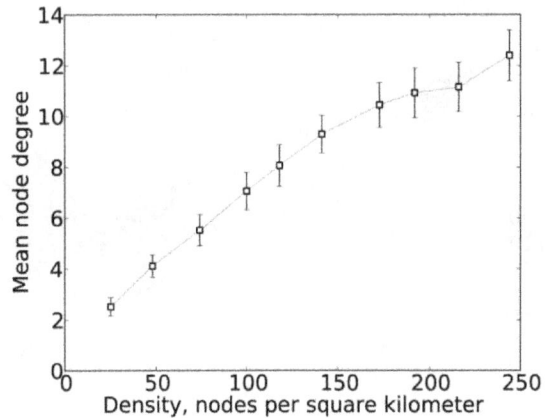

Figure 2. Mean node degree versus density of nodes in simulated environment.

on line-of-sight transmission [19]. The foliage model and density of obstacles are intended to represent an outdoor estuarial environment similar to that encountered by the FSN project. The multipath effect of the signal reflecting off the ground is modeled. The reflection is assumed to result in a 180° phase change and no attenuation.

The size of the simulated environment is modified as needed to yield desired node densities. The minimum and maximum densities are 25 and 250 nodes per square kilometer. Figure 2 shows the mean node degree (number of neighbors in the physical connectivity graph) for the different simulated densities. The bars represent the 95% confidence interval for the estimate of the mean.

4.2. Co-channel interference model

The interfering effect of two nodes transmitting on the same channel at the same time (usually called a 'collision') is one of the main constraints on the decentralized schedule.

The IEEE802.15.4-2006 standard specifies required jamming resistance for interference coming from an adjacent channel (1 channel away) or an alternate channel (2 channels away), but does not specify a required resistance to interference on the same channel. The Texas Instruments CC2420 2.4 GHz IEEE802.15.4-2006 compliant transceiver [20] has a specified co-channel rejection of -3 dB; in other words, if node A receives a transmission from node B with p dBm power, and a simultaneous transmission from node C with $(p-3)$ dBm power, the transmission for B will be received correctly and the transmission from C rejected. We use this model for our simulation. Adjacent and alternate channel interference are not modeled in this simulation.

4.3. Node mobility model

Each node is modeled as a mobile device moving at a constant speed in the environment described above. The speed of each node is drawn from a uniform distribution

---

[1]As an on-line addition to this paper, the source code of the simulator is made freely available at http://float.berkeley.edu.

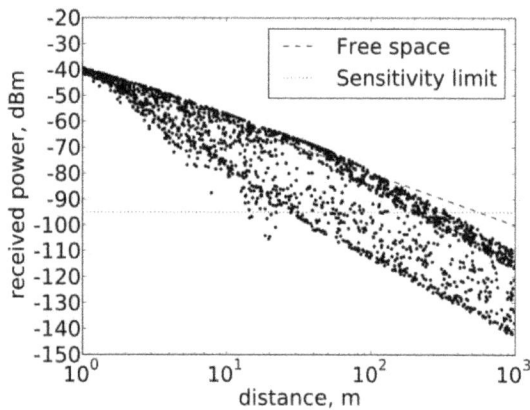

Figure 3. Received power from randomly chosen locations in simulated environment.

over $[0.8, 1.2]$ ms^{-1}. Each node transmits at 0 dBm (1 mW) using an isotropic antenna. The height of the antenna from the ground (used for the multipath calculations) is drawn from a uniform distribution over $[0.7,1.3]$ m for each node. Node motion is controlled by a random waypoint procedure: nodes select a cardinal direction randomly, then a distance to move in that direction. When they reach their destination, they repeat the selection process. The nodes are confined to a square area with dimensions determined by the desired node density.

Figure 3 shows the received power for randomly located transmitter and receiver nodes in the simulated environment.

5. Experimental set-up

On 19 November 2010, an implementation of the TSCH algorithms presented in Section 3 was tested using 10 Berkeley FSN drifters in the Grant Line Canal near Tracy, California.

The algorithms were implemented on Texas Instruments eZ430-RF2500 platforms, which consist of an MSP430 16-bit 16-MHz micro-controller and a CC2500 radio chip. The radio chip was programmed to communicate on the frequencies of the IEEE802.15.4-2006 standard, on the 2.4 GHz frequency band.

Each drifter was equipped with an eZ430-RF2500 platform. Distributed synchronization of those nodes was facilitated by a *pulse per second* (PPS) signal generated by the GPS unit on board the drifter, which provides a 1 Hz synchronization pulse with 25 ns jitter. The memory footprint of the implemented algorithms is 6 kB of flash memory and 500 B of RAM memory[2].

Both synchronization algorithms as well as a physical connectivity discovery mechanism were executed concurrently by the nodes using a 'master' superframe of 100

Table 2. Superframe structure.

Slot	Function
0–2	ASN synchronization
3–58	Physical graph discovery
63–79	Aloha algorithm
82–98	Reservation algorithm

frames, and scheduling various operations within that framework, as shown in Table 2. The idea is to gather baseline physical connectivity data and to run both algorithms simultaneously to allow for fair comparison of their performance. Each slot is 10 ms long; the superframe repeats every second.

(i) The *physical graph discovery* phase consists of each node deterministically broadcasting on each channel in sequence (i.e. there are no collisions). When not transmitting, a node listens for its peers and records which node was heard, on what slot, and on what frequency channel. Because there are 10 drifters and 16 channels, it takes 160 physical graph discovery slots to completely survey the connectivity. With 56 slots per superframe dedicated to physical discovery, we obtain a full image of the physical connectivity every three superframes, that is, every 3 s.

(ii) During the *Aloha algorithm* phase, the nodes execute the scheduling algorithm presented in Section 3.2. During the *reservation algorithm* phase, the nodes execute the scheduling algorithm presented in Section 3.3. These algorithms are executed independently from each other and from the physical graph discovery phase. As in the simulation-based study, both phases are 17 slots long.

We use the results of the physical graph discovery as an estimate of the instantaneous connectivity in order to evaluate the algorithmic performance of the Aloha and reservation algorithms.

The results of the discovery phase, and the state/neighbor/channel tables from each TSCH algorithm, were output from the eZ430-RF2500 motes and recorded using the data logging capabilities of the FSN drifter. Seven hours of data were recorded, resulting in over 250000 records of connectivity and algorithm state[3].

The Berkeley FSN drifters acted as passive floating sensors, being carried by the water current at approximately 0.3 ms^{-1}. Variations in the channel velocity profile caused their relative positions to change during the experiment. Overall, connectivity was not as highly dynamic as the simulation environment.

---

[2]The firmware source code is available at http://wsn.eecs.berkeley.edu/svn/ezwsn/.

[3]The gathered traces are made freely available at http://wsn.eecs.berkeley.edu/connectivity/.

6. Results

6.1. Static metric: relative connectivity

The static connectivity test in the simulated environment proceeds as follows:

(i) simulate 25 mobile nodes for 60 s;

(ii) pick a node and a superframe at random;

(iii) from the physical connectivity graph, count the number of unique edges incident to that node over the superframe (i.e. the number of one-hop neighbors connected for at least 1 slot during the superframe); this is the degree of the node;

(iv) from the logical connectivity graph, find the number of outbound edges (for the unidirectional test) or find the number of neighbors with both an outbound and inbound edge (the bidirectional test);

(v) the ratio of the logical connection count to the node degree is the *connectivity ratio* for the node. A connectivity ratio of 0.8 indicates that a logical link is present 80% of the cases a physical link is. A connectivity ratio of 1 is the best possible case.

To process the experimental results, the procedure was similar: a node and superframe were picked at random from the experimental logs, and the calculation of the connectivity ratio proceeded as in the simulation case.

Figures 4 and 5 show the mean connectivity ratio versus the node degree for 1250 simulations, for both unidirectional and bidirectional connections. Error bars represent the 95% confidence interval in the estimate of the mean.

In simulation, the reservation-based algorithm outperforms the Aloha-based algorithm at almost all node degrees (the confidence intervals overlap for degree 1). The reservation-based algorithm has more resources

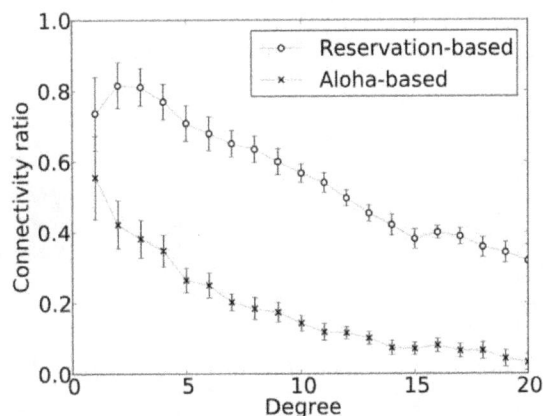

Figure 5. Mean connectivity ratio by degree for *bidirectional* links in simulated environment.

allocated to neighbor discovery, and a successful advertisement/connection request exchange results in a bidirectional connection. For both algorithms, increased node degree results in a decreased relative connectivity ratio. More local nodes mean more collisions between Aloha advertisements, which reduces the effectiveness of neighbor discovery, and more cases of multiple nodes responding to an advertisement, resulting in collisions and lost connectivity. The superframes also fill up when more neighbors are present; since the superframe size is 17 slots, a node cannot have bidirectional links with more than eight neighbors. The difference between the Aloha-based and reservation-based algorithm performance at high node degrees, however, demonstrates that both collisions and saturation must be significant.

In the experimental results, shown in Figures 8 and 9, a different relationship between the Aloha and reservation performance is observed. For the unidirectional case, the reservation algorithm dominates at lower network degrees, as in the simulation results, but under more connected conditions, the reservation algorithm performance

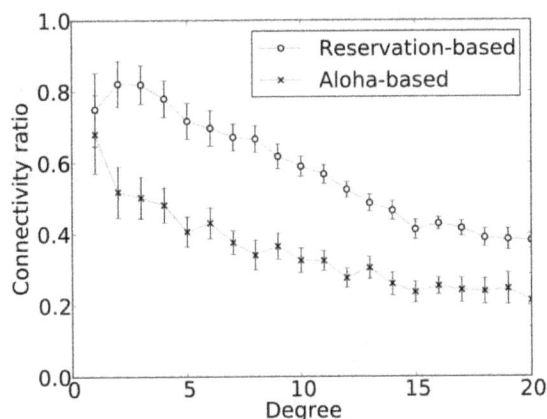

Figure 4. Mean connectivity ratio by degree for *unidirectional* links in simulated environment.

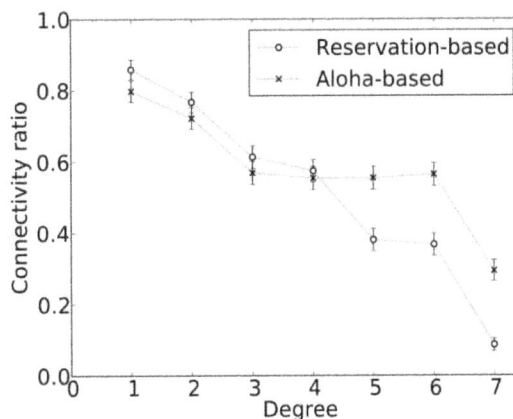

Figure 6. Mean connectivity ratio by degree for *unidirectional* links: experimental data.

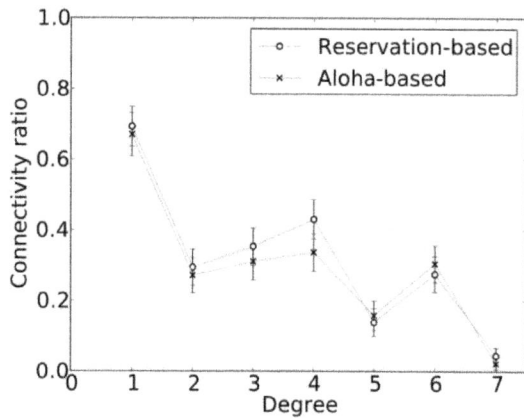

Figure 7. Mean connectivity ratio by degree for *bidirectional* links: experimental data.

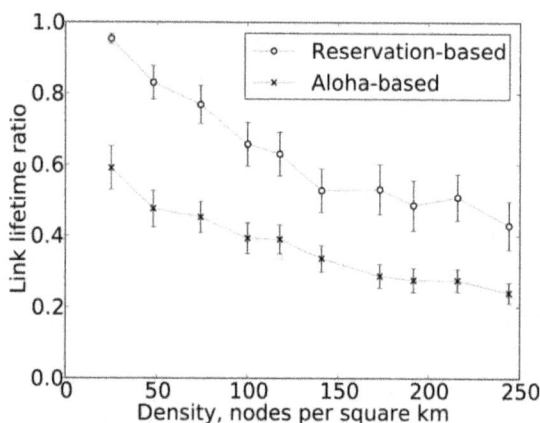

Figure 9. Mean *bidirectional* link lifetime ratio versus density in simulated environment.

suffers. This phenomenon is not well explained by the analysis applied to the simulation results. In the bidirectional case, we see a change in the performance of both algorithms at different network densities, but the results are too close to judge that one algorithm is outperforming the other. In both cases, the overall trend (higher density leading to lower connectivity ratio) is consistent with the simulation results. The regime where the Aloha algorithm outperforms the reservation algorithm in the unidirectional case remains unexplained.

6.2. Dynamic metric: link durations

The dynamic link duration test proceeds as follows:

(i) simulate 25 nodes for 60 s;

(ii) pick a node and a superframe at random;

(iii) pick one of the edges on the physical connectivity graph incident to that node at random; this is the link we will test;

(iv) count the number of consecutive superframes (forward and backward in time) in which this link is in the physical connectivity graph; this is the *physical link duration*;

(v) count the number of superframes in which the link exists in the logical connectivity graph, either as a unidirectional link (the original node to the destination) or as a bidirectional link; this is the *logical link duration*;

(vi) the ratio of the logical link duration to the physical link duration is the *link lifetime ratio*. A link lifetime ratio of 0.8 indicates that the algorithm has scheduled a logical link 80% of the time a physical link is present. That is, if two nodes are within radio range for 10 s, they can exchange data for 8 s. A link lifetime ratio of 1 is the best possible case.

For the experimental results, the procedure is the same, with random node and superframe drawn from the experimental logs.

Figures 8 and 9 show the mean link lifetime ratio versus the density of the nodes in the simulated environment for 1250 simulations. The bars represent the 95% confidence interval for the estimate of the mean. The degree of the node is not well defined over many superframes, as the physical and logical connectivity change. While the static connectivity test could use the node degree as the independent variable, for the dynamic link duration test we use the node density as a surrogate. See Figure 2 for the relationship between the mean node degree and node density.

The dynamic performance in simulation also shows that the reservation-based algorithm outperforms the Aloha-based algorithm. Again, the Aloha-based algorithm is at a disadvantage, because its advertisement/connection request transactions build unidirectional links, not bidirectional links. At low densities, the ratio between the

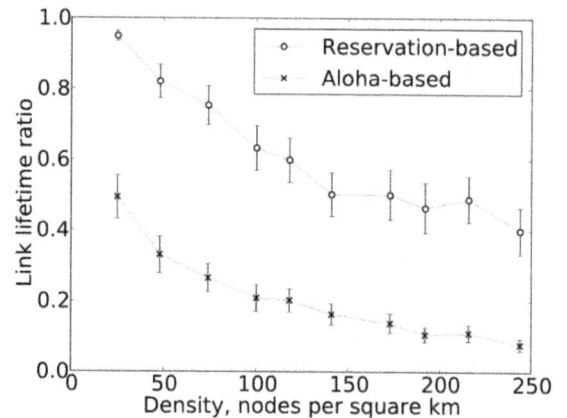

Figure 8. Mean *unidirectional* link lifetime ratio versus density in simulated environment.

algorithms' performances for bidirectional links is roughly 2, which suggests that the unidirectional/bidirectional allocation difference dominates in this regime. But at higher densities, the difference between the two algorithms widens, which means other effects must be significant as well.

The saturation effects at work in the connectivity tests are also significant in the dynamic case. Links can be broken by co-channel interference, if another pair of nodes begins transmitting at the same channel/slot as an existing link. Nodes that have many active links also have less vacant slots available to form new links. Saturation effects alone cannot explain the decreased performance at high density, however, since the Aloha-based algorithm's performance decreases significantly more than the reservation-based algorithm's performance.

The reservation-based algorithm benefits when advertisements are exchanged frequently, because information about connected neighbors is carried by the advertisement packets. The reactivity of the reservation-based algorithm therefore increases at higher densities, as nodes learn about possible new neighbors more quickly. Because the advertisements in the reservation-based algorithm carry more information than the advertisements in the Aloha-based protocol, the reservation-based algorithm gains relative performance at higher node densities.

Although the dynamic link survival time test can be applied to the experimental data, the experimental was conducted at essentially a single density condition. We therefore do not have values of the dynamic test at different densities, and cannot explore the density–link time relationship as in Figures 8 and 9. The results of the dynamic link survival time test are summarized in Table 3.

The dynamic lifetime test shows strong performance for both algorithms, under either the unidirectional or bidirectional case, with no statistically significant difference in the mean lifetime ratios. Although the value of the mean lifetime ratio is consistent with those observed in Figures 8 and 9, having both the Aloha and reservation algorithms perform (practically) identically is inconsistent with our observations in the simulated system. A major difference between the two scenarios is the distribution of link lifetimes in the physical connectivity graph. In the simulated environment, the connectivity is highly dynamic, and the short simulation time (60 s) places an upper bound on the link lifetime. In the experimental set-up, link lifetimes ranged from as short as 1 s to several

hours long. When the connectivity is not as dynamic, the increased reactivity of the reservation-based algorithm is not an advantage, and the algorithms have similar performance.

7. Conclusions and future work

In this article, we present what is, to our knowledge, the first scheduling algorithm for TSCH networks which both is distributed and which copes with mobile networks. The two variant algorithms are based on an advertisement and rendezvous scheme: nodes continuously advertise their presence to allow neighbor nodes to discover and contact one another. An inactivity threshold mechanism is used to tear down previously established links.

The algorithms are tuned for a network of 25 drifter nodes randomly moving inside a lake or river. Simulation results show, under realistic propagation and mobility models, the efficiency of the algorithms. The simulation results in Figures 5, 6, 9, and 10 support the conclusion that the reservation-based algorithm outperforms the Aloha-based algorithm in practically all density conditions. Experimental results (Figures 7 and 8; Table 4) do not show a significant advantage to one algorithm or the other; the major difference between the experimental set-up and the simulated system was the rate at which links formed and dropped, which suggests that in an environment with highly dynamic connectivity, including networks of mobile nodes, devoting additional resources to neighbor discovery and coordination pays off.

The goal of the scheduling algorithms presented in this article is to establish two-way connections between neighbor nodes, subject to the constraints of the superframe structure and the physical connectivity. We did not make assumptions about what kind of data is sent over the links; the latency, throughput, and reliability requirements are not specified. These scheduling algorithms could be adapted to meet either predetermined or dynamic provisioning requirements. For example, a pair of nodes that need to exchange a large amount of data might wish to schedule more than one transmission slot per superframe.

Many WSN applications are highly energy-constrained. Our scheduling algorithms, as described here, require the radios to constantly either receive or transmit. This may consume too much power for some applications. An obvious modification is to reduce the duty cycle of the Aloha coordination activities; the algorithms could be implemented exactly as written, while only performing Aloha listen/transmit actions on a subset of the idle slots. The obvious trade-off is between the energy consumed for Aloha coordination versus the reactivity of the network to changes in the physical connectivity graph. Further work will focus on characterizing the rate of change of the connectivity graph and determining a method for balancing power consumption and reactivity. Comparing the

Table 3. Link lifetime results from experimental data.

Algorithm	Link type	Mean lifetime ratio and 95% confidence interval
Aloha	Unidirectional	0.80 ± 0.03
Reservation	Unidirectional	0.79 ± 0.03
Aloha	Bidirectional	0.80 ± 0.03
Reservation	Bidirectional	0.82 ± 0.03

performance of these algorithms to that of previously proposed algorithms like TRAMA will also yield insight into the trade-offs made when designing algorithms for static versus mobile connectivity.

Acknowledgements. The authors thank Kevin Weekly and Carlos Oroza for their assistance during the experiment. Andrew Tinka acknowledges the support of NSERC.

References

[1] TINKA, A., STRUB, I., WU, Q.F. and BAYEN, A.M. (2010, August) Quadratic programming based data assimilation with passive drifting sensors for shallow water flows. *Int. J. Control* 83: 1686–1700.

[2] IEEE (2006). *IEEE Standard for Information Technology—Telecommunications and Information Exchange Between Systems—Local and Metropolitan Area Networks—Specific Requirements. Part 15.4: Wireless Medium Access Control (MAC) and Physical Layer (PHY) Specifications.*

[3] IEEE P802.15 Working Group for Wireless Personal Area Networks (WPANs) (2009). *IEEE P802.15.4e/D0.01 Draft Standard for Information Technology—Telecommunications and Information Exchange Between Systems— Local and Metropolitan Area Networks—Specific Requirements—Part 15.4: Wireless Medium Access Control (MAC) and Physical Lay.*

[4] POLASTRE, J., HILL, J. and CULLER, D. (2004) Versatile low power media access for wireless sensor networks. In *Second ACM Conference on Embedded Networked Sensor Systems (SenSys)* (ACM Press), 95–107.

[5] BUETTNER, M., YEE, G.V., ANDERSON, E. and HAN, R. (2006) X-MAC: a short preamble MAC Protocol for duty-cycled wireless sensor networks. In *4th International Conference on Embedded Networked Sensor Systems (SenSys)* (ACM Press).

[6] YE, W., SILVA, F. and HEIDEMANN, J. (2006) Ultra-low duty cycle MAC with scheduled channel polling. In *4th ACM Conference on Embedded Networked Sensor Systems (SenSys)* (ACM Press), 321–334.

[7] ERGEN, S.C. and VARAIYA, P. (2006) PEDAMACS: power efficient and delay aware medium access protocol for sensor networks. *IEEE Trans. Mob. Comput.* 5: 7.

[8] RAJENDRAN, V., OBRACZKA, K. and GARCIA-LUNA-ACEVES, J.J. (1995) Energy-efficient, collision-free medium access control for wireless sensor networks. *Wireless Networks* 12: 1.

[9] BURRI, N., VON RICKENBACH, P. and WATTENHOFER, R. (2007) Dozer: Ultra-low power data gathering in sensor networks. In *International Conference on Information Processing in Sensor Networks* (New York: ACM).

[10] SOHRABI, K., GAO, J., AILAWADHI, V. and POTTIE, G.J. (2000) Protocols for self-organization of a wireless sensor network. *IEEE Pers. Commun.* 7: 5.

[11] WATTEYNE, T., MEHTA, A. and PISTER, K. (2009) Reliability through frequency diversity: why channel hopping makes sense. In *6th ACM International Symposium on Performance Evaluation of Wireless Ad Hoc, Sensor, and Ubiquitous Networks (PE-WASUN)* (ACM Press).

[12] WATTEYNE, T., LANZISERA, S., MEHTA, A. and PISTER, K.S.J. (2010) Mitigating multipath fading through channel hopping in wireless sensor networks. In *IEEE International Conference on Communications (ICC)*.

[13] DOHERTY, L. and PISTER, K. (2008) TSMP: Time Synchronized Mesh Protocol. In *Parallel and Distributed Computing and Systems (PDCS)* (Orlando, FL).

[14] DOHERTY, L., LINDSAY, W. and SIMON, J. (2007) Channel-specific wireless sensor network path data. In *16th International Conference on Computer Communications and Networks (ICCCN)* (New York: IEEE), 89–94.

[15] GUSTAFSSON, D. (2009) *WirelessHART—Implementation and Evaluation on Wireless Sensors.* Master's Thesis, Kungliga Tekniska hogskolan.

[16] SONG, J., HAN, S., MOK, A.K., CHEN, D., LUCAS, M., NIXON, M. and PRATT, W. (2008) WirelessHART: applying wireless technology in real-time industrial process control. In *IEEE Real-Time and Embedded Technology and Applications Symposium* (Los Alamitos, CA: IEEE Computer Society), 377–386.

[17] HART COMMUNICATION FOUNDATION (2008) *HART Field Communication Protocol Specifications.* Standard Revision 7.1.

[18] ISA (2009) *Wireless Systems for Industrial Automation: Process Control and Related Applications.* ISA-100.11a-2009.

[19] OESTGES, C., VOLLACIEROS, B.M. and VANHOENACKER-JANVIER, D. (2009) Radio channel characterization for moderate antenna heights in forest areas. *IEEE Trans. Veh. Technol.* 58: 4031–4035.

[20] TEXAS INSTRUMENTS (2007) *CC2420, 2.4 GHz IEEE 802.15.4 / ZigBee-Ready RF Transceiver (Rev. B).* Data Sheet SWRS041B.

Appendix

Pseudocode listings for the two proposed algorithms are given below. The Aloha-based algorithm is described in Algorithm 1. The reservation-based algorithm has different behaviors during timeslot 0 and other slots; its slot 0 behavior is given in Algorithm 2, while the behavior at other times is given in Algorithm 3.

Algorithm 1: Aloha-Based Scheduling

```
for each slot i in 0..L-1
    S[i] = Aloha
    N[i] = NULL
    C[i] = NULL
end for
loop
    Go to the next slot i
    if S[i] == Aloha
        if uniform(0,1) < 0.1
            Find the set {j} of all other slots with state S[j] ==
            Aloha
            for each of these slots
                S[j] = Receive Connection Request
                C [j] = uniform(1,15)
            end for
            Send Advertisement with slots and channels
            {(j,C[j])}, on channel 0
        else
            Listen for an Advertisement on channel 0
            if Advertisement {(j,C[j])} received
                Find own set of slots {k} which are of state S[k]
                ==   Aloha
                if {k} ∩ {j} is not empty
                    Choose common slot n in {k} ∩ {j} randomly
                    S[n]= Transmit Connection Request
                    C[n] set to the receiving channel, read from
                    Advertisement
                    N[n] set to the node that sent the
                    Advertisement
                end if
            end if
        end if
    else if S[i]==Receive Connection Request
        Listen for a Connection Request to self on channel
        C[i]
        if valid Connection Request received
            Send Acknowledgment
            S[i] = Receive Data
            N[i] set to the ID of the requesting node
        else
            S[i] = Aloha
        end if
    else if S[i] == Transmit Connection Request
```

```
        Send Connection Request on channel C[i] to node
        N[i]
        if Acknowledgment received
            S[i] = Transmit Data
        else
            S[i] = Aloha
            N[i] = NULL
        end if
    else if S[i] == Receive Data or S[i] == Transmit Data
        if no successful communication for 5 consecutive su-
        perframes
            S[i] = Aloha
            N[i] = NULL
        end if
    end if
end loop
```

Algorithm 2: Reservation-Based Scheduling, Initialization, and Slot 0 Behavior

```
for each slot i in 0..L-1
    S[i] = Aloha
    N[i] = NULL
end for
C[0] = uniform_integer(0,15)
P = {}
D = {}
loop
    Go to the next slot i 10
    if i == 0
        if P is not empty and uniform(0,1) < 0.1
            Choose (j,c) randomly from neighbors of interest
            in P
            Transmit Advertisement to node j on channel c
            if Acknowledgment received
                set state of all advertised slots to S[k] = Receive
                Connection Request
            end if
        else
            Listen for an Advertisement on channel C[0]
            if Advertisement received
                Send Acknowledgment
                If neighbor of interest, choose common slot n
                (similar to Algorithm 1)
                S[n] = Transmit Connection Request
                N[n] = the ID of the node that sent the
                Advertisement
                C[n] = the receiving channel for that slot in
                the Advertisement
            end if
        end if
    else
        execute Algorithm 3
    end if
end loop
```

Algorithm 3: Reservation-Based Scheduling, Behavior for Slots other Than 0

```
if S[i] == Aloha
    if uniform(0,1) < 0.1
        Find the set {j} of all other even slots with S[j] ==
        Aloha
        for each j
            S[j] = Receive Connection Request
            C[j] = uniform_integer(1,15)
        end for
        Send Advertisement listing {(j,C[j])} and all tuples in
        D on channel 0
    else
        Listen for an Advertisement on channel 0
        if Advertisement {(j,C[j])} received
            Add new possible neighbors to P using the infor-
            mation in the Advertisement
            Find own set of slots {k} with S[k] == Aloha
            if {j} ∩ {k} is not empty
                Choose common slot n in {j} ∩ {k} randomly
                    S[n] = Transmit Connection Request
                    N[n] = the ID of the node that sent the
                    Advertisement
                    C[n] = the receiving channel for that slot in
                    the Advertisement
                end if
            end if
        end if
else if S[i] == Receive Connection Request
        Listen for a Connection Request for self on channel
        C[i]
        if valid Connection Request received
            Send Acknowledgment
            S[i] = Receive Data; S[i+1] = Transmit Data
            N[i] and N[i+1] = the ID of the requesting node
        else
            S[i] = Aloha
        end if
else if S[i] == Transmit Connection Request
        Send Connection Request on channel C[i] to node
        N[i]
        if Acknowledgment received
            S[i] = Transmit Data; S[i+1] = Receive Data
            Put (N[i], C[i]) in D
            Remove N[i] from P if present
    else
        S[i] = Aloha
    end if
else if S[i] == Receive Data or S[i] == Transmit Data
        if no successful communication for 5 consecutive
        superframes
        S[i] = Aloha
        move N[i] from D to P
        N[i] = NULL
    end if
end if
```

15

Middleware-based Security for Hyperconnected Applications in Future In-Car Networks

Alexandre Bouard[1,*], Dennis Burgkhardt[1], Claudia Eckert[2]

[1]BMW Forschung und Technik GmbH, Munich, Germany
[2]Technische Universität München, Garching near Munich, Germany

Abstract

Today's cars take advantage of powerful electronic platforms and provide more and more sophisticated connected services. More than just ensuring the role of a safe transportation mean, they process private information, industrial secrets, communicate with our smartphones, Internet and will soon host third-party applications. Their pervasive computerization makes them vulnerable to common security attacks, against which automotive technologies cannot protect. The transition toward Ethernet/IP-based on-board communication could be a first step to respond to these security and privacy issues. In this paper, we present a security framework leveraging local and distributed information flow techniques in order to secure the on-board network against internal and external untrusted components. We describe the implementation and integration of such a framework within an IP-based automotive middleware and provide its evaluation.

Keywords: Security & Privacy, Access Control, Middleware, CE Device, Third-Party Application, Automotive Application, Car-to-X Communication, Decentralized Information Flow Control, Dynamic Data Flow Tracking

1. Introduction

During the last two decades, vehicles evolved into very complex systems embedding powerful electronic platforms for various purposes, e.g., safety, infotainment. While still fulfilling their primary goal of transportation means, cars are now offering a plethora of new connectivity interfaces and communicate with numerous external communications partners: the Internet, Consumer Electronic (CE) devices, road-side units and other cars [1]. Like smartphones, the car will soon host Third-Party Applications (TPAs) [2]. Such a connectivity and new features will obviously allow a better customization of the car and a stronger tethering between all on-board and external communication partners. On the other hand, this may raise the threat level and increase the attack likeliness through these newly extended communication interfaces.

Recently, cars have been shown to be vulnerable against simple attacks involving packet sniffing/injection and more complex ones, like buffer overflows [3]. These attacks were performed by attackers having physical access to the car and its on-board network, but later work have show the feasibility to compromise the car through most of its external communication interfaces[4, 5]. In addition, today's automotive applications are mostly developed for a specific platform and for a precise car model. The car manufacturer knows the developer and can therefore set contractually certain responsibilities and testing processes. While not providing a complete security, such a strategy allows the car maker to keep the application integration process under its control. Loadable and on-the-fly installable applications have revolutionized the CE world but may shake up the static architecture of the car. While being mostly foreseen for the infotainment purpose, such applications will get access to Internet, several on-board functions and may secretly compromise the integrity and data confidentiality of the car [6].

At a functional level, limited communications technologies (e.g., Controller area network (CAN),

---

*Corresponding author. Email: alexandre.bouard@bmw.de

Media oriented systems transport (MOST)) and drastic requirements for low latency and high robustness let only very little space to security. Part of the solution seems to lie in the use of Ethernet and the Internet Protocol (IP) as standard for the on-board communications [7]. A larger bandwidth and mature security protocols will allow to secure the communications between two on-board platforms, but may remain insufficient in order to achieve a holistic solution. Future automotive applications will become more and more complex and partly designed by third parties. They will simultaneously trigger critical functionalities of the car and handle large amounts of data presenting different level of sensitivity. Not considering the whole information security problem, i.e., how information travels through the system, may lead to privacy breaches and, even worse, to safety malfunctionings, which could endanger the passengers' life. In order to keep on producing safe and secure vehicles, car manufacturers need to secure the on-board architecture accordingly.

Information flow control (IFC) is about controlling how information spread into a system and has been successfully applied for distributed systems. Our approach proposes to make use of the middleware layer to enforce the security. Through this layer, on-board applications exchange security metadata expressing their security concerns/requirements and can enforce policies relevant for the decentralized information flow control model (DIFC) we developed. Dynamic data flow tracking (DDFT) engines allow to taint data of interest and to follow their propagation within a running application in order to monitor and control its potentially malicious behavior. We chose to use such techniques to secure the integration of TPAs and couple them to our DIFC model via the middleware. For comparison, we design and evaluate a second security solution for integration of TPAs, which makes use of isolation/virtualization techniques and a DIFC-based network input/output monitoring. At the edge of the on-board network, a security communication proxy filters inbound and outbound communications based on pre-defined DDFT and DIFC policies and allows a secure and privacy-aware tethering of online services and external devices.

The main contributions presented in this paper are:

- a DIFC authorization model regulating on-board communications and integration within the middleware logic;

- a customized DDFT environment based on *libdft* [8], which locally monitors TPAs and is coupled to the DIFC framework;

- a security architecture for the communication proxy, extended from our previous work [9] and complying with the enforcement of DIFC and DDFT rules;

- a prototype implementation of our security framework integrated within our automotive adaptation of the IP-based middleware *Etch* [10].

The rest of the paper proceeds as follows. After having given a brief overview of today's automotive security and threats to it, we introduce the main security and privacy goals of our work as well as related work in Section 2. Section 3 describes our architecture for on-board security middleware and communication proxy as well as introducing exemplary *Security & Trust Levels*. Then Section 4 introduces our model for DIFC and DDFT techniques in detail, after which Section 5 describes our implementation. Section 6 proposes an evaluation of our prototype and security concepts and finally Section 7 concludes this article.

2. Background and Related Work

In this section we provide some background information about the automotive on-board architecture and its security shortcomings. We then define the threats and goals we consider in this work, as well as some attack scenarios.

2.1. Current & Future On–board Network

Today, the on-board network of a premium vehicle includes up to 70 interconnected electronic control units (ECUs). The ECU network is organized around specific domains, e.g., infotainment, power train, and is interlinked via several communication bus technologies, e.g., CAN, MOST, which necessitates complex application gateways for interoperability. On-board automotive applications are divided in elementary blocks over diverse ECUs and exchange braodcasted signal-based messages. Recently, some research work highlighted numerous security issues due to a lack of protection on the communication bus and poor ECU implementations. Common attacks have been successfully performed on both local [3, 11] and remote interfaces [4, 5]. Plaintext communications without authentication mechanisms allow an attacker with access to the on-board network to easily sniff and inject packets in order to misuse internal protocols. A lack of input validation of some ECUs allows to bypass authentication mechanisms via buffer overflow techniques in order to reprogram the platform. In some cases, the poor software implementation does not reflect the published standards and allows to directly activate the reprogramming mode of the platform [3]. Remote attackers may also be able to compromise communication interfaces like the Bluetooth interface via a brute force attack or the GSM access gateway via buffer overflow attack in order to gain access to the on-board network [4].

For tomorrow, the use of Ethernet/IP as on-board communication standard has been strongly investigated

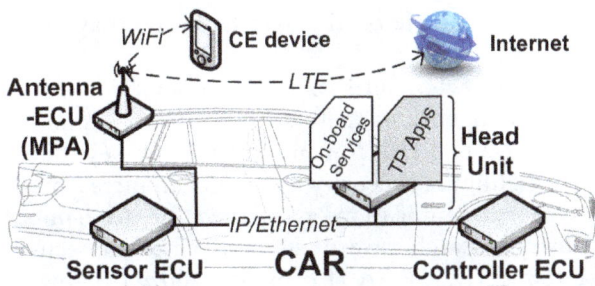

Figure 1. On-board Network Architecture and considered scenarios. Solid right-angle lines represent the wired on-board network. The dashed arrows represent external communications over different wireless networks.

and could be part of the security answer [7]. First a larger bandwidth will allow to comply with the requirements of future automotive applications [12]. It will allow to exchange large objects like environment model for driving assistance or infotainment content for audio and video purposes. Secondly, mature security protocols, developed specifically for the Internet world, will be instantly applicable and should provide a suitable protection for all bus communications. With Ethernet/IP, automotive applications will remain complex and distributed, but the design of engineering-driven middleware will greatly simplify their management. Such middleware will abstract and automate the network addressing and security enforcement [13, 14]. The clear separation between middleware and application logic will allow car manufacturers to separate the security logic from the application part and to significantly decrease the risk of buffer overflows thanks to security programming and code validation techniques [15]. In addition, car manufacturers will centralize most external communication interfaces (e.g., LTE, WiFi) around a multiplatform antenna-ECU (MPA) [16]. The MPA design offers them the opportunity to setup a single security gateway for all Car-2-X (C2X) communications, easy to verify and maintain.

2.2. Threats and Goals

Today's car are facing several challenges. Their functional behavior relies on complex software, which are optimized to run on resource limited-platforms, rarely updated and process a significant amount of data of different sensitivities. Attackers are already taking advantage of defects in the application logic and in weak security mechanisms. This usually results in privacy breaches (vehicle tracking [5]), in endangering the car's integrity (unauthorized reprogramming of an ECU [3]), and even worse in threatening the life of the its occupants (partial brake disabling [3]). If nothing is done, the emphasis of the use of C2X communications, more and more complex on-board applications and the

integration of TPAs will only increase the security risk. For the moment these attacks are mostly performed by the research community, but this could quickly change. The use of Ethernet/IP and a security middleware solves some of our issues but do not cover the whole information security problem. These do not address the threats related to unintentional bugs of the on-board applications. Besides, they do not consider unfair authorized parties, internal (e.g., TPA) or external (e.g., CE device), trying to bypass security policies, i.e., by trying to access or leak information they have no authorization for. This work aims at improving the information security in cars and at addressing the threats we just mentioned.

Attacker model: In this paper, we consider an attacker both acting internally and externally. Externally the attacker has access to all communication interfaces, can potentially get authenticated, and have her messages forwarded by the proxy to the inside. In addition, she can get access to the on-board network, sniff the traffic and inject new packets. She may as well write a TPA and have it installed in the car through a legitimate channel, e.g., the application store of the car manufacturer. We assume the attacker to be bounded only polynomially in computational power and storage, so that the current cryptographic primitives, e.g., AES, RSA, can be assumed to be secure, since there are no known algorithms able to break them in polynomial time. Thus she cannot break strong cryptographic protocols or successfully guess random numbers. We restrict the attacker to software-based attacks so that she cannot physically tamper ECUs, e.g., read or flash memory content. This work does not consider denial-of-service attacks.

Use Cases & Attack scenarios: Our use cases are depicted in Figure 1 and consider both internal and external untrusted components, over which the car manufacturer has no control. We define as service a group of on-board application sharing the same middleware layer. We take the example of a TPA running on the Head Unit (HU). It communicates with several on-board services and with the Internet and some CE devices over the MPA. Internet services and CE devices are authorized to get access to some on-board services and can get authenticated. The TPA is considered as conform with the internal application programming interface (API) of the car. However it may present a poor implementation exploitable by an attacker. We here mainly focus on attack scenarios leveraging the TPA to (1) compromise the integrity of the car or (2) leak confidential information.

1. **Integrity Attack Scenario:** A malicious TPA may send bogus packets on the on-board network (e.g., a shellcode) or access/modify locally resources of the HU (e.g., filesystem) in order to disturb the

car functioning. Secondly, a malicious external communication partner may send bogus messages to the car, forwarded by the MPA and try to disturb the car as well.

2. **Confidentiality Attack Scenario:** A malicious TPA may get access to sensitive data, stored on the HU (e.g., the home address of the driver in the navigation module) or received from another service (e.g., preference settings of a user from the seat controller). Even without the permission, the TPA may try to send them outside, either directly over the proxy or through an intermediate step, for example a buggy service communicating with the outside. Otherwise as previously, an external communication partner may leverage a bug of an on-board application and try to retrieve confidential information it should not get access to.

Assumptions: Next-generation ECUs will take advantage of security middleware able to establish communications channels over strong security protocols like IPsec [14]. Each of them will be equipped with a hardware security extension providing key storage and secure boot [17]. Consequently, we assume that after ignition of the car, the middleware, the operating system (OS) and the hardware are not compromised and stay so during the runtime. Thus, we trust the middleware of every service to establish secure communication channels with each others and to enforce suitable security mechanisms when it is required and expected.

2.3. Security Architecture requirements

With regard to our automotive context, we define the following additional security challenges/ requirements. Even when enforcing security, the car should provide performances (for high throughput and large bandwidth) and a robustness at least equivalent as they are currently. Security solutions should be optimally performing on all platforms, even on the resource-limited ones. Our solutions should not require regular updates or financial extra cost. The security should be easy to manage and should not increase the application complexity for all application developers and end-users.

2.4. Related Work

Cars already communicate with our smartphones via USB, 3G or Bluetooth. Depending on the standard, traditional challenge/response schemes ensure the access to basic web browsing, car information, phone- and audio-functionalities. More critical features like remote door opening/locking are preformed via GSM or 3G through a server of the car manufacturer. The server acts like a firewall. However this solution is

expensive, not scalable on the long term and may not be secure [18].

Industry projects: Until recently, automotive security has focused on anti-theft devices such as immobilizer and secure RFID transponder for car key. But the newly highlighted security issues and an increasing use of C2X communications reoriented the academic and industry research toward automotive holistic security solutions. The EVITA [19] project and its follow-up SEIS [7] aimed at securing the on-board network. They both proposed a modular framework establishing internal secure communication channels and leveraging secure hardware platforms. On the other hand, a project like SeVeCom [20] addressed the security issues of future vehicle communication networks. They designed C2X protocols using encryption and authentication mechanisms, which got implemented on the V2X platforms of the sim^{TD} project [21]. But none of these projects really formalize the transition of data between outside and inside or consider the damages that external data could cause on the inside. They all rely on strong security components on the edge of the on-board network performing the enforcement of static access control lists (ACLs).

Securing a corporate network presents some similarities with the automotive context, e.g., when integrating mobile devices. Their approaches make use of strong authentication mechanisms and device integrity measurements in order to establish network connections and VPN tunnels [22]. However they usually lack specifications for a secure resource- and data-management. In the context of SEIS, [9] proposes a proxy-based architecture for a secure CE-device integration. The proxy evaluates the security level of the device and communication and shares it with the ECUs. We chose to extend these concepts to our architecture and to complete it with a more formal security model.

About IFC: IFC is a form of mandatory access control. Resources (e.g., documents) and principals (e.g., persons) having access to them are given a label, i.e., a clearance level. A label-based partial order defines whether the access is authorized or not [23]. DIFC extends the IFC concepts [24]. A resource (e.g., an application) can be allowed to divide its access rights and create new labels to manage with more flexibility its access control. DIFC was adapted at the granularity of a process: processes are separated between trusted and untrusted during runtime. Label-based rules are enforced locally by a customized OS [25, 26]. But for distributed applications, OSs can exchange their labels through the network as well [27]. However these approaches are too fine granular and suffer from a too significant performance overhead. For a lighter approach, we chose to enforce DIFC labels only on on-board communications between services. Exchanging labels to enforce IFC is not new. For Pedigree, a central server and customized network switches distribute and enforce IFC policies on every network communication [28]. However in-car applications are distributed over ECUs with different OSs/hardwares. In order to reduce the risk of errors, latency and maintenance complexity, the DIFC cannot rely on any central entity and cannot be enforced in the OS or its hardware. We therefore chose to enforce DIFC at the application level, especially in the middleware layer.

About DDFT: For more security, smartphone applications are tested before their release on an online application store and are usually isolated from each other thanks to the sandboxing mechanisms of the mobile OS. However it is not flawless [6, 29]. We therefore consider 2 other runtime options. Either (1) we isolate the TPA from other on-board services and monitor its inputs/outputs, like in Section 4.1. Otherwise (2) we monitor the TPA itself and what happens during runtime, for example by using DDFT techniques, like in in Section 4.2. DDFT allows to taint and track data of interest within a running application/system and have been successfully applied for various purposes, e.g., malware monitoring [30] or privacy-aware smartphone monitoring [31]. DDFT offers two approaches: monitoring the whole host [31] or just one process/application [8]. Considering our requirements for robustness and low latency, we orient our work toward a lighter approach, the second one. This solution causes some overhead but does not require any OS or source code modification and has been already used for distributed [32] and automotive [33] environments. But monitoring or isolating TPAs will not be sufficient. Thus, we decide to combine the isolation/monitoring techniques to our DIFC model via the middleware.

3. On-board Security Architecture: Secure Middleware and Communication Proxy

As mentioned in the introduction, Ethernet/IP will be intensively used by car manufacturers as standard for the on-board communications. The rest of this section provides an overview about our secure middleware layer (Section 3.1), the architecture of our security proxy (Section 3.2), and) a taxonomy for external untrusted communication partners (Section 3.3). Finally, Section 3.4 discusses the benefits of such a security architecture and pinpoints its shortcomings.

3.1. A Security Middleware Extension

By definition, the middleware abstracts the communication interfaces and hides the network complexity from the application logic. It may as well automate the security enforcement and therefore allow the application developer to be completely security-unaware. We present here an architecture for a security middleware

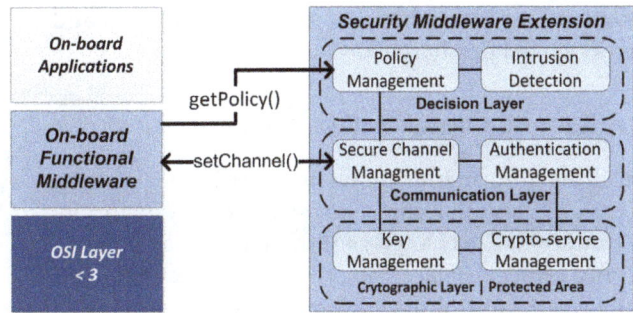

Figure 2. Architecture of the security middeware extension and its association to a functional middleware.

extension (SME) [14], that can be easily coupled to any middleware layer.

Figure 2 presents the three-layer SME architecture we are considering here. Such a modularization offers enough adaptivity and combination possibilities to comply with all the different security levels required by our use cases. For example, a simple temperature sensor, which only sends information to the engine controller, will not provide as much security features as the HU which deals with multiple communication partners and very untrusted data. The security layers are organized as follows:

- *The decision layer* provides security decisions by means of static policies. These policies mostly regulate the establishment of a communication channel between 2 on-board platforms and the access to all resources present on an ECU. It provides a direct API *getPolicy()*, which is available from the functional middleware level and returns the policy decision. In addition it may include functionalities for network monitoring and intrusion detection.

- *The communication layer* provides the security plug-ins for the communication protocol implementation and associated filtering mechanisms. It is accessible from the middleware through the interface *setChannel()* which opens a network socket and allows to specify the chosen security protocol, authentication scheme and encryption strength.

- *The cryptographic layer* is in charge of the key management and the cryptographic processing, i.e., the data encryption/decryption and signature generation/verification. For more security, this layer may be included in a hardware security module for protecting the key material from an attacker and providing integrity mechanisms such as remote attestation.

Due to the risk of latency and errors, the configuration of the middleware and its SME is statically set up during the vehicle assembly or during periodic system updates. It mostly concerns the definition of security associations for IPsec channels between two ECUs and their associated preshared keys.

3.2. A Security Communication Proxy

Controlling information flows in distributed systems like cars is essential for a holistic security solution. ECUs internally exchange genuine packets and therefore only require secure communication channels and simple access control mechanisms. But the integration of untrusted devices and online services, over which the car manufacturer has no control, necessitates a more complex authorization model. In order to get the best user experience, the car manufacturer cannot ban any type of CE device or trendy online community service, only because they do not comply with certain security prerequisites. It can only restrict their access to the car and adapt the security mechanisms on a case by case basis. In addition, ECUs communicate with the outside behind the MPA. Being more than just a super-antenna, the MPA is a complete ECU and decouples the communication between on-board and outside like a NAT router. Such decoupling allows the car to use a unique optimal security protocol for the on-board network, while letting the choice of the outside protocol to CE- and web-application developers. It also requires the MPA to be able to support the "blind" ECU, e.g., by providing it information about the external device or service, the identity of the concerned user, the used wireless protocols or the likeliness of a security threat, so that the ECU can enforce the right security decision.

For this purpose, we developed an application-independent in-band middleware protocol, allowing internal exchanges of security metadata. Concretely, the middleware header is extended with a new field, specifying the security and trust context in which data are or may be exchanged with an external peer. Instead of directly considering the privacy aspect of any single piece of information, we focus on the trust we grant the peer and quantify that. The security aspect characterizes how secure the external communication protocols are. On the other hand, the trust aspect characterizes how trustworthy the remote device or online service is considered to be. We call this context *Security & Trust Level (STL)* and provide its precise evaluation in Section 3.3. In order to distinguish whether the STL defines the current communication situation or whether it defines a required situation to send a message out, we define the STL_{status} and the STL_{req}, respectively. The life cycle of the STL as represented in Figure 3 is explained in the following:

The communication security proxy: The proxy is implemented on the MPA and stands in the middle of every communication happening between an internal entity and the outside. Unlike traditional NAT routers simply forwarding IP packets, the proxy

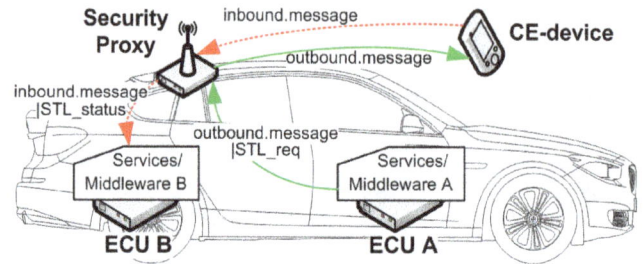

Figure 3. STL life cycle.

really decouples these communications and acts like a translation interface between external and internal middleware-based protocols. It dynamically manages all external communication channels and their security features, i.e., encryption process, key management and authentication schemes. Internally, the proxy communicates with ECUs over static IPsec channels. Future C2X use cases foresee exchanges of big objects at a high frequency. The proxy will not be able to perform deep packet inspection and is as a consequence application-unaware. The proxy evaluates for each external entity a STL_{status} and adds it in the middleware header of every inbound packets. In addition, the proxy enforces on them a first coarse domain-based filtering, for example an online social network service will not get access to any service of the power train management domain. Inversely, for every outbound message, the proxy, before forwarding it, makes sure that every received STL_{req} is conform to the actual STL of the communication situation. Section 3.3 provides more information about STL-based policies.

Middleware & STL: The middleware of every on-board service intensively uses the STL concept. All STL-based policies are managed in its *Decision Layer* and are invoked when receiving or emitting a packet. Based on the received STL_{status}, the security middleware decides whether it is safe and authorized to process such a packet. Depending on the ECU capacity and the STL, the middleware can adapt its security processing and pass the data through a security parser, e.g., access request to a SQL database, or run the data in an isolated environment, e.g., JavaScript code in an isolated web-browser. A received STL_{req} determines the data sensitivity and requires the middleware to decide whether its applications are allowed to receive such data. Inversely, when sending a packet to the on-board network, the middleware automatically extends the message header with a STL_{req} reflecting the sensitivity of the payload information, i.e, industrial secret or private information. Any communication is concerned, since a multicast address could inadvertently forward a packet with private data to the proxy, i.e., to the outside. The applications on top of the middleware are totally STL-transparent, the STL enforcement happens in the middleware. Like most policies on the ECUs, the STL-based policies are defined by the car manufacturer at design time and are statically set up in the SME.

3.3. The STL Taxonomy

Section 3.2 introduced the concept of STL. It defined it as the security and trust context in which data are ("$_{status}$") or should be ("$_{req}$") exchanged with the outside. The rest of the section proposes an evaluation of (1) its security aspects, (2) its trust aspect and (3) its enforcement.

1) SL definition: We define the SL as a qualitative characterization of the security strength of an external communication protocol. The SL is characterized as follows:

- **SL=0** Communication providing no security or presenting exploitable design flaws.
  Example: Plaintext; WEP encryption; TLS+DES or RC4 with a 56-bits key;

- **SL=1** Communication providing strong authentication of the external peers and data integrity (i.e., against unauthorized modifications).
  Example: WPA2 encryption; Message in plaintext protected by HMAC-SHA1;

- **SL=2** Communication as secure as SL=1 and, in addition, providing strong confidentiality (i.e., one secret key per user, no shared key between users).
  Example: TLS+AES; IPsec+AES;

- **SL=3** Communication as secure as SL=2 and assuring the presence of a secure hardware element protecting the cryptographic materials of the external peer.
  Example: SL2-protocol + remote attestation.

2) TL definition: We define the TL as an abstract representation of how trustworthy the data sender and receiver are. The notion of trust is usually defined as a mix between 3 components: reputation, reliability and security [34]. The security has already been considered, thus the TL focuses on the 2 remaining ones. The evaluation criteria of the TL should be clear and easy to assess. We consider that data may only be misused, if they are (1) physically and (2) juridically accessible, i.e., (1) if the data leave the car and (2) if the receiver is legally allowed to endanger the user's privacy (e.g., data selling/forwarding, data stored on an unprotected server). The TL should reflect such risks and is evaluated based on the following criteria:

- **Criterion 1 (Cr.1) "Local Usage":** determines whether the data are limited to an on-board usage only.

- **Criterion 2 (Cr.2) "Anonymization":** determines whether data have to be anonymized, when released out, i.e., whether an external receiver may be able to trace back the identity of the car or of the user.

Table 1. Binary decision tree used for TL evaluation.

	$Cr.\,1 \rightarrow$	$Cr.\,2 \rightarrow$	$Cr.\,3 \rightarrow$	TL
Case 1	true	-	-	3
Case 2	false	true	-	2
Case 3	false	false	true	1
Case 4	false	false	false	0

- **Criterion 3 (Cr.3) "Jurisdiction":** determines whether the external receiver is considered as a safe "place of jurisdiction" (POJ), i.e., whether the servers hosting the online service are located in a country imposing a regulation protecting the user's privacy.

In order to determine the TL of an external peer, we use the simple binary decision tree of Table 1. Every criterion is evaluated iteratively, a "true" answer stops the process and sets the TL value. Highly sensitive data, like industrial secrets, should never be released (Cr. 1=true) and thus are assigned TL=3. Very sensitive data, like the car position, should be able to leave the car but not endanger the driver's privacy (Cr. 2=true) and therefore should be anonymized (TL=1). An application for local hazard warning, broadcasting the position of an accident (and also the position of the car) should be able to do so, only if the emitted packets do not include traceable information about the user's or car identity. Data with a low sensitivity, like the driver's username and settings, can be released without anonymization but only to services presenting a safe POJ (Cr.3=true, TL=1), for example a banking service, whose servers are in Germany. A service presenting an unsafe POJ, like Facebook in USA, should only be able to receive nonsensitive data with TL=0. While Cr. 1 and Cr. 2 are easy to assess and enforce, Cr. 3 needs to be determined based on recommendations from international privacy experts [35].

The TL taxonomy is quite simple and provides an efficient way to control the data release with the outside. But further tests with more use cases should be performed. The TL criteria are very coarse, but give to the car manufacturer a simple way to configure a "by default" privacy/trust-aware behavior. For a more flexible usage, the user should be able to change the assigned TL of an online service, like a social network of her choice, and allows it to receive some data with TL=1 as well.

3) STL enforcement: We consider security and trust as two independent variables necessitating separated enforcements and evaluations. Indeed, anonymized data with a TL=2 may be sent with a SL=1 in plaintext (e.g., local hazard warning scenario), while data with a TL=1 may be sent with a SL=2 (e.g., banking scenario), because the user does not want such information to be eavesdropped. As a consequence, we define the STL as the concatenation of the SL and the TL, i.e., STL=(SL,

TL). For an easy management, we limit ourselves to 4 SL values and 4 TL values, coded over 4 bits in the middleware header.

Concretely, data with a STL_{req}, which arrives on the proxy will be allowed to be released to an external service or device X: 1) if X complies with the conditions of the received TL and is authorized to receive data with such a TL and 2) if the communication with X provides a SL greater than or equal to the received SL. However, such conditions may be too constraining and may never allow certain data to leave the car. Declassification methods allowing to assign a lower STL to some data or to just add an exception on the proxy should be possible. But those methods should only be part of use cases predefined by the car manufacturer and if necessary should involve the driver's decision, e.g., if it is her private data. Further considerations about declassification methods are not provided in this work.

STL-based polices are statically implemented in the ECUs and do not require any update. Either the on-board service generates the data to be sent and associates its own STL_{req} depending on the appropriate policy, or the ECU received the data from another ECU and before forwarding them, labels them with the received STL_{req}. The proxy should regularly receive notifications to update the TL of new external services and the SL of new or flawed communication protocols. The CE device case is bit particular, as it gets authenticated by the proxy and is assigned a STL_{status}. This STL depends on the used connection protocol for the SL and is assigned a TL=1, since we assume that the user's device is under her control and is therefore safely handling her private data.

3.4. Intermediary Discussion

Enforcing the security in the middleware provides a clear separation between security/networking management and application logic. Such an approach abstracts the security model and makes it more efficient to enforce and easier to verify. Security programming methods can be easier to apply and can solve many security flaws related to stack pointers overwriting attacks, e.g., buffer overflows. However our STL model has shown its limits: it implicitly considers a unique user in the car and cannot handle the information of more than one simultaneous passenger or driver, respectively. Adding the unique ID of a user to the STL label may not be sufficient to secure on-board information flows. For the moment our approach do not consider any TPA or any solution for constraining it to respect authorized information flows and to not act maliciously. The following Section 4 provides a more formal security model completing our architecture and STL-based enforcement.

4. Controlling Information Flows in Cars

IFC is about monitoring the in-car propagation of data defined as data of interest. Such data may be interesting to track within the car either because they are sensitive data and their release should be controlled, or because their integrity is essential to preserve, e.g., when processed by a safety-critical application. We propose to monitor the information flow at two different levels: at the network level between on-board applications in Section 4.1 and within the TPA in Section 4.2. Like for our previous approach, we chose to enforce security and IFC via the middleware.

4.1. Decentralized Information Flow Control

We defined the term services as a group of on-board applications running on top of a same middleware layer. The applications belonging to a same service share the same security concerns for confidentiality (e.g., because they share data of same sensitivity) and integrity (e.g., because they trigger the same critical mechanisms). For this reason, a security label, characterizing such concerns, is assigned to every service perpetrating network exchanges. The middleware, independent from the on-board applications, is in charge of monitoring and labeling the on-board network communications. Comparisons between the labels of 2 services allow to protect the integrity and the confidentiality of the information they process, e.g., by isolating corrupted data from a critical application or preventing an unauthorized information disclosure. The rest of the section presents our formal DIFC model, inspired from [27].

Security Labels. One label is assigned to each service. A label includes two subcomponents: a secrecy label S and an integrity label I. S and I are two sets of tags. A tag is defined as a security concern of an individual about the secrecy (in S) or the integrity (in I) of the information they process. For the ECUs, a tag is a unique value implemented as a bit-string. We refer to it with a symbolic name like b_s or b_i, where the subscripts s and i designate the concerns for respectively the secrecy and the integrity and b the principal (e.g., the service b or the CE device b), whose concern is characterized. We call service tag and user tag a tag which designates the security concerns of respectively an on-board service or a user and her CE device. The secrecy tags are "sticky", i.e., information from a service labeled with b_s cannot flow to a service lacking it. The integrity tags are "fragile", i.e., information from a service labeled with b_i can flow to a service lacking but will then lose its label.

The labels establish a lattice enforcing a form of mandatory access control, as shown in Figure 4. Formally, information from a service A labeled with S_A and I_A can flow to a service B labeled with S_B and

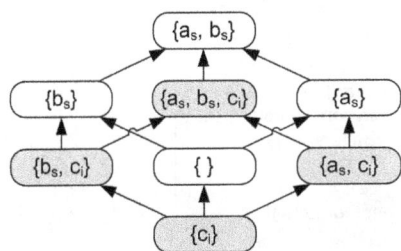

Figure 4. Label-based lattice. This example includes 2 secrecy tags a_s and b_s and one integrity tag c_i. Boxes represent service labels. Grey boxes show labels including c_i. Arrows link labels between which information can flow.

I_B if and only if the tags of S_A are included in S_B and I_A contains the tag of I_B. The partial order "$\prec$" (pronounced "can flow to") is defined as follows:

$$L_A \prec L_B \text{ iff } S_A \subseteq S_B \text{ and } I_A \supseteq I_B,$$

where $L_A = (S_A, I_A)$ and $L_B = (S_B, I_B)$

However the on-board services are distributed over several ECUs and do not know each other's label. We therefore chose to label the messages as well. When service A with label L_A sends a message M to service B with label L_B, A assigns to M the label L_M such as $L_A \prec L_M \prec L_B$. Assigning such a label to M allows A to disclose the information first to M ($L_A \prec L_M$) and to make sure that B will receive it if it fulfills the condition $L_M \prec L_B$, i.e., $L_A \prec L_B$ by transitivity of the partial order $\prec$.

Concretely in our scenario, the HU stores data belonging to different users with different secrecy tags, so that only the appropriate TPA and CE device can receive the data of a particular user.

Tag Ownership. If an information could only follow the partial order $\prec$, labeled messages could only travel to services with a greater or equal secrecy level and some information would never be able to leave the car. Our DIFC model decentralizes the exception management. In addition to its label, each service is assigned a set of tags called ownership O. A tag t included in the ownership O of a service S allows S to derogate to the restriction imposed by t. We say that S owns t. Obviously, no service should own all the tags of the system. Each service should rather own a minimal set of tags in order to remain functional.

We note $\prec_O$ (pronounced "can flow to, given O") the new partial order including the concept of ownership. So that an information can flow from A labeled with L_A to B labeled with L_B given the ownership O, given (1) the secrecy tags of S_A should be included in S_B except for the secrecy tags of O; and (2) the integrity tags of I_B should be included in I_A except for the integrity tags of O. We formally define $\prec_O$ as:

$$L_A \prec_O L_B \text{ iff } S_A - O \subseteq S_B - O \text{ and } I_A - O \supseteq I_B - O,$$

where $L_A = (S_A, I_A)$ and $L_B = (S_B, I_B)$

Practically, a service A with ownership O_A can send a message to a service B with ownership O_B if and only if $L_A \prec_{O_A} L_M \prec_{O_B} L_B$. For example, a TPA will not be given any ownership and as a consequence will not be able to omit restrictions imposed by its label, e.g., it will not be able to bypass the secrecy tag of a user in order to leak her private data. On contrary, the proxy will be given ownership of the secrecy user tag in order to be able to communicate simultaneously with several CE devices requesting data labeled with different secrecy tags. We consider the proxy as providing a high security level and therefore we trust it to use its ownership in a secure manner.

In order to express a new security concern, during runtime a service can create and own a new tag. At its discretion, it can grant the ownership to other services. For each new user U, the proxy generates a new secrecy tag u_s and grants it to the HU, so that the HU can label and protect the private data of U.

Dynamic Label Assignment (DLA): A DLA is an explicit request from a service A to another service B to increase the label of B with a new tag. As mentioned in the Section 2.4, we consider here a TPA enclosed in an isolated cell on the HU. Like for a "black box", the HU can only monitor the inputs and outputs of the isolated cell and therefore of the TPA. At first the TPA is empty labeled without any ownership and thus cannot receive any sensitive, i.e., secrecy-labeled, information or contact integrity-critical functions. For example, in order to exchange private data of the driver d, the HU, which owns d_s, imposes per DLA the TPA to extend its label with d_s. The TPA cannot take the tag d_s out of its label later and is therefore only able to send messages to services including d_s in their label or ownership. The TPA is label unaware and does not manage its own label. Instead, a trusted dedicated service of the HU is in charge of it and filters all inputs/outputs of that TPA.

Automotive DIFC Architecture. We chose to monitor network communications and not every process of the ECU in order to not suffer from a too big overhead and to limit the risk of errors. Services are isolated from each other in their own address space or are on different ECUs. Applications of a same service share the same security concerns and the same middleware layer. We therefore chose to label them together and to rely on the middleware layer to enforce the label-based conditions. As shown in Figure 5, applications in different services communicate through their respective middleware. The Secure Channel Manager and the labeler are part of the *Communication Layer* of the SME. The Secure Connection Manager provides the logic and protocol implementation to establish secure communication channels. The labeler extends the header of each message with a field for the label and enforces the

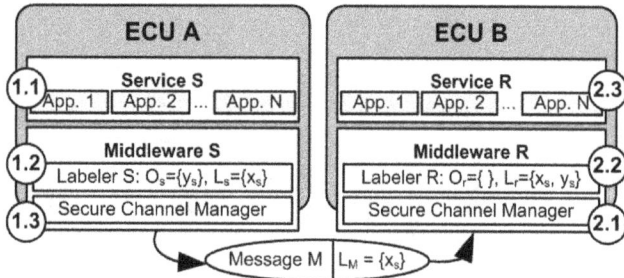

Figure 5. DIFC-based communication between 2 services R and S. L and O are the label and ownership of a service. x_s, y_s, and x_i are secrecy and integrity tags.

1.1 - App.1 of service S sends the message M.

1.2 - Middleware S labels M such as $L_S < L_M$.

1.3 - Middleware S sends the labeled message M.

2.1 - Middleware R receives the labeled message M.

2.2 - Middleware R checks whether $L_M < L_R$.

2.3 - App. N of service R receives the message M.

	Shadow memory, taint of		
	x	y	z
0: init();	(0,0)	(0,0)	(0,0)
1: buffer **x** = receiveBuffer(from_a_ECU);	(1,1)	(0,0)	(0,0)
2: buffer **y** = readBuffer(from_sensitive_file);	(1,1)	(2,2)	(0,0)
3: buffer **z** = proccessBuffers(**x**, **y**);	(1,1)	(2,2)	(1,1)or(2,2)
4:	(1,1)	(2,2)	(1,1)or(2,2)
5: write(**z**, in_output_file);	(1,1)	(2,2)	(1,1)or(2,2)
6: sendBuffer(**z**, to_another ECU);	(1,1)	(2,2)	(1,1)or(2,2)

Figure 6. Example of code with data dependencies (left side – in bold, the data to taint) and its taint propagation (right side).

partial order $<_O$. The applications of a service are totally label-transparent and are not involved in the DIFC security process. The rest of this subsection provides more information about label assignment and policies.

Label assignment: At first, the label of each service S includes the integrity and secrecy tag (s_i and s_s) characterizing its own security concerns. The assignment of additional tags in the label or ownership are then specified during design phase by the car manufacturer and depend on the use cases the service is involved in. An example of assignment is provided for evaluation in Section 6.1. During runtime, the proxy is the only service able to generate new tags related to new users and to grant them to the relevant ECUs. We do not consider the addition of any new on-board service, after the car left the assembly line.

DIFC label-based policies: These are managed by the *Decision Layer* of the SME. They provide the decisions to enforce in the labeler, e.g., whether to pass a received labeled message to the applications or which label to add to the output messages. Like for the label assignment, these policies are defined by the car manufacturer during the design phase.

4.2. Dynamic Data Flow Tracking

Our second approach for TPA integration makes use of DDFT techniques. Like for a "grey box", this allows us to get an insight of what happens during the execution. These techniques provides an efficient way of detecting data of interest and to track their propagation within a running application.

Tracking and Controlling the Execution. DDFT tools monitor each machine instruction performed by the TPA and detect every system call and every data flow

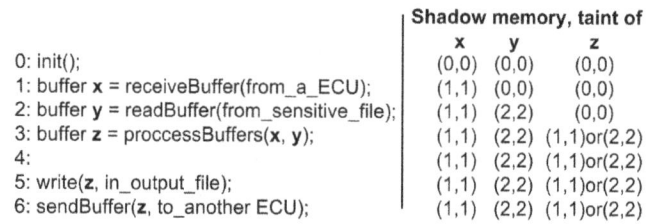

between registers and memory. They usually make use of dynamic binary instrumentation (DBI) frameworks like Intel's Pin [36] in order to inject custom code within the execution for a policy enforcement. They can therefore raise a warning or stop the execution in case of an application behaving in contradiction with one of their policies. When they are configured properly, they allow to eliminate numerous attacks related to stack pointers overwriting, e.g., buffer overflow [37], string format [38] and return-oriented programming [39] attacks. In order to explain how DDFT tools proceed, we focus on the following 3 points: taint sources, intra-taint propagation and taint sinks. For the rest of this section we consider the pseudocode provided in the left part of Figure 6.

Taint sources are programs or memory location, through which data enter the application. If recognized as data of interest, they are tainted in the shadow memory. The shadow memory is a mapping between the actual memory of the application and its taints. In our scenario, we identify as sources every traditional input/output channel potentially used by the TPA, e.g., inter-process communications (like pipes), filesytem and network socket. Concretely, the DDFT tool monitors the functions "receiveBuffer()" (line 1) and "readBuffer()" (line 2) and taints their returned buffers "x" and "y" accordingly.

Taint propagation: All along the application execution, tainted data are tracked, while they are altered and processed. An example of taint propagation is given in the right part of Figure 6. In our example, the function "processBuffers()" (line 3) produces out of two tainted buffers the buffer "z" that should be tainted as well. Originally, DDFT tools were about detecting security attacks and a simple binary tainting (i.e., one bit of the shadow memory tainting a byte of the real application memory) was enough to detect whether untrusted data were overwriting critical parts of the stack. However, our scenarios consider data of different sensitivity that require several taint values. We therefore propose to taint the data according to the STL taxonomy introduced earlier. For reminder, the STL does not consider data integrity, but only considers the security of the

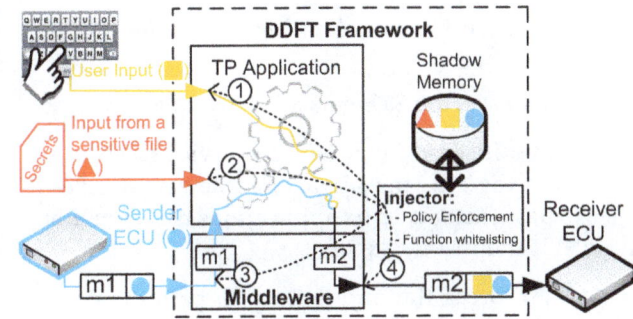

Figure 7. Overview of the DDFT framework in the on-board network. The colored shapes represent different levels of sensitivity, that are expressed by the taint values (i.e., yellow square (1), red triangle (2), blue round(3)). These taints are injected using binary instrumentation (*Injector*). The *Injector* monitors the execution, especially system calls (dotted lines) and the taint propagation between memory and registers. $m1$ and $m2$ are tainted messages sent respectively to and from the TPA. The TPA output $m2$ shows a combination of the sources "round" and "square" but not "triangle" and is therefore tainted accordingly.

communication and the trust of the external communicating entity. Since the car manufacturer does not control the development of the TPA, every output of the TPA has to be considered as potentially dangerous to process.

Taint sinks are programs and memory locations, where the presence of a taint is checked and where a policy may be enforced. The policies mostly concern the decision about passing the data to a function or using the data as program control data, e.g., return address. For our example, it concerns the fact of writing "z" in a file (line 5) or sending it (line 6).

Car-wide Taint Propagation. DDFT locally ensures security and can detect a security attack or a privacy leakage. For its integration in our car-wide security framework, our DDFT tool instruments the middleware layer as well and takes an active part in the in-band middleware protocol.

The Middleware-based taint propagation is presented in Figure 7. System calls related to the network socket management are monitored and intercepted by the Injector. The header of the incoming message is scanned, the STL field is extracted and its value is used as taint value for the data payload in the shadow memory (bullet 3). On the other hand, before being sent out, the outgoing message gets injected a new STL value relative to the sensitivity of the whole payload (bullet 4). The STL value is determined from the shadow memory by the Injector and carries the most relevant STL found, i.e., the highest SL and TL found. In addition the DDFT tool checks if the TPA is authorized to communicate with the remote service over a dedicated secure channel.

Middleware enforcement: After receiving a TPA message, the middleware layer of the on-board service extracts the STL field and enforces the appropriate policy. The middleware implementation of the service is static and cannot enforce a different policy for each new TPA. Instead, the service middleware relies on the DDFT framework and trusts it to have authorized the communication and to have provided an accurate STL value. Then based on DDFT/DIFC generic policies, applicable to all TPAs, the middleware decides to pass the payload to the aimed application. More details about the DDFT/DIFC policies, linking the STL of the DDFT engine to the DIFC model, are provided in Section 4.3.

DDFT Policies: We identify 2 types of DDFT policies, both of them enforced in the DDFT framework:

a) *Static polices:* Such rules are embedded in the DDFT framework and enforced for all TPAs. They list the authorized services, with which a TPA can communicate. They provide the taint propagation rules (intra- and middleware-based) and the rules related to the taint sinks and taint sources (e.g., how to taint keyboard inputs, which tainted files can be read/overwritten). These rules are static and defined by the car manufacturer at design time.

b) *Dynamic policies:* These rules allow a better customization of the permission given to a TPA and specify additional services or files that can be accessed. They are specific for one application and loaded in the DDFT framework like a Android rule set during the TPA installation. The rule set is evaluated against the existing static rules and needs to be approved and signed by the car manufacturer. Other privacy-relevant policies and exceptions may be directly specified by the driver thanks to on-board configuration interfaces and displays of pop-up windows.

4.3. Coupling DIFC and DDFT

The DDFT/DIFC interface concerns the middleware layers of the services having direct communications with the TPA. It gave them a way to interpret the received STL taint from the TPA based on their DIFC label. Like the DIFC approach, the applications of a service are unaware of this interface.

The TPA and the DDFT framework are not part of the DIFC model and therefore not assigned any label. It allows the TPA to receive information from the whole car without any constraint. The DDFT engine provides accurate STL taints, which gives a precise idea of the output sensitivity. For example in Figure 7, even if the TPA gets as inputs highly sensitive data from a file, the output STL indicates that the payload was processed only from data of lower sensitivity and should be processed accordingly. In comparison to the

DIFC approach, DDFT allows the TPA to be more functional, even when handling critical data.

For this DIFC/DDFT approach, the DDFT is DIFC label-unaware and only receives messages with STL taints. It allows us to keep the DDFT tool simple, efficient and generic and not to worry about the different specificities of all car models. Then, the TPAs are most likely to receive information from on-board services and the outside, process them and directly communicate with the outside. The amount of traffic from the TPA to on-board services will remain minor. As a consequence, the security should be based on a taxonomy oriented towards a secure information release with the outside, like the STL.

Due to the limited number of taints and the privacy risk, the data of only one user should reach the TPA. Therefore all monitored TPA are assigned one user identity (ID). Like for the user tags, these IDs are defined and distributed by the proxy. For each message exchanged between a service and a TPA, a STL and a user ID are added to its header. The DDFT tool filters inbound messages based on the provided ID. The middleware of the concerned service can easily characterize whose privacy is concerned.

Proxy interface: Most sensitive outputs of the TPA will aim toward the outside. In a same manner as explained in Section 3.3, the proxy enforces a STL-based filtering. In addition to the STL considerations, the proxy makes sure that the ID joint to the STL is appropriate for the communication, e.g., the ID of a user U communicating with U's CE device. The proxy provides therefore 2 types of filtering: (1) STL-based for communications with the TPA and (2) DIFC-based for communications with on-board services.

Service interface: All on-board services can send data to the TPA. Their middleware just provide the right user ID (if these data are private) and a suitable STL value. The DIFC labels are enforced to create information flows respecting their integrity and confidentiality. Since the integrity of the TPA outputs cannot be assessed, the transition between DIFC label and STL taint only focus on the information confidentiality.

For messages flowing from an on-board service to a TPA, the sent STL value depends on the secrecy labels of the service:

- for a label involving tags expressing a high secrecy for the car manufacturer, the STL gets a TL=3. The SL is not relevant since the data will not leave the car. The list of high-secrecy tags is defined by the car manufacturer and available in each service interface.

- for a label involving tags expressing the secrecy of a user, but not expressing a high secrecy for the car manufacturer, the STL getss a TL=2 or 1 depending on their privacy level. The SL depends on the user's

settings since it is her own data, but as a default value a SL=2 is advised.

- in any other case, the STL gets a TL=0 and a SL=0.

For the opposite case, i.e., when the service receives a message from the TPA, the service middleware decides whether to pass the data to its applications based on the received STL and its own label:

- a STL=(*,3) forces the middleware to pass the data to an application handling highly confidential data that cannot leave the car. Thus an application having a user secrecy tag in its label is not able to receive such data. An authorized service should also include high secrecy tag of the car manufacturer in its label.

- a STL=(*,1) or (*,2) forces the middleware to pass the data to an application handling the private data of a user. Therefore a service with a user's secrecy tag corresponding to the received ID field should be able to get the data.

- a STL=(*,0) indicates that the data are not sensitive and can be passed to all kind of applications.

These last rules do not really consider the SL part of the STL. If sent to the outside, a communication from a service has to go through a DIFC-based enforcement at the proxy level, which is already statically configured. The SL is more relevant for unknown and dynamic cases where security has to be evaluated and configured on-the-fly, like with the TPA.

5. Implementation

This section describes our prototypical implementation combining a middleware-based DIFC enforcement to a DDFT engine.

5.1. The Middleware

As basis of our implementation, we chose to make use of the C-version of the middleware *Etch* [10]. *Etch* is an open-source software project under the Apache 2.0 licence and is considered as a serious candidate for the automotive purpose [13]. Our middleware copes with two types of enforcement: the first one related to the DIFC model and the second one to the DDFT monitoring. We therefore developed two middleware versions. The DIFC version extends the serialization of the middleware header with two fields of 15 bytes, one for the secrecy label, the other for the integrity label. For the DDFT version we extended the header with a 2 integer fields including the 2 values of the STL , i.e., the SL and the TL. The DIFC version is used between all on-board services, while the DDFT one is only used for communication involving the TPA. As a consequence, services communicating with the TPA are aware of the 2 types of header serialization. Like any traditional middleware, the payload serialization and security enforcement are separated from the application logic. *Etch* allows to precise the label of a service and the authorized taints through an adapted interface description language (IDL) and provides an automatic code generation for the enforcement of DIFC and DDFT policies.

We developed a communication proxy similar to the one presented in our previous work [9]. The proxy provides two secure communication interfaces: an external one using the TLS protocol with CE devices and online services and an internal one for on-board services establishing IPsec communication channels. Internal and external communication partners communicate over a mirror-service of the proxy, which makes the communication decoupling totally transparent for both of them. The proxy is application unaware. It enforces a message filtering based on the labels or taints present in the middleware header of all outbound messages. For inbound message, the proxy adds in the middleware header the corresponding labels or taints, depending on the communication target, i.e., an on-board service or a TPA. The proxy determines the user identity for the establishment of the related user tags and ID values based on the TLS certificate provided by the user's CE device or based on other security credentials that an online service can provide.

5.2. Isolation Cell

Regarding our first approach in Section 4.1 for integration of TPAs, we make use of the XEN® hypervisor 4.2 [40], that we set up on the HU. We run the trusted HU middleware and applications, which are developed by the car manufacturer, in the most privileged domain, called Dom0. The untrusted TPA runs in an unprivileged cell, called DomU. Communications between Dom0 and the DomU occur over a virtualized bridge. The XEN environment enforces a complete isolation of the DomU otherwise. The TPA runs on top of a label-unaware *Etch* middleware. The middleware of the HU service in Dom0 is therefore able to receive both labeled and unlabeled middleware header. The HU service acts like a forwarder and enforces DIFC policies for all traffic going to and coming from the TPA.

5.3. The DDFT Engine

Regarding our second approach in Section 4.2, we make use of the DDFT framework *libdft* [8]. *Libdft* relies on the Intel's Pin for DBI. This tool provides relatively good performance in comparison to other DDFT engines [33] and a well-defined API for a customizable security enforcement. More than just using this framework, we extended its expressiveness and its taint propagation mechanisms in order to deal with the 16 values of

the STL. Originally, one byte of memory was tagged with one bit in the shadow memory, it is now one byte tagged by 4 bits. We limited our choice to 16 values in order to keep the size of the shadow memory reasonable and the taint propagation mechanisms efficient. We extended the *libdft* framework with the possibility to differentiate user inputs, i.e., from the keyboard, from file inputs. For the file management, we implemented in *libdft* a system of whitelist, which specifies which file can be accessed by the TPA in reading or writing and for which taint values. We developed a network system call monitoring able to scan every incoming message, extract the taints from the middleware header and taint the data accordingly. In a same manner, the framework can now detect the function calls which send network messages and automatically inject them with the suitable taint values of their payload.

5.4. Testing Environment

We performed this prototypical implementation and the experiments presented in Section 6.2 on three computers: the CE device, the proxy and the HU. They are interlinked with a Gigabit Ethernet and are running a standard 32-bit Fedora Linux on an Intel Atom N270 (1.6 GHz) with 1 GB or RAM. The DomU runs a Debian 6.0 Linux with 256 MB of allocated RAM. While being more resourceful than most embedded platforms of the car, our platforms provide a performance similar to a current HU [41]. Our *Etch* middleware presents a suitable performance, when tested on resource limited microcontroller [13]. Our implementation does not perform extensive modifications of the middleware and therefore should not significantly impact the middleware performance. However this last point should be verified for a more rigorous validation.

6. Evaluation

In order to evaluate our system, we first discuss the security of our concepts and how our system would react during the attack scenarios presented in Section 2.2. Then in the second part, we quantify the overhead of our implementation and discuss the functional requirements presented in Section 2.3.

6.1. Security Evaluation

For this section, we refer to the two attack scenarios defined in Section 2.2. We describe for our two security approaches – isolation and DDFC – how our system would react and which threats can be stopped.

First Approach – DIFC & Isolation. Figure 8 presents an example of label distribution and helps us to understand how DIFC can secure on-board communications and the integration of TPAs and CE devices. The CE device connects to the proxy and gets a direct access to

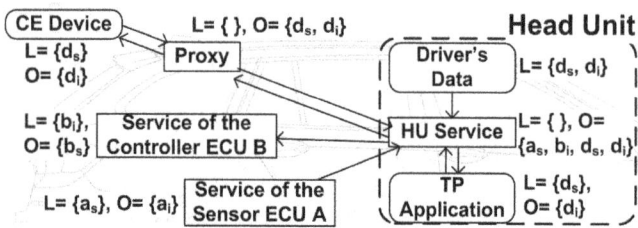

Figure 8. Security scenario. Rectangular boxes represent services running on an independent IFC aware middleware. Round boxes represent IFC unaware applications, devices and files. Solid arrows represent middleware-based communications.

on-board services and an access to the TPA through the intermediary of a HU service. The TPA, always through the intermediary of a HU service, can access the driver's data stored on the HU, receives data from ECU A and triggers mechanisms of ECU B.

Our DIFC model does not propose any service hierarchy, instead all services are distrustful with each other. As a consequence, a successful attack or bug will have a limited impact and will compromise only the tags of the affected services. In our scenario, the CE device of the driver d is authenticated by the proxy, which binds the device to the tags d_s and d_i. The proxy afterwards grants the HU with the ownership of the new tags. The HU performs a DLA of these tags with the TPA, so that the latter can access the driver's data. CE device and TPA cannot be trusted to enforce DIFC rules, that's why their middleware is DIFC unaware and we rely on the HU service and proxy to enforce security.

The TPA is confined to the two driver's tags d_s and d_i. The presence of d_s in its label allows the TPA to get access to the driver d's data on the HU. The ownership of d_i allows the TPA to write on the d's data. The presence of d_i in its label would constrain it to receive data labeled with d_i and would prevent it from accessing for example non-sensitive configuration files. A label with d_s but without d_i would limit it to a "read-only" access. The d_s labeling forces a malicious TPA to only communicate with the d_s-labeled CE device (i.e., belonging to the driver d) and prevents it from communicating with other peers. Because the CE device is bound to one person and labeled with the tags of one user, we do not allow the TPA to be labeled with several user tags.

In a same manner, a CE device is limited to one user and gets access to the data of one user and to non sensitive data. Sensitive data, i.e., from a service labeled with a secrecy service tag, like a_s for ECU A, are blocked by the proxy and are not forwarded to the CE device. Like the TPA, the CE device cannot trigger mechanisms of services with an integrity service tag in their label, like b_i for ECU B. The proxy is the only on-board entity which generates new user tags and grants

their ownerships. The proxy is empty labeled in order to always be able to communicate with several CE devices and other online services.

The labels provide also an efficient way to constrain information flows between on-board services. The HU service can receive information from ECU A if and only if it has the tag a_s in its label or ownership. As a consequence, even if a message from ECU A, labeled with a_s, is forwarded by a multicast address. The proxy will never send such message out and no unauthorized service, i.e., without the tag a_s, will process it. On the other hand, since the HU service owns the tags a_s, it may occasionally forward a_s-labeled data to the TPA. The (safety) mechanisms of ECU B will only be triggered by authorized services, i.e., services having b_i in their label or ownership.

Approach conclusion: No service is fully trusted and owns all the tags of the system. Label, ownership and DLA allow services to express their security concerns for integrity and secrecy The on-board services rely on their remote on-board communication partners to enforce the right DIFC policy. If we consider our integrity attack scenarios, the TPA is isolated in a cell and cannot disturb the HU functioning. The label enforcement of the TPA and of the CE device are performed by the HU service and the proxy, respectively. The exchange of labeled messages allows the communicating services to determine if the message/its source have a sufficient integrity level in order to be processed or trigger a mechanism. Regarding the confidentiality scenario, the labeling of the CE device and the TPA prevents them from accessing highly sensitive data. In case of a bug causing a privacy issue, the labeling of the messages allows the proxy to detect an unauthorized communication and to block it.

Second Approach - DIFC & DDFT. This second approach differs only in the way we monitor the TPA. Communications between services and with CE devices are secured by the same methods as previously, except that the communications with the TPA benefit from a new security enforcement. Both our attack scenarios feature a TPA presenting exploitable vulnerabilities. A well configured DDFT can detect such attacks and stop the application before it harms the car or a user's privacy. The rest of this section reasons directly with our attack scenarios.

Integrity attacks: The DDFT framework is configured to detect every system call which involves exchanges of information with the outside. It detects in particular inter-process communications, e.g., with critical HU processes, shared memory and filesystem access and can block all of them. The DDFT framework only authorizes certain network communications and file accesses. Authorized and non critical files are whitelisted by the DDFT and their access is monitored. Therefore the HU functioning cannot locally be disturbed, the DDFT framework can even enforce policies limiting its resource consumption, e.g., against denial of service attacks. At a remote level, a communication with another service is only possible if it is authorized by the rule set of the TPA and if the service has a policy authorizing the communications with a TPA. Therefore a TPA will not be able to reach critical functionalities, e.g., from the brake controller, and disturb the car functioning.

Confidentiality attacks: These attacks mostly concern the release of sensitive information to the outside and have to go through the proxy. The DDFT framework whitelist the file that the TPA can access and makes sure that it does not access files containing the private information of a user it is not assigned to. The services which communicates with the TPA specify in the middleware header the STL of the data and the ID of the user, whose privacy is concerned. A TPA can be sent all information, the DDFT is trusted to propagate the STL taint and to check that the ID suits the user it was assigned. If the TPA tries to directly send some data through the proxy, the DDFT engine injects a STL and ID field in the payload. The proxy can then ensure that the addressee is the one specified in the header and that the STL condition is fulfilled. The situation is a bit more complex if the TPA attempts to release the data through another service communicating with the outside. That is why a service dealing with the private data of a user, i.e., having a user's tag in its label or ownership, will refuse a message with a TL=3. Only services not communicating with the outside may receive data with TL=3, i.e., services with secrecy service tags reflecting a high secrecy for the car manufacturer. The processing of data tainted with a TL=1 or 2 will depends whether the service can handle private information, i.e., whether the service has the user's secrecy tag in its label or ownership. The decisions about whether to process and which SL to use later are defined by the car manufacturer and statically set up in the middleware of the service. As a consequence, sensitive data coming from the TPA are processed by services respecting the concerns expressed by the STL and allowing a privacy-aware release with the outside.

Approach conclusion: This second approach combines the advantages of DIFC for on-board and CE device-based communications and provides more granularity as for the integration of TPA. The TPA accesses more data, even the confidential ones and is still able to communicate with untrusted partners if the message does not involve sensitive information.

Limitations and potential solutions. In Section 2.2, we assumed the integrity of the middleware/OS of every ECU and the proxy. However secure boot and remote attestation do not protect or detect runtime attacks, which can be very harmful if they compromise critical ECUs like the proxy or the HU. Other runtime intrusion detection systems have to be considered as well [42]. Such security tools perform scans of critical data structures and recognition of instruction patterns within a running platform. They generally significantly degrade the system performance and should be used in a carefully selected manner. Successful attacks on the proxy could be mitigated by its compartmentalization thanks to hypervisor or microkernel techniques. Even if an isolated cell of the proxy gets compromised, the proxy detects it and shuts down the cell without impacting the other genuine cells.

Regarding the integration of a TPA in an isolated cell, we mentioned that they can only be used by a unique user. Several users, willing to use simultaneously a TPA, may require the car to assign a virtual machine per user in order to preserve their privacy. This solution may be too resource costly. The DDFT approach, on the other hand is less heavy. It can monitor several applications for different users and isolate them from each other.

Partial security conclusion. Unlike OSs like Android, which control their applications with a set of coarse permissions, DDFT allows a fine security granularity. More than just isolating, DDFT stops the TPA, before the attacker takes control of it. Unlike the isolation solution, TPAs monitored by DDFT are more functional and several applications can be used by several users. DIFC/DDFT is therefore a better solution from a pure security point of view.

6.2. Functional Evaluation

Section 6.1 justifies that the DDFT approach is providing a deeper monitoring and a better flexibility than the isolation approach. This section assesses our implementation overhead and determines whether this choice can be corroborated from a functional point of view.

We measure the middleware throughput (in call/sec) between an application of the CE device and an on-board TPA for a scenario similar to the one presented in Section 6.1. We performed our tests with different security features and in various situations in order to demonstrate the overhead caused by our two approaches for TPA integration. Benchmarks are run on three separated machines as described in Section 5: the HU, the proxy and the CE device. The CE device application sends a simple *Etch* message including one integer to the TPA. Before arriving to the TPA, the message goes through the intermediary of the proxy and then a service of the HU, where DIFC can be enforced. After reception, the TPA retrieves a series of integers from a file on the HU. Based on the received and read integers, the TPA generates a buffer and sends

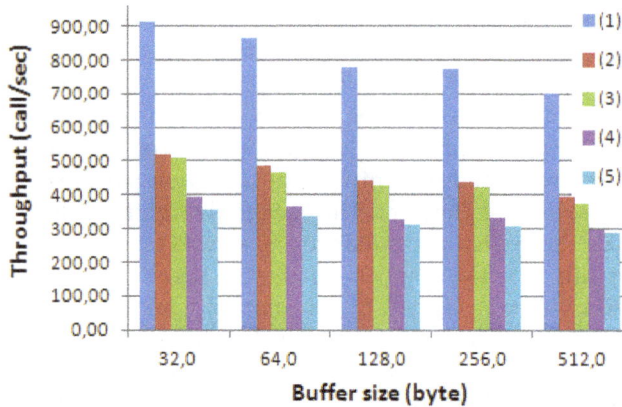

Figure 9. Middleware throughput average for various buffer sizes and security modes enabled.

Mode Description:
(1) No security feature is enabled
(2) The encryption between CE device, proxy and HU is enabled.
(3) DIFC is enforced between proxy and HU + (2).
(4) The TPA runs in a Virtualization cell + (3)
(5) The TPA is DDFT monitored + (3)

Table 2. Normalized middleware throughput performance of the scenario presented in Section 6.2. These results are computed from the throughput averages of the different modes. The mode of this table are similar to the used for Figure 9. Factor (i) presents the normalized performance with (1) as reference, while factor (ii) takes (3) as reference.

Enabled Mode	(1)	(2)	(3)	(4)	(5)
Factor (i)	1	0.57	0.55	0.43	0,40
Factor (ii)	-	-	1	0.78	0.73

it to the CE device through the same inverse path, i.e., through the HU service and then the proxy. We vary the size of the buffer in order to stress the middleware and the taint propagation mechanisms. In this scenario the TPA plays the role of an infotainment server, e.g., for music or picture. Our first set of measurements is performed without any security features enabled (1) and we use this case as reference. We then did the same while encrypting the communication channel (2), i.e., by using TLS for the link CE device–proxy and IPsec for the proxy–HU. After which, in addition, we enforced DIFC between proxy and HU (3). The measurements of (3) gives a lower bound overhead caused by the communication security without any TPA consideration. The case (4) uses the same security features as (3) and monitors the TPA with our DDFT framework and the HU service, which enforces the DDFT/DIFC coupling rules. The measurements of (4) allow to determine the overhead imposed by a DIFC/DDFT framework and our custom taint propagation mechanisms. Finally the case (5), like (4), uses the security of (3), but instead of DDFT, the TPA runs in an isolated XEN cell and the HU service enforces DIFC with the TPA. The measurements of (5) allow to evaluate the impact caused by DIFC and the isolation cell. The throughput results of Figure 9 present the average of ten measurements of 5000 calls each.

Discussion: After normalization of the results of Figure 9, we realized that the performances for different enabled security features and regardless of the buffer size are proportionally similar. The normalized

average throughput can be found in Table 2. First this table shows that the use of security protocols like TLS and IPsec is responsible for the biggest impact (~43%). The processes for encryption/decryption and generation/verification of security fields like HMAC are extremely costly, when used with a high throughput system. A second less consequent performance decrease (~4%) is due to the enforcement of DIFC between proxy and HU. For each packet, proxy and HU service checks whether the integrity and secrecy labels are valid for the invoked function. The last performance impact is caused by the security feature securing the TPA. For the isolation case, the performance decreases by 27% and for the DDFT by 22%. In the first case, it is due by the virtualization of the network bridge and the OS, in the second case by the function call monitoring and the taint propagation mechanisms. As a conclusion, for our scenario isolation and DDFT presents a similar performance and could be used for infotainment use cases involving a CE device and requiring a moderate bandwidth (1,18 Mbit/sec for the DDFT and 1,23 Mbit/sec for the isolation). Our experimentation involves a communication link between HU service and proxy, this link is used here to show the impact of DIFC on our scenario and because the isolation approach requires an intermediary step on a HU service. However with our DDFT approach, a TPA can directly communicate with the proxy and reaches a bandwidth of 2,14 Mbit/sec when using DDFT and encryption.

However, our evaluation is only focused on the middleware level in a small 3-node network and for a specific scenario involving simple TPAs and CE-based applications. Tests performed with *libdft* for bigger applications such as a web-browser [8] or a MP3-player [33] have shown a more significant latency (up to 28 times slower), whereas the isolation cell tends to provide a constant performance, independently from the application complexity. As a consequence, DDFT should only be applied for simple application handling private data and requiring a lot of interaction with external partners. DDFT should besides optimize its use of trusted libraries, i.e., libraries, which are not

Table 3. Summary of our system evaluation with respect to the attack scenarios of Section 2.2 and the requirements of Section 2.3 (More details are provided in Section 6).

	1 - Approach DIFC/Isolation	2 - Approach DIFC/DDFT
Monitoring paradigm	system oriented	single-process oriented
Integrity scenario		
• Protection of the HU from a TPA	Yes, by isolation	Yes, by monitoring
• Protection of other ECUs from a TPA	Yes	Yes
• Protection of the ECUs from external attackers	Yes, same DIFC-based mechanisms for both approaches	
Secrecy scenario		
• Protection of the user's data from a TPA	Yes	Yes, more fine-granular
• Privacy-aware data release, with a TPA	Yes	Yes, more dynamic/flexible
• Protection of the user's data from external attackers	Yes, same DIFC-based mechanisms for both approaches	
Requirements		
• Performance	Partially, independent of the application complexity	Partially, dependent on the application complexity
• Solutions for all platforms	No	Yes
• No regular updates necessary	Yes	Yes
• No financial extra cost	Yes, software-based solution	Yes, software-based solution
• No complexity increase for developers and users	Yes, the communication decoupling of the proxy makes users and developers security-unaware	

monitored. The use of the isolation approach can be used for a whole system and its TPAs, e.g., an Android partition. Additional investigations involving larger networks generating more traffic are strongly recommended for real-world validation.

About some functional requirements: This paragraph offers a brief evaluation of our system based on the requirements defined in Section 2.3. Our implementation partly fulfills our performance requirement, it does better than low-speed communication like CAN but is far from the high-speed one like MOST. For bandwidth demanding use case, like for audio/video streaming, other protocols like IEEE1722 [43], which skips the use of UDP and IP, are more efficient. Our security framework has been implemented with an efficient automotive middleware, but the testing with resource limited platform should be performed. Unlike the DDFT engine, the security solution involving isolation and virtualization is resource-costly and therefore may not be suitable for resource-limited platforms. Then security was designed in an engineering-driven way. Enforcing security in the middleware allows the security engineering team to not know in detail the application logic and to focus on abstracted flows of information. Finally with our proxy-based architecture, developers of CE device application are totally security-unaware and are not imposed any (security) protocol, as far as the proxy provides a suitable interface and can compute the STL.

Before giving our conclusion, Table 3 proposes a short summary of the evaluation of our two approaches.

7. Conclusion

In this paper, we presented a security architecture for next-generation automotive on-board networks, which combines different information flow control techniques. Our system proposes two levels of security. Locally very untrusted components, like TPAs, are monitored thanks to a custom DDFT engine. DDFT allows to fully control the TPA behavior without modifying the binary executable: At the network level, our DIFC model allows every on-board service to independently express its security concerns and to trust the remote on-board communication partner to respect it. For this purpose, the security is ensured by the middleware. This layer is in charge of the network communications, is strictly separated from the application logic and allows the on-board exchange of security metadata. The middleware is also used as a glue between DIFC and DDFT and provides the translation interfaces between the two models. A Proxy-based architecture, on the edge of the embedded network, filters inbound and outbound messages and allows an on-the-fly evaluation of the security and trust level of all external communications. The taxonomy used to evaluate this security and trust level is also directly used by the DDFT engine and by the service middleware. More than just providing the monitoring of one application, our security framework proposes a DIFC-based monitoring approach of a whole system thanks to isolation and virtualization techniques. With respect to the integrity and confidentiality attack scenarios in Section 2.2, our proposed framework successfully prevents those attacks and offers a light-weight and easy-to manage middleware-based security and privacy framework by using the aforementioned techniques. It also meets almost all architecture requirements from Section 2.3. More than giving a recommendation about which solution to use, we see our 2 approaches for TPA integration as complementary; the choice for an approach should be carefully evaluated and dependent on the application complexity and the requirements for performance and security flexibility. However, while enhancing the in-car security and privacy, these solutions have shown some performance limitations. More rigorous validations therefore require additional investigations in order to determine a suitable tradeoff between acceptable performance and security enforcements on resource-limited hardware.

References

[1] Elliott, M.-A. (2011) The Future of Connected Cars. http://mashable.com/2011/02/26/connected-car/ (accessed on 14 march 2013).

[2] Lutz, Z. (2011) Renault debuts R-Link, an in-dash Android system with app market http://www.engadget.com/2011/12/09/renault-debuts-r-link-an-in-dash-android-system-with-app-market/ (accessed on 14 march 2013).

[3] Koscher, K., Czeskis, A.,Roesner, F., Patel, S., Kohno, T., Checkoway, S., McCoy, D., Kantor, B., Anderson, D., Shacham, H. and Savage S. (2010) Experimental Security Analysis of a Modern Automobile. In *Proceedings of the 2010 IEEE Symposium on Security and Privacy* (Washington DC: IEEE Computer Society), 447–462. doi:10.1109/SP.2010.34

[4] Checkoway, S., McCoy, D., Kantor, B., Anderson, D., Shacham, H., Savage, S., Koscher, K., Czeskis, A., Roesner, F.and Kohno, T. (2011) Comprehensive Experimental Analyses of Automotive Attack Surfaces. In *Proceedings of the 20th USENIX conference on Security* (Berkeley: USENIX Association), 6–6.

[5] Rouf, I., Miller, R., Mustafa, H., Taylor, T., Oh, S., Xu, W., Gruteser, M., Trappe, W. and Seskar, I. (2010) Security and Privacy Vulnerabilities of In-car Wireless Networks: a Tire Pressure Monitoring System Case Study. In *Proceedings of the 19th USENIX conference on Security* (Berkeley: USENIX Association), 21-21.

[6] Slivka, E. (2012) Apple Pulls Russian SMS Spam App from App Store. http://www.macrumors.com /2012/07/05/apple-pulls-russian-sms-spam-app-from-app-store/ (accessed on 15 March 2013).

[7] Glass, M., Herrscher, D., Meier, H., Piastowski, M. and Shoo, P. (2010) SEIS - Security in Embedded IP-based Systems. In *ATZelektronik worldwide*, 2010-01, 36–40.

[8] Kemerlis, V., Portokalidis, G., Jee, K. and Keromytis, A. (2012) libdft: Practical Dynamic Data Flow Tracking for Commodity Systems. In *Proceedings of the 8th ACM SIGPLAN/SIGOPS conference on Virtual Execution Environments* (New York: ACM), 121–132. doi:10.1145/2151024.2151042

[9] Bouard, A., Schanda, J., Herrscher, D., Eckert, E. (2012) Automotive Proxy-based Security Architecture for CE Device Integration. In *Procceedings of 5th International Conference Mobilware on Mobile Wireless Middleware, Operating Systems, and Applications* (Heidelberg:Springer-Verlag), 62–76. doi:10.1007/978-3-642-36660-4_5

[10] Homepage of *Etch*. http://incubator.apache.org/etch/ (accessed on 15 March 2013).

[11] Hoppe, T., Kiltz, S., Dittmann, J. (2008) Security Threats to Automotive CAN Networks & Practical Examples and Selected Short-term Countermeasures. In *Proceedings of the 27th international conference on Computer Safety, Reliability, and Security* (Heidelberg: Springer-Verlag), 235–248. doi:10.1007/978-3-540-87698-4_21

[12] Maier. A. (2012) Ethernet - The Standard for In-car Communication. In *2nd Ethernet & IP @ Automotive Technology Day.* http://www.ethernettechnology day.com/downloads/18 _Alexander_Maier_-_BMW.pdf (accessed on 15 March 2013).

[13] Weckemann, K., Satzger, F., Stolz, L., Herrscher, D., and Linnhoff-Popien, C. (2012) Lessons from a Minimal Middleware for IP-based In-car Communication. In *Proceedings of the IEEE Intelligent Vehicles Symposium 2012* (Washington DC: IEEE Computer Society), 686–691.

[14] Bouard, A., Glas, B., Jentzsch, A., Kiening, A., Kittel, T., tadler, F., and Weyl, B. (2012) Driving Automotive Middleware Towards a Secure IP-based Future. In *Proceedings of the 10th ESCAR Embedded Security in Cars Conference.*

[15] Clarke, E. M., Grumberg, O. and Peled, D. (1999) Model Checking. MIT Press.

[16] Mecklenbrauker, C. F., Molisch, A. F., Karedal, J., Tufvesson, F., Paier, A., Bernado, L., Zemen, T., Klemp, O., Czink, N. (2011) Vehicular Channel Characterization and Its Implications for Wireless System Design and Performance. In *Proceedings of The IEEE Special Issue on Vehicular Communications* (Washington DC: IEEE Computer Society), 99(7): 3646–3657.

[17] Fujitsu Semiconductor Europe (2012) Fujitsu Announces Powerful MCU with Secure Hardware Extension (SHE) for Automotive Instrument Clusters. In Fujitsu Press Release. http://www.fujitsu.com/emea/news/pr/fseu-en_20121129-1044-fujitsu-mcu-secure-hardware-extension-atlas-l.html (accessed on 15 March 2013).

[18] McMillan, R. (2011) 'War Texting' Lets Hackers Unlock Car Doors via SMS. http://www.pcworld.com/article/236678/War_Texting _Lets_Hackers_Unlock_Car_Doors_via_SMS.html (accessed on 15 March 2013).

[19] Homepage of the EVITA project. http://evita-project.org/ (accessed on 15 March 2013).

[20] Homepage of the SeVeCom project. http://www.seve com.org/ (accessed on 15 March 2013).

[21] Homepage of the sim^{TD} project. http://www.simtd.org/ (accessed on 15 March 2013).

[22] Detken, K.-O., Fhom, H. S., Stehman, R., Dietrich, G. (2010) Leveraging Trusted Network Connected for Secure Connection of Mobile Devices to Corporate Networks. In Pont, A., Pujolle, G. and Raghavan, S.V.

[eds] *Communications: Wireless in Developing Countries and Networks of the Future* (Heidelberg:Springer-Verlag). doi:10.1007/978-3-642-15476-8_16

[23] Department of Defense (1983)   Trusted Computer System Evaluation Criteria In *Orange Book*

[24] Myers, A. C., Liskov, B. (2000) Protecting Privacy Using the Decentralized Label Model. In *ACM Transactions on Software Engineering and Methodology* (New York:ACM), 9:410–442.

[25] Efstathopoulos, P., Krohn, M., VanDeBogart, S., Frey, C., Ziegler, D., Kohler, E., Mazières, D., Kaashoek, F., Morris R. (2005) Labels and Event Processes in the Asbestos Operating System. In *Proceedings of the 20th ACM symposium on Operating systems principles* (New York:ACM), 17–30. doi:10.1145/1095810.1095813

[26] Zeldovich, N., Boyd-Wickizer, S., Kohler, E., Mazières, D. (2006) Making Information Flow Explicit in Histar. In *Proceedings of the 7th symposium on Operating systems design and implementation* (Berkeley:USENIX Association), 263–278.

[27] Zeldovich, N., Boyd-Wickizer, S. and Mazières, D. (2008) Securing Distributed Systems with Information Flow Control. In *Proceedings of the 5th USENIX Symposium on Networked Systems Design and Implementation* (Berkeley:USENIX Association), 293–308.

[28] Ramachandran, A., Mundada, Y., Tariq, M., and Feamster, N. (2008) Securing Enterprise Networks Using Traffic Tainting. In *Special Interest Group on Data Communication.*

[29] Zdziarski, J. (2012) Hacking and Securing iOS Applicaitions. (O'Reilly Media, Inc.), chap 13.1 Sandbox Integrity Check. http://my.safaribooksonline.com/book/-/97814493252 13/jailbreak-detection/sandbox_integrity_check (accessed 15 March 2013).

[30] Yin, H., Song, D., Egele, M., Kruegel, C. and Kirda E. (2007) Panorama: Capturing Systemwide Information Flow for Malware Detection and Analysis. In *Proceedings of the 14th ACM conference on Computer and communications security* (New York:ACM), 116–127. doi:10.1145/1315245.1315261

[31] Enck, W., Gilbert, P., Chun, B., Cox, L., Jung, J., McDaniel, P. and Sheth, A. (2010) Taintdroid: An information-flow tracking system for realtime privacy monitoring on smartphones. In *Proceedings of the 9th USENIX conference on Operating systems design and implementation* (Berkeley:USENIX Association), 393–407.

[32] Zavou, A., Portokalidis, G., Keromyitis, A. (2011) aint-Exchange: A Generic System for Cross-Process and Cross-Host Taint Tracking. In *Proceedings of the 6th International conference on Advances in information and computer security* (Heidelberg:Springer-Verlag), 113–128.

[33] Schweppe, H. and Roudier, Y. (2012) Security and Privacy for In-vehicle Networks. In *1st IEEE International Workshop on Vehicular Communications, Sensing, and Computing* (Washington DC: IEEE Computer Society).

[34] Shankar, V., Urbam, G., Sultan, F. (2002) Online trust: a stakeholder perspective, concepts, implications, an future directions.   In *Journal of Strategic Information Systems*, 11(3):325–344.

[35] Ling T. C. et al: Baker & McKenzie - Global Privacy Handbook. IACCM (2012)

[36] Luk, C.-K., Cohn, R., Muth, R., Patil, H., Klauser, A., Lowney, G., Wallace, S., Reddi, V.J. and Hazelwood, K. (2005) Pin: building customized program analysis tools with dynamic instrumentation. In *Proceedings of the 2005 ACM SIGPLAN conference on Programming language design and implementation* (New York:ACM), 190–200. doi:10.1145/1065010.1065034

[37] Levy, E. (Aleph One). (1996) Smashing the Stack for Fun and Profit. In *the Phrack Magazine*, 7(49), chap 14.

[38] Scut, team teso (2001) Exploiting Format String Vulnerabilities. Technical Report, http://julianor.tripod.com/bc/formatstring-1.2.pdf.

[39] Shacham, H., Page, M. and Pfaff, B. (2004) On the Effectiveness of Address-space Randomization. In *Proceedings of the 11th ACM conference on Computer and communications security* (New York:ACM), 298–307. doi:10.1145/1030083.1030124

[40] Barham, P., Dragovic, B., Fraser, K., Hand, S., Harris, T., Ho, A., Neugebauer, R., Pratt, I. and Warfield, A. (2003) Xen and the art of virtualization. In *Proceedings of the 19th ACM symposium on Operating systems principles* (New York:ACM), 164–177. doi:10.1145/945445.945462

[41] BMW AG. Navigation System Professional, http://www.bmw.com/com/en/insights/technology/tec hnology_guide/articles/navigation_system.html (accessed 15 March 2013).

[42] Garfinkel, T., Rosenblum, M. (2003) A Virtual Machine Introspection Based Architecture for Intrusion Detection. In *Proceedings of NDSS Symposium 2003*, Internet Society

[43] IEEE Standards Association (2011) IEEE 1722 - Layer 2 Transport Protocol Working Group for Time-sensitive streams http://grouper.ieee.org/groups/1722/ (accessed 15 March 2013).

16

An efficient geo-routing aware MAC protocol for underwater acoustic networks

Yibo Zhu*, Zhong Zhou, Zheng Peng, Michael Zuba, Jun-Hong Cui

Department of Computer Science and Engineering, University of Connecticut, Storrs, CT 06269, USA

Abstract

In this paper, we propose an efficient geo-routing aware MAC protocol (GOAL) for underwater acoustic networks. It smoothly integrates self-adaptation based REQ/REP handshake, geographic cyber carrier sensing, and implicit ACK to perform combined channel reservation and next-hop selection. As a result, it incorporates the advantages of both a geo-routing protocol and a reservation-based medium access control (MAC) protocol. Specifically, with its self-adaptation based REQ/REP, nodes can dynamically detect the best next-hop with low route discovery cost. In addition, through geographic cyber carrier sensing, a node can map its neighbors' time slots for sending/receiving DATA packets to its own time line, which allows the collision among data packets to be greatly reduced. With these features, GOAL outperforms geo-routing protocols coupling with broadcast MAC. Simulation results show that GOAL provides much higher end-to-end reliability with lower energy consumptions than existing Vector-Based Forwarding (VBF) routing with use of a broadcast MAC protocol. Moreover, we develop a theoretical model for the probability of a successful handshake, which coincides well with the simulation results.

Keywords: geographic cyber carrier sensing, geo-routing, MAC, self-adaptation, underwater sensor network

1. Introduction

Underwater acoustic network is a promising technique that could connect underwater vehicles, sensor nodes, and other devices working in an underwater environment *via* acoustic channels. It can be used to collect oceanographic data and monitor oceanic volcano activity or oil/gas fields [1–3]. Although it is a class of *ad hoc* networks, the routing and medium access control (MAC) protocols for terrestrial *ad hoc* networks cannot serve it. This is because of its long signal propagation delay, narrow channel bandwidth, and high node mobility. These issues also provide challenges in designing efficient routing and MAC protocols for underwater acoustic networks [1–5].

In underwater acoustic networks, traditional routing protocols such as AODV [6] do not work because their costly route discovery process is unsuitable in long-delay underwater environments. Geo-routing protocols, such

as VBF [7], VBVA [8], and DBR [9], are preferred here. These protocols do not need dedicated route discovery and forward packets directly based on the nodes locations. Since location information is indispensable for many aquatic applications [10–14], these protocols do not cause much extra cost and are very efficient from the routing perspective.

However, geo-routing protocols [7, 9] are usually based on the broadcast nature of the underlying acoustic channel. It is highly possible that multiple nodes are selected as the next-hop, which can lead to collisions if all of these next-hop candidates relay the packet. Although the self-adaptation methods such as those in [7, 9] narrow down the size of the candidate set to some extent, the collision probability is still very high without proper MAC design and optimization.

Existing MAC protocols for underwater acoustic networks, such as R-MAC [15], UWAN-MAC [16], and T-Lohi [17], are usually based on channel reservations.

*Corresponding author. Email: yibo.zhu@engr.uconn.edu

In these protocols, senders and receivers interact with each other to reserve the channel for data communications. Before the channel reservation process, sender must know the exact receiver. Unfortunately, it cannot be satisfied by current geo-routing protocols since a node cannot know its next-hop node beforehand in the stateless routing protocol. For example, in R-MAC, a node reserves a channel by measuring the propagation delay and mapping the slot at the sender side to the receiver side, which is not compatible with the geo-routing protocols that cannot provide the next-hop information. Therefore, a new MAC protocol which can effectively suppress collisions and can be smoothly combined with geo-routing protocol is highly desirable.

In this paper, we propose an efficient Geo-rOuting Aware MAC protocoL (GOAL) for underwater acoustic networks which smoothly integrates self-adaptation based REQ/REP, geographic cyber carrier sensing, and implicit ACK to find the next-hop node and perform channel reservation at the receiver side. Utilizing self-adaptation based REQ/REP, a forwarder can determine the best next-hop with little route discovery cost. By adopting geographic cyber carrier sensing, collisions among the data packets are almost eliminated. With implicit ACK strategy, control messages are significantly reduced and thus fewer collisions occur among control packets. With these techniques, GOAL is energy-efficient and provides high end-to-end reliability. Another remarkable feature is that GOAL can work in mobile underwater acoustic networks with localization services such as SLMP [14].

The rest of this paper is organized as follows. Section 2 briefly discusses the related works. Then, GOAL is presented in detail in Section 3. After that, a theoretical model is developed for the probability of a successful handshake in Section 4. Subsequently, simulation results and discussion are shown in Section 5. At last, Section 6 provides our concluding remarks and future work.

2. Background and related work

In this section, we will first review related works on geo-routing protocols in underwater acoustic networks and demonstrate the disadvantages in collision resolutions. We will then review MAC protocols for underwater acoustic networks and their differences from our work.

In underwater acoustic networks, nodes communicate *via* acoustic channels with long propagation delay and therefore take more time and consume more energy to perform route discovery. Luckily, a couple of localization algorithms [11–14] have been designed for underwater acoustic networks, which make geo-routing possible. Geo-routing protocols which are based on the nodes location have been gaining significant attention because they perform stateless routing which allows their routing cost to be very low. In VBF [7], for example, packets are forwarded along the routing pipe from the source to the

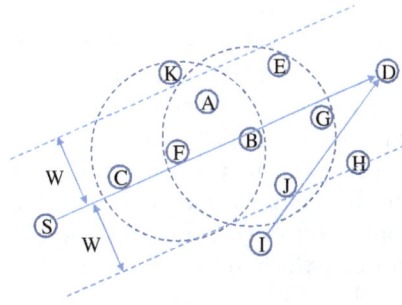

Figure 1. VBF: a self-adaptation based geo-routing protocol.

destination. All nodes within the routing pipe will participate in the packet forwarding process. In DBR [9], a depth-based geo-routing protocol, packets are forwarded to nodes with less depth and finally arrive at the sink nodes deployed on the sea surface. A more recent routing protocol is VBVA [8], which is based on VBF and incorporates the void avoidance capability. In all these protocols, a node does not explicitly choose the next-hop but cooperates with its neighbors to determine the best relay node(s) according to some self-adaptation schemes.

The basic idea of self-adaptation is as follows. When an eligible node, one that is nearer to the destination than the previous hop, gets a data packet, it starts to back off according to its location before forwarding the packet. The self-adaptation scheme tries to guarantee that a better relay node backs off for a shorter time so that the best relay node ends backoff and then forwards data packet first. For example, as shown in Figure 1, node C and node K are ineligible because the former's position is worse than that of node F and the latter lays out of the pipe area from the source node S to the sink node D. Then, only node A and node B are eligible to forward. Comparing with node A, node B is closer to the vector from source node S to sink node D, and node B is closer to the sink node D than other neighbors of forwarder F. Thus, according to the self-adaptation scheme, the backoff time of node B is shorter than that of node A after receiving the DATA packet from node F. As a result, node B first forwards the DATA packet. By overhearing the forwarding, other nodes, such as node A, cancel the backoff and do not forward the packet any further. In this procedure, several optimal relay nodes[1] can forward first and other nodes are suppressed by the overheard forwarding. As this procedure is repeated, the packet will get closer and finally arrive at the sink node.

Although such a self-adaptation scheme can improve the system's performance to some extent, it cannot prevent MAC collision when there are two or more adjacent nodes forwarding packets at the same time. As shown in Figure 1, if node J happens to forward the DATA packet

---

[1]Some nodes might have already forwarded the data packet before overhearing the forwarding by the best relay node.

for node I when node B also relays the DATA packet, collision might occur at the common neighbors of nodes B and J, and therefore the DATA packet might not be further forwarded. This harms the end-to-end reliability of the routing protocol. Additionally, the dropped packets waste plenty of energy, i.e. they are not energy-efficient.

In order to further improve the performance of geo-routing protocols, effective collision resolution schemes should be employed, which are usually implemented in the MAC protocols. MAC protocols have been widely investigated for underwater acoustic networks in the last few years. In FAMA [18], RTS/CTS and carrier sensing are combined to avoid collision. However, it is not energy-efficient because the REQ/REP packet is very long and consumes lots of energy to transmit. Slotted-FAMA [19], a modified FAMA, tries to improve the energy-efficiency problem by slotting the time and sending both control and data packets at the beginning of a slot. In this way, the length of an RTS/CTS packet is not determined by the maximal propagation delay as that in FAMA and therefore is much more energy-efficient. However, the RTS/CTS handshake requires the routing protocol to explicitly provide the next-hop, i.e. it cannot act as the MAC protocol for self-adaptation based geo-routing protocols.

COPE-MAC [20], a novel RTS/CTS-based MAC protocol, enables nodes to perform reservations in parallel such that nodes can send multiple DATA packets in one RTS/CTS/DATA/ACK round. For long-delay featured underwater acoustic networks, COPE-MAC is energy-efficient and can significantly improve the throughput. Unfortunately, similar to Slotted-FAMA, it also requires the explicit next-hop and therefore cannot be coupled with self-adaptation based geo-routing protocols.

In T-Lohi [17], a short tone message is used to reserve the channel to send data. However, even though a node does not receive any tone during a contention period, it cannot ensure that there is no collision at the receiver side. In other words, it still suffers from the hidden terminal problem and cannot effectively avoid the collisions.

R-MAC [15] consists of three phases. In the first phase, each node measures the propagation delay to its neighbors. In the second phase, each node reserves a receiving slot at the receiver side and then the receiver confirms if the reservation is collision-free. This phase can make sure that there is no collision at the receiver side for the data packet. In the last phase, each node follows the reservation in the second phase to transmit the data packets. An explicit receiver address is needed in phase two for the channel reservation, and therefore R-MAC cannot work with self-adaptation based geo-routing protocols as well.

Unlike other reservation-based MAC protocols, UWAN-MAC [16] does reservation *via* one-way communication. Assuming the delay between neighbors does not vary, each node piggybacks the relative sending time of the next packet in current packet. As a result, a node knows when it will receive the next packet. However, such a one-way handshake cannot solve the hidden terminal problem. Therefore, collision is still heavy in multi-hop networks. In addition, UWAN-MAC requires nodes to foresee the exact sending time of the next packet, which is unpractical in self-adaptation based geo-routing protocols.

Different from above works, GOAL, the new approach in this paper, smoothly integrates the self-adaptation scheme and MAC reservation techniques. It first employs the self-adaptation scheme to do handshaking and finds the next-hop. Similar to implicitly finding the best relay in self-adaptation based geo-routing protocols, the cost of selecting the next-hop is low in our protocol. Additionally, the receiving slot is reserved based on the geographic information during the handshake, and then the DATA packet can be forwarded without collision. Thus, GOAL finally avoids more collisions while keeping a low routing cost.

3. Description of GOAL

In this section, we will discuss our new GOAL, which is reservation-based and can smoothly integrate with any known geo-routing protocols with self-adaptation. For instance, if GOAL adopts the self-adaptation scheme of VBF, its functionality is equivalent to that of a reservation-based MAC protocol coupling with VBF. We first present the basic idea of GOAL. Then, we describe its three key components, self-adaptation based REQ/REP handshake, geographic cyber carrier sensing, and implicit acknowledgement. Specifically, we apply the self-adaptation scheme of VBF to GOAL as a special case in the description. (Note that GOAL can be used with any self-adaptation scheme.) After that, we will provide an example to show the overall working process of GOAL with detailed analysis.

3.1. Basic idea

Geo-routing aware MAC protocol is a reservation-based MAC protocol. In GOAL, each node maintains a time schedule, which records the time slot corresponding to its neighbors' packet sending/receiving time. Whenever a node wants to send a packet, it should make sure that the selected sending time does not overlap with any existing time slot in the time schedule line. In this way, the DATA packet can be sent collision-free. In order to map the sending/receiving slot to a node's own time schedule line, the self-adaptation based REQ/REP handshake is employed where only a few qualified neighbors are allowed to reply a REP packet for the REQ packet received, which will reduce the collisions. The REQ/REP handshake process in GOAL is used to implement twofold functionalities: determining the next-hop and

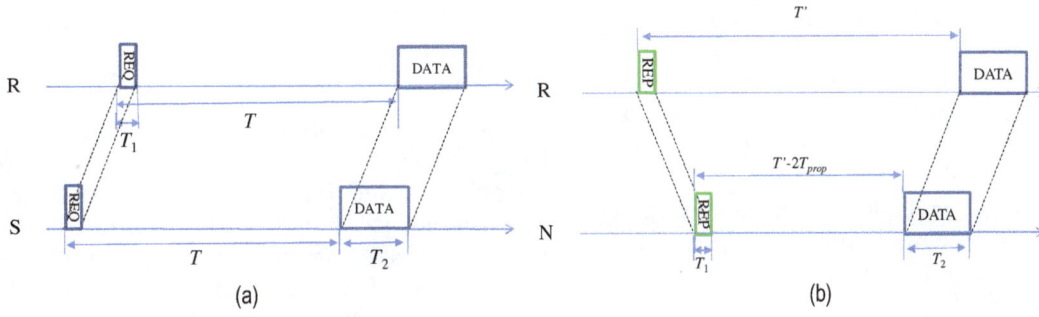

Figure 2. (a) Mapping neighbor's sending slot and (b) mapping neighbor's receiving time slot.

mapping the neighbors' sending and receiving slots to a node's own time schedule.

As shown in Figure 2(a), when sending a packet, node S piggybacks the transmission time T_2 and the relative sending time T of the next packet. After receiving this packet, node R can calculate the receiving time of the packet, i.e. it maps the sending time of the packet at node S to its time schedule line. Specifically, the interval between the time when node S sends the first and second packet is T. Assuming that the propagation delay between nodes R and S does not change much during the coming time period T, the interval between the time points when node R begins to receive these two packets is still T. Therefore, after node R finishes receiving the first packet, it knows that it will receive next packet during time slot $[T - T_1, \; T - T_1 + T_2]$, where T_1 and T_2 are the transmission times of the first and second packet, respectively. Note that the time slot is expressed by relative time and can be easily converted to absolute time.

The method of mapping the receiving time is illustrated in Figure 2(b). When node R sends the first packet, which has a transmission time of T_1, it then notifies its neighbors including node N that it will receive the next packet after a time period of T'. Suppose that node N knows that the propagation delay between itself and node R is T_{prop}. After completely receiving the first packet, node R knows that there will be a collision at node R if it emits any packet signal during $[T' - T_1 - 2T_{\text{prop}}, \; T' - T_1 - 2T_{\text{prop}} + T_2]$, where T_2 is the transmission time of the packet which node R will receive. Therefore, in order to avoid the collision at node R, node N must make sure that the sending interval does not overlap with time slot $[T' - T_1 - 2T_{\text{prop}}, \; T' - T_1 - 2T_{\text{prop}} + T_2]$ when it sends out any packet.

Applying these two mapping schemes, a node can map its neighbor's sending and receiving time period to its own time schedule line to avoid collisions when transmitting a DATA packet.

3.2. The GOAL protocol

The GOAL protocol consists of three parts: self-adaptation based REQ/REP handshake, geographic cyber carrier sensing, and implicit acknowledgement. As described in Section 3.1, self-adaptation based REQ/REP handshake and geographic cyber carrier sensing are used to determine the optimal next-hop and reserve channel for DATA packets. In addition, implicit acknowledgement is imported to reduce the number of control messages. The details of these parts are provided as follows based on the network topology shown in Figure 1.

Self-adaptation based REQ/REP handshake. When the current forwarder F intends to send out a DATA packet, it first selects a qualified sending time to broadcast a REQ{P_S, P_F, P_D, T, T_{DATA}} packet. Via the REQ packet, node F tells its neighbors that it will send the DATA packet T time later[2] and the corresponding transmission time is T_{DATA}. It also provides the location of the source, current forwarder, and the destination, which are P_S, P_F, and P_D, respectively.

After receiving REQ, neighbors of node F then know that they will receive the DATA packet $T - T_{\text{REQ}}$ time later (by applying the method of mapping neighbor's sending slot), where T_{REQ} is the transmission time of the REQ packet. Then, the neighbors who have a better location than that of node F start to back off according to the self-adaptation scheme of VBF. Once a node terminates the backoff process, it will send node F a REP{P_{thisnode}, T', T_{DATA}} packet, where P_{thisnode} is the location of this node and T' is the relative time that it will send a DATA packet. Due to the broadcast feature of acoustic medium, part of node F's neighbors, which might still be in the backoff state, can overhear the REP packet. Then, they cancel the backoff process because the overheard REP implies that there is a better relay. Finally, node F decides the next-hop according to the received REP packets. If it does not receive any REP, it waits for a random time period and tries to resend the REQ. Otherwise node F must receive at least

---

[2]Note that T must be bigger than double the maximum propagation delay plus the maximum backoff time, which is decided by the self-adaptation scheme. Otherwise, node F cannot determine the next-hop before it sends a DATA packet. In this case, node F will issue a new REQ for this DATA packet later.

one REP packet from its neighbors. In this case, it sets the next-hop as the one with the shortest adaptation time, which node F can for use the location information in REPs to calculate again. Once the pre-scheduled DATA sending time comes, node F sends the DATA packet to the selected next-hop.

As an improvement, multiple DATA packets to the same sink node can be transmitted in one packet train [19], and therefore REQ/REP handshake could perform reservation for multiple DATA packets in one round. This strategy can clearly enhance the efficiency of handshake.

Geographic cyber carrier sensing.   In underwater acoustic networks, it is difficult for nodes to completely avoid collisions due to the long propagation delay. In order to address this issue, nodes in GOAL apply the geographic information and the two mapping methods in Section 3.1 to map neighbors' packet sending and receiving slots to their time schedule line. In this way, nodes can figure out when they can send a packet and when they cannot. This is similar to carrier sensing, and therefore we have named it geographic cyber carrier sensing. With geographic cyber carrier sensing, the collisions at neighbors can be greatly alleviated if the selected packet sending time does not overlap with any reserved slot in its time schedule line.

Specifically, after receiving the REQ packet, a node knows that it will receive the DATA packet during $[T - T_{\mathrm{REQ}}, T - T_{\mathrm{REQ}} + T_{\mathrm{DATA}}]$ by applying the method of mapping neighbor's sending time slot, where T_{REQ} stands for the transmission time of REQ packet. Then, this node converts the time slot to absolute time and inserts it into its time schedule line.

The REP packet has twofold functionalities: responding the REQ packet and notifying neighbors to avoid collisions. On one side, with the REP packet from node B, the sender of REQ knows that node B is a potential next-hop. On the other side, based on the information in REP packet, other neighbors of node B can evaluate the propagation delay T_{prop} between themselves and node B. The evaluation method is to use propagation speed to divide the Euclidean distance. Then by applying the method of mapping neighbors' receiving time slot, this node will know that there will be a collision at node B if it sends packet   during $[T - T_{\mathrm{REP}} - 2T_{\mathrm{prop}},$ $T - T_{\mathrm{REP}} - 2T_{\mathrm{prop}} + T_2]$, where T_{REP} is the transmission time of REP packet. To avoid collisions, this node should not emit any packet signal during this period. Note that the propagation delay measure method might introduce an error because the acoustic signal is transmitted along a bent path and the nodes are mobile. In order to tolerate this error, guard time T_{guard} is introduced, i.e. the propagation delay is in range $[T_{\mathrm{prop}} - T_{\mathrm{guard}}, T_{\mathrm{prop}} + T_{\mathrm{guard}}]$. Thus, the time period becomes $[T - T_{\mathrm{REP}} - 2T_{\mathrm{prop}} - 2T_{\mathrm{guard}}, T - T_{\mathrm{REP}} - 2T_{\mathrm{prop}} + T_{\mathrm{DATA}} + 2T_{\mathrm{guard}}]$.

Based on geographic cyber carrier sensing, nodes can obtain their neighbors' sending and receiving schedule after the REQ/REP handshake. By recording the schedules in their time schedule line, nodes can conveniently choose a qualified time to send packet.

Implicit acknowledgement.   In terrestrial *ad hoc* networks, REQ/REP/DATA/ACK can substantially improve the reliability of one-hop transmissions. However, if this scheme is applied in underwater acoustic networks, there are more collisions among control packets because of low bandwidth and long propagation delay. A possible way to address these challenges is to adopt an implicit acknowledgement scheme to reduce the number of control packets. Specifically, if the node that receives the DATA packet is not the destination, it must send REQ to determine the next-hop within a certain time. Because the previous hop is still within the one-hop range with a high probability, it can also overhear the REQ. Based on this heuristic rule, REQ is revised to include the packet identifier (PID) of the DATA packet. As a result, the previous hop can confirm that the DATA packet is successfully forwarded.

For the destination node, it explicitly acknowledges the DATA packet using an ACK packet. In addition, a node will send an explicit ACK packet without a backoff if it receives a REQ when both of the following conditions are met: (i) the location of this node is better than the sender of REQ; (ii) this node has received the DATA packet for which the REQ packet requested.

For any node, if it does not receive an implicit acknowledgement or ACK packet within certain time after sending out the DATA packet, it will initiate a new REQ/REP/DATA round to retransmit the DATA packet. Although retransmission can improve the transmission reliability, the maximum number of times that retransmission can occur should not be infinite. This is because it will introduce more delay and energy consumption. Therefore, we define the number of maximum retransmission times as a tradeoff. Specifically, one node can transmit and retransmit a DATA packet at most the number of maximum retransmission times, where a retransmission is caused by having failed to overhear the corresponding REQ packet or an ACK packet. If the number of maximum retransmission times is exceeded, the node should give up trying to resend the DATA packet.

3.3. An example of GOAL

In the example, the network topology is shown as Figure 3(a). Node F tries to forward the DATA packet from source node S to destination node D. Following the GOAL protocol, node F selects a qualified sending time to broadcast a REQ$\{P_{\mathrm{S}}, P_{\mathrm{F}}, P_{\mathrm{D}}, T, T_{\mathrm{DATA}}, \mathrm{PID}\}$ packet. By means of the REQ packet, node F notifies its neighbors that it will send the DATA packet T time later and the corresponding transmission time is T_{DATA}. With the information in the REQ packet, nodes C, A, and B figure out that they will receive the DATA packet $T - T_{\mathrm{REQ}}$

time later. Note that node C will discard this REQ because its location is worse than that of current forwarder node F. Therefore node F's neighbors, except node C, start to back off according to the self-adaptation scheme in VBF.

Similar to VBF, when node B first exits the backoff state, it sends a REP$\{P_B, T', T_{DATA}\}$ packet to node F. By over-hearing the REP packet from node B, node A realizes that there is a better relay and cancels the backoff. Additionally, based on the information in the REP packet, node A can evaluate T_{BA}, which denotes the propagation delay between node B and node A. Thus, node A will not send any packet during time interval $[T' - T_{REP} - 2T_{BA} - 2T_{guard}, \ T' - T_{REP} - 2T_{BA} + T_{DATA} + 2T_{guard}]$. As well, node F finds out that the next-hop could be node B after receiving the REP packet. When the scheduled DATA packet sending time comes, node F sends the DATA packet to node B since node B is the optimal one. Later on, node B tries to forward the DATA packet and sends a REQ$\{P_S, P_B, P_D, T, T_{DATA}, PID\}$ packet to do the handshake. After receiving this REQ packet, node F knows that the forward-ing is successful and then prepares to forward the other DATA packets.

3.4. Properties of GOAL

In GOAL, nodes apply the self-adaptation scheme in the REQ/REP handshake process to determine the next-hop. This procedure is similar to the general self-adapta-tion based geo-routing protocol for data packets. Since a REQ/REP packet is much shorter than a DATA packet, the probability of collision among REQ/REP packets in GOAL is much lower than that among DATA packets in self-adaptation based geo-routing protocol. Note that the use of geographic cyber carrier sensing allows DATA packets in the GOAL protocol to be almost collision-free, and the entire collision probability is accordingly lower than that in self-adaptation based geo-routing protocols. As a result, GOAL provides a higher end-to-end reliability than self-adaptation based geo-routing protocol coupling with broadcast MAC.

As discussed above, GOAL introduces MAC collision among short REQ/REP packets while avoiding collision among long data packets. As a result, the collision prob-ability is reduced. It is clear that the collision among long DATA packets wastes more energy than the collision among short ones, and thus GOAL requires less energy consumption for packet delivery than self-adaptation based geo-routing protocols plus broadcast MAC.

However, in order to achieve the above desirable features, GOAL incurs a longer delay. As explained in Section 3.2, nodes schedule the sending time of a DATA packet T time later after sending the REQ packet, where T is at least the maximum backoff time plus double the maximum propagation delay. Moreover, due to implicit acknowledgement strategy, nodes will also wait for more than one round trip time. In addition, nodes in GOAL will perform retransmission if any failure occurs during the forwarding procedure, which will also increase the delivery delay. Therefore, the delivery delay in GOAL is higher than self-adaptation based geo-routing protocols plus broadcast MAC.

4. Analysis of probability of successful handshake

In this section, we theoretically explore the probability of a successful handshake in GOAL.

4.1. Notations

In this section, we define the terms in the analysis as follows.

T_I: The given node maps the arrival time points of all neighbors' DATA packets to its own time schedule line. Then, T_I is the length of the interval between the arrival time points of two successive DATA pack-ets on the time schedule line.

T_I': The given node maps the scheduled sending time of all neighbors' DATA packets to its own time sche-dule line. Then, T_I' is the length of interval between the sending time points of two successive DATA packets on the time schedule line.

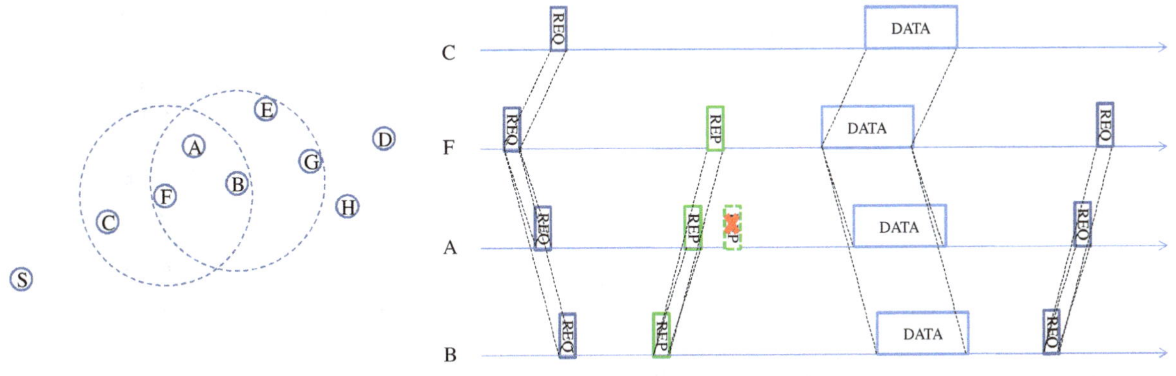

Figure 3. (a) Network topology and (b) one hop forwarding.

T_D: Transmission time of the DATA packet.

T_c: Transmission time of the control packet (both REQ packet and REP packet).

n: The average number of neighbors of each node.

k: The average number of neighbors which reply a REP packet to the sender of a REQ packet based on the self-adaptation algorithm used by GOAL.

λ: The traffic rate of DATA packets generated by each node.

λ_D: The real traffic rate of DATA packets that each node sends to the channel.

P_c: The probability that the control packet is transmitted successfully.

P_r: The probability that the requested time slot does not overlap with any existing reserved slot on the time schedule line.

$\tilde{P}_H$: The probability of the successful handshake.

4.2. Protocol simplification and assumptions

In GOAL, whenever the MAC layer gets a DATA packet from the upper layer, it will cache this DATA packet if it is forwarding another DATA packet in a REQ/REP/DATA round. After transmitting the current DATA packet successfully, the node starts a new round to transmit the next DATA packet. To simplify the analysis, we revised GOAL such that a REQ/REP/DATA round for different DATA packets can be overlapped. Specifically, when a node is forwarding a DATA packet in a REQ/REP/DATA round, it will cache the DATA packet from the upper layer and send a REQ to perform reservation for the DATA packet in the coming interval between DATA packets. Additionally, retransmissions will no longer happen. In other words, if REQ/REP handshake fails or a DATA packet is transmitted unsuccessfully, the DATA packet will be dropped. We also make the following assumptions throughout the analysis part of this paper.

(i) Similar to existing works [21–23], packet error rate is not considered in the model. In other words, a packet is dropped only if there is a collision.

(ii) The traffic at every node follows a Poisson process with same λ parameter and the DATA packets are of the same length.

(iii) The length of the REQ packet is same as that of the REP packet.

(iv) Relay traffic is not considered, i.e. all nodes only do one-hop transmission.

(v) As described before, GOAL employs implicit ACK and the ACK packet is only sent by the sink node. Without relay traffic, there are few ACK packets in the networks. Therefore, we reasonably ignore ACK packets in the analysis.

(vi) A multi-hop network scenario is considered in the analysis. Accordingly, the analysis also involves the interferences incurred by hidden terminals, which is an important issue but ignored by most of the existing analysis works on MAC protocols.

4.3. Analysis

In this part, we deduce $\tilde{P}_H$ by analyzing the handshake process between sender S and its neighbors $R_i (1 \leq i \leq n)$, where node S is an arbitrary node in the network.

In GOAL, as illustrated in Figure 4, every node records the time slots when its neighbors receive or send DATA packets. To avoid the collision with DATA packets, control packets (both REQ and REP) are transmitted during the interval between these time slots.

Although the arrival time of the DATA packet follows a Poisson process, the sending time of the DATA packet and the control packet no longer follows the Poisson process due to the schedule strategy in GOAL. However, because every DATA packet is preceded by a REQ packet and the REQ is sent immediately during the next available interval after a DATA packet arrives, the probability that N_c control packets are sent by node R_i and its neighbors, except node S, during a given interval of length T_I' equals the probability that N_c DATA packets arrive during the same time period. Thus, this probability still can be evaluated by utilizing the Poisson process. Therefore, if T_I' and N_c are fixed, we can obtain P_{N_c} as

$$P(N_c = n_c) = \frac{\lambda_c^{n_c} e^{-\lambda_c T_I'}}{n_c!}, \tag{1}$$

where λ_c is the traffic rate of control packets sent by node R_i and its neighbors, except node S, total n nodes. Since the traffic of control packets consists of REQ traffic and REP traffic, we can obtain

$$\lambda_c = n(\lambda_{REQ} + \lambda_{REP}). \tag{2}$$

Note that every DATA packet is preceded by a REQ packet, which implies REQ packets are subject to the same Poisson process as the input traffic. Therefore, we have $\lambda_{REQ} = \lambda$. Regarding λ_{REP}, it is related to λ_{REQ}. Specifically, due to collisions, a REQ packet correctly arrives at node R_i with probability P_c. After getting the REQ packet, node R_i checks if the requested time slot overlaps with any existing reserved time slot for DATA packets. If there is no overlap, it will reply a REP packet. Otherwise, it will discard the packet. According to the self-adaptation scheme, k out of the n neighbors of node S will reply the REP packet on average, i.e. each neighbor replies with probability k/n. Additionally, considering that each node is a potential REP replier of its n neighbors, we can get

$$\lambda_{REP} = n\frac{k}{n} P_c P_r \lambda_{REQ} = k P_c P_r \lambda_{REQ}, \tag{3}$$

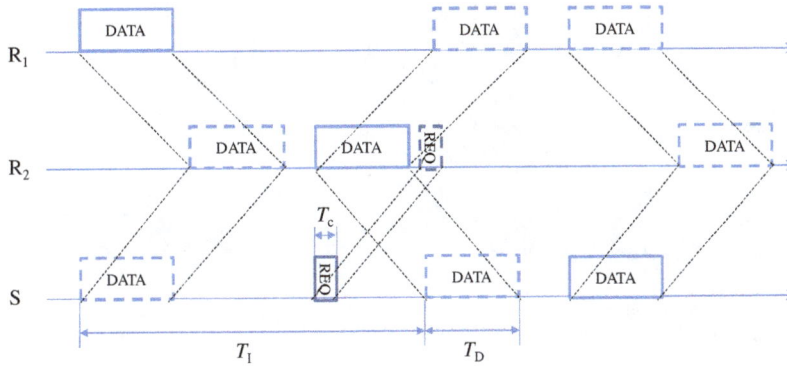

Figure 4. An example of reserved time slots on nodes' time schedule line.

where P_r is the probability that the requested time slot does not overlap with any existing time slot for DATA packets. It can be evaluated as

$$P_r = e^{-2T_D n \lambda}. \quad (4)$$

Substituting equations (4) and (3) into equation (2), we have

$$\lambda_c = n(1 + k P_c P_r)\lambda_{REQ}. \quad (5)$$

Based on P_{N_c}, we can calculate the probability of event A, which denotes the event that a REQ packet sent by node S collides with any control packets at node R_i. According to the strategy in GOAL, the sending time of control packets approximately follows a uniform distribution. If there are N_c control packets sent or received by R_i in the interval of length T_I', excluding the control packets from node S because there is no collision among the control packets sent by node S, then the collision probability of event A is

$$P\left(A | N_c = n_c, T_I' = T\right) = \left(1 - \frac{2 T_c}{T - T_D}\right)^{n_c}. \quad (6)$$

Note that the handshake is successful as long as node S receives a REP packet from any neighbor. Recall that on average there are k potential repliers according to the self-adaptation scheme. Accordingly, we can calculate the probability of the event H, which represents the successful handshake, *via* equation (7).

$$P\left(H | N_c = n_c, T_I' = T\right) = 1 - \left(1 - P\left(A | N_c = n_c, T_I' = T\right)\right)^k. \quad (7)$$

Then, considering the distribution of N_c, we can get

$$P\left(H | T_I' = T\right) = \sum_{n_c=0}^{\infty} \left(1 - \left(1 - P\left(A | N_c = n_c, T_I' = T\right)\right)^k\right). \quad (8)$$

From equation (8), we can see that we still need to figure out the distribution of T_I' to get $\tilde{P}_H$. Although T_I follows an exponential distribution, T_I' does not follow the exponential distribution anymore due to the schedule strategy in GOAL. As shown in Figure 5, for example, the ith packet arrives at t_i, and the duration of interval $[t_i, t_{i+1}]$ is T_i. After being scheduled, the sending time of the ith packet becomes t_i', and the interval between t_i' and t_{i+1}' is T_i'. During this procedure, the interval length between DATA packets is adjusted to be at least $T_m(T_m \geq T_D)$.

For a given interval sequence $T_i(0 \leq i \leq l + 1)$, there are two conditions for that l successive intervals $T_1 \sim T_l$ are adjusted to T_m but T_{l+1} is reduced to an interval longer than T_m. One is that T_0 should be long enough to allow $t_1' = t_1$. Here, $t_1' = t_1$ means the scheduled time t_1' is not affected by schedule scheme in GOAL. And the other is

$$\sum_{i=1}^{j} T_i \leq j T_m \quad \text{for } 1 \leq j \leq l$$
$$\sum_{i=1}^{l+1} T_i > (l+1) T_m \quad (9)$$

Note that all T_i follow an exponential distribution with the same λ_{dc} parameter, where λ_{dc} is the total traffic rate of DATA packets sent by any given node, its neighbors, and hidden terminals. Since the time slots confirmed by REP packets might overlap with the slots requested by REQ packets, some slots are not successfully reserved. Specifically, as long as overhearing a REP packet confirms an overlapped slot, the sender of the REP packet will remove the slot it intends to reserve on its time schedule time and try to reserve later. Therefore, we will overestimate the traffic rate of DATA packet if counting all time slots reserved by control packets in. To avoid this case, we approximate λ_{dc} as the sum of the traffic rate of DATA packets sent by this node, the traffic rate of DATA packets sent by all neighbors, and the traffic rate of non-overlapped DATA packets sent by hidden terminals. These three parts correspond to the three items in equation

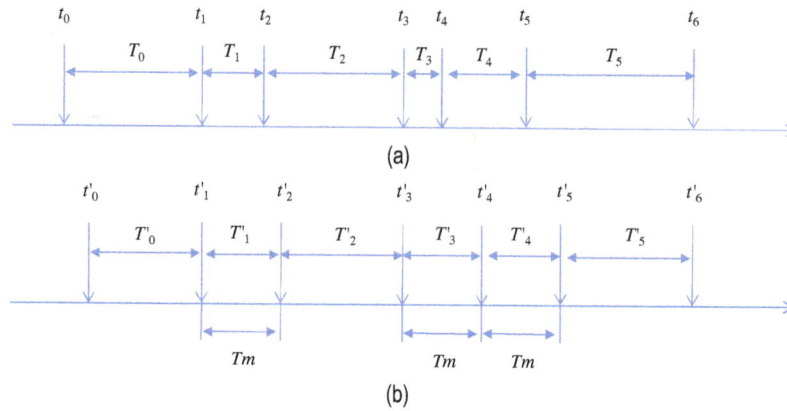

Figure 5. (a) Sequence of DATA packet arrival time and (b) Sequence of DATA packet sending time after being scheduled $(t'_1 = t_1, \; t'_3 = t_3, \; t'_6 = t_6)$.

(10), respectively. For the last item, we resort to a constant factor α, which can be trained with the simulation data, to obtain the traffic rate of non-overlapped DATA packets since it is complicated to model the probability of the overlapped packets. Therefore, λ_{dc} can be expressed as

$$\lambda_{dc} = \lambda_D + n\lambda_D + \alpha nk\lambda_D = (1 + n + \alpha nk)\lambda_D. \quad (10)$$

In equation (10), since a node will send out a DATA packet only if the handshake is successful, λ_D can be evaluated by the following equation.

$$\lambda_D = \tilde{P}_H \lambda. \quad (11)$$

Based on the distribution of T_i and the two conditions, we can calculate the probability $P(L = l | t_1 = t'_1)$ of l successive intervals are all adjusted to intervals of length T_m as

$$P(L = l | t_1 = t'_1) = \int_0^{T_m} \int_0^{2T_m - T_1} \cdots \int_0^{lT_m - \sum_{j=1}^{l-1} T_j}$$
$$\times \int_{(l+1)T_m - \sum_{j=1}^{l} T_j}^{\infty} \left(\lambda_{dc} e^{-\lambda_{dc} T_m}\right)^{l+1} dT_{l+1} dT_l \cdots dT_2 dT_1$$
$$= \lambda_{dc}^l e^{-(l+1)\lambda_{dc} T_m} \int_0^{T_m} \int_0^{2T_m - T_1} \cdots \int_0^{lT_m - \sum_{j=1}^{l-1} T_j} \quad (12)$$
$$\times \, dT_l \cdots dT_2 dT_1.$$

Note that equation (12) also denotes the probability that the $(l+1)^{\text{th}}$ interval is long enough such that the arrival time of the next packet will not be adjusted. Thus, each T_{l+1} is actually one case of T_0. Then, we can get

$$P(t_1 = t'_1) = \sum_{l=1}^{\infty} P_l. \quad (13)$$

With $P(t_1 = t'_1)$ and equation (12), we can calculate $\delta_l(T)$, the probability that an interval of length $T > T_m$ is reduced to the interval of length T_m among a given l successive adjusted interval.

$$\delta_l(T) = P(t_1 = t'_1) \sum_{i=c}^{l} \lambda^l e^{-(l+1)\lambda T_m} \int_0^{L_{U_1}}$$
$$\cdots \int_0^{L_{U_j}} \cdots \int_{T_i = T}^{T+\epsilon} \cdots \int_0^{L_{U_b}} \cdots \int_0^{L_{U_l}}$$
$$dT_l \cdots dT_b \cdots dT_i \cdots dT_j \cdots dT_1, \quad (14)$$

where $\epsilon \to 0^+$ and

$$c = \left\lceil \frac{T}{T_m} \right\rceil$$
$$L_{U_b} = bT_m - T - \sum_{b=1}^{b-2} T_b \qquad . \quad (15)$$
$$L_{U_j} = \begin{cases} jT_m - \sum_{b=1}^{j-1} T_b & \text{if } c \le i - j \\ iT_m - T - \sum_{b=1}^{j-1} T_b & \text{otherwise} \end{cases}$$

Furthermore, we can calculate the probability of interval of length $T(T > T_m)$ is adjusted to T_m as

$$\delta(T) = P(t_1 = t'_1) \sum_{l=2}^{\infty} \delta_l(T). \quad (16)$$

With $\delta(T)^3$, we can calculate the distribution of T'_I as follows

$$P(T'_I = T)$$
$$= \begin{cases} \int_0^{T_m} \lambda e^{-\lambda T'} dT' + \int_{T_m}^{\infty} \delta(T') dT' & \text{if } T = T_m \\ \lambda e^{-\lambda T} \epsilon - \delta(T) & \text{if } T > T_m \\ 0 & \text{otherwise} \end{cases} \quad (17)$$

Finally, with the distribution of T'_I, we can calculate $\tilde{P}_H$ as

$$\tilde{P}_H = \sum_{T'_I} \left(P\left(H | T'_I = T\right) P\left(T'_I = T\right) \right). \quad (18)$$

In equation (18), since both sides contain $\tilde{P}_H$, it is not easy to get a closed form of $\tilde{P}_H$. Therefore, we resort to the numerical method (iteration) to calculate $\tilde{P}_H$.

5. Performance evaluation

In this section, we use simulations to evaluate the performance of GOAL. Aqua-Sim [24], a NS-2 based underwater acoustic network simulator developed by the UWSN lab at the University of Connecticut, has been used for our simulations.

5.1. Simulation settings

In the simulation, nodes are randomly deployed within a 300 m × 300 m × 500 m area. When a node detects an event, it will send the data collected to the sink node. To simplify the simulations, we make two assumptions: (i) a node can detect the event occurring within its sensing range and (ii) event lasts for a long period of time[4], such that nodes send data to the sink node periodically as long as it can sense the event. This period is defined as sensing interval.

All nodes can move freely in horizontal two-dimensional space, i.e. in the X–Y plane. The speed of a node follows a uniform distribution between 0.2 and 1.5 m s^{-1}. The transmission range is set to 120 m. The sink node, which is the destination for all data packets, is fixed at (250, 250, 0). The sensing range of nodes is 80 m. The number of maximum retransmission times is set to be 6. Each simulation lasts for 5000 s. The energy consumption parameters are based on a commercial underwater acoustic modem, UMW 1000, from Link-Quest [25]: the power consumption on transmission mode is 2 W; the power consumption on receive mode is 0.75 W; and the power consumption on sleep mode is 8 mW.

Three metrics are used to quantify the performance: packet delivery ratio, energy consumption per byte, and delivery delay. Specifically, the packet delivery ratio is the ratio of the total number of packets sent by source nodes to the number of packets received by the sink node. The energy consumption per byte is to divide the total network energy consumption by the number of data bytes successfully received by the sink. The delivery delay is the average end-to-end delay of each packet received by the sink. We compare the performance of GOAL with VBF

coupling with that of broadcast MAC (we use VBF for short in the rest of this work) [7].

5.2. Simulation results

Impacts of data sensing interval. In this set of simulations, the number of nodes in the network is fixed to be 100 and the size of the DATA packet is set to 300 bytes. Then, we change the data sensing interval of every node from 20 to 70 s.

As shown in Figures 6, 7, and 8, GOAL can provide a high end-to-end reliability. Figure 6 shows that GOAL can provide a much higher packet delivery ratio than VBF. This is because GOAL can greatly reduce collisions by its REQ/REP handshake process and its channel reservation mechanism. Additionally, we can see that the packet delivery ratio of GOAL increases while the sensing interval becomes larger. This is because nodes with a larger sensing interval generate fewer packets, which causes fewer collisions. Since the number of maximum retransmission times is fixed, the packet delivery ratio is improved when there are fewer collisions. We can also observe that the packet delivery ratio of VBF does not vary much while the sensing interval increases. This is because VBF is a best-effort protocol and the collision probability of VBF mainly depends on the self-adaptation scheme, which is highly related to the node distribution. Note that the size of network is fixed and nodes are uniformly deployed. Hence, the node distribution is decided by node density. In this simulation set, node density is fixed and therefore the packet delivery ratio keeps nearly the same value.

GOAL can also achieve high energy efficiency. From Figure 7, we can observe that GOAL is more energy-efficient than VBF, especially when the sensing interval becomes larger. This is because in GOAL, when the sensing interval is shorter, multiple packets can be sent together with just one REQ/REP handshake, which can improve the system's energy efficiency. In addition, as the sensing interval becomes larger, less data packets

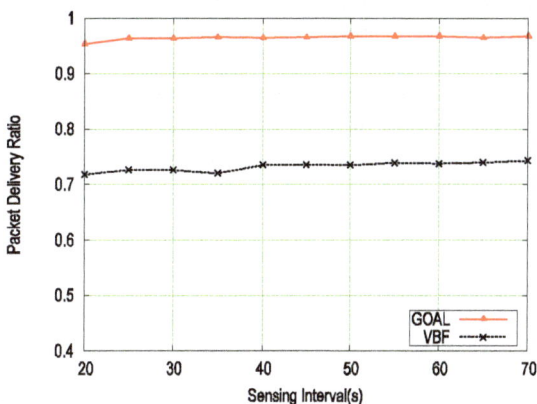

Figure 6. Packet delivery ratio with varying sensing interval.

---

[4]This is practical. For example, oceanic volcano usually belches slight smoke and ashes for a long time before it finally erupts.

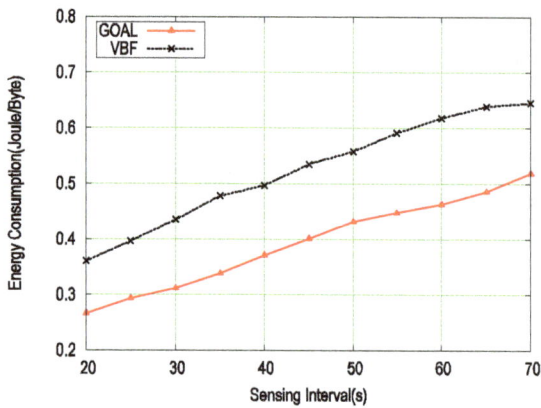

Figure 7. Energy consumption with varying sensing interval.

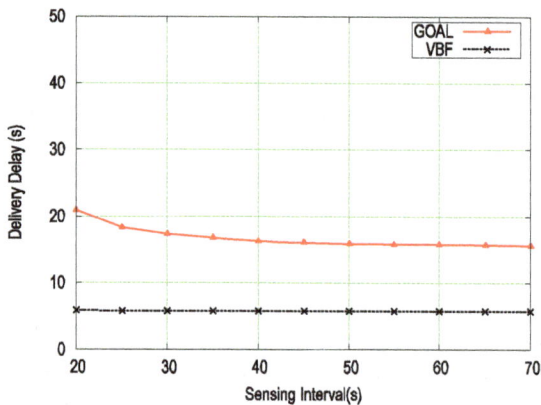

Figure 8. Delivery delay with varying sensing interval.

are sent in the network. Therefore, most nodes will waste energy in the idle state with a constant rate (8 mW). Additionally, this increases the energy consumption when the sensing interval is larger.

Considering the reliability requirement, the energy consumption in VBF is much higher than that in GOAL. For example, let us set P_G as the delivery ratio of GOAL and P_V as the delivery ratio of VBF, and set E_G and E_V as the energy consumption of GOAL and VBF, respectively. To achieve the same packet delivery ratio, VBF should perform retransmission for N times on average and thus the energy consumption is NE_V, where N satisfies

$$1 - (1 - P_V)^N = P_G. \tag{19}$$

Hence, N can be expressed as

$$N = log_{1-P_V}(1 - P_G). \tag{20}$$

In Figure 6, for example, the packet delivery ratio of GOAL and VBF is approximately 0.97 and 0.73, respec-

tively, when the sensing interval is 50 s. Applying the above equation, we can obtain the average number of times that VBF should transmit each packet to reach the same packet delivery ratio as GOAL, which is as follows:

$$N = log_{1-0.73}(1 - 0.93) = 2.67. \tag{21}$$

Therefore, the energy consumption of VBF should be at least doubled. In other words, the energy consumption in GOAL is less than half of that in VBF, which indicates that GOAL is more energy-efficient.

Figure 8 shows us that the end-to-end delay of GOAL decreases with the increasing sensing interval. This is because collisions increase when the sensing interval is shorter. With collisions, nodes have to initiate a new REQ/REP/DATA round to do retransmission, which introduces extra delay. As the sensing interval becomes larger, fewer collisions and retransmission appear. Therefore, the delay decreases while the sensing interval increases. For VBF, which is a best-effort protocol, the delivery delay has almost nothing to do with traffic rate, but is mainly decided by the backoff time in the self-adaptation scheme. Thereby, the delivery delay in VBF does not change significantly in Figure 8.

Impacts of node density.   In this set of simulations, we set the sensing interval of every node to be 50 s and change the number of nodes in the network from 70 to 120. The size of the DATA packet is fixed to be 300 bytes.

The impact of node density is shown in the next three figures. In Figure 9, we can see that the packet delivery ratio of GOAL is much higher than that of VBF. Again, this is because GOAL reduces more collision than VBF and VBF is a best-effort protocol. Also, we can see that the packet delivery ratio of both GOAL and VBF increases while there are more nodes in the network. One reason as mentioned before is that GOAL largely reduces the MAC collision by performing reservation for DATA packets. The other reason is that it is related to the self-adaptation scheme. Specifically, when the node density is lower, there are fewer qualified next-hops according to the self-adaptation scheme. Particularly, some forwarders do not have a qualified next-hop. In VBF, the DATA packet is dropped in such a case. In GOAL, forwarding failure can be detected by missing the implicit acknowledgement, and therefore a retransmission is issued.

From Figure 10, we can observe that GOAL consumes less energy than VBF for transmitting every unit data from source to sink. The reason is similar to that of Figure 7. In VBF, the collision probability is higher than that in GOAL. Moreover, each collided packet in VBF wastes more energy than that in GOAL because the packet in VBF is much longer. As a result, GOAL saves more energy. Similar to the analysis for Figure 7, if we analyze the energy consumption with the same packet

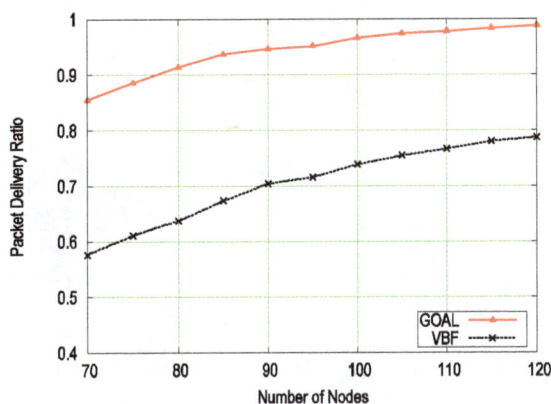

Figure 9. Packet delivery ratio with varying number of nodes.

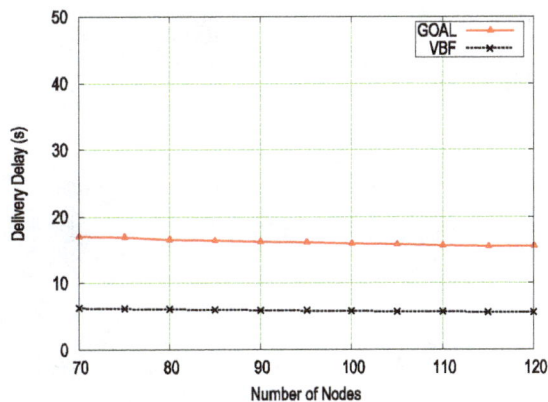

Figure 11. Delivery delay with varying number of nodes.

delivery ratio, we can see that the energy consumption in GOAL is much less than that in VBF, especially when there are less nodes within the network.

In Figure 11, the delivery delay of GOAL is higher than that of VBF. The reason has been mentioned before: the handshake and implicit acknowledgement in GOAL introduce more delay while VBF is a best-effort protocol which does not care whether the transmission to next-hop is successful. Due to the same reason, the delivery delay of VBF is almost a constant in Figure 11. Additionally, we can observe that the delivery delay of GOAL slightly decreases while the node deployment becomes dense. This is because dense deployment improves the probability that the next-hop with a better location can be found. According to the self-adaptation algorithm, a better location implies a shorter backoff time. Therefore, the total delivery delay is reduced.

Impacts of DATA packet size. In this set of simulations, we set the sensing interval and number of nodes as 50 s and 100, respectively. Then, we compare the performance of GOAL and VBF plus broadcast MAC by varying the size of DATA packet from 200 to 400 bytes with step 20.

From Figure 12, as the size of the DATA packet increases, the packet delivery ratio of VBF decreases. This is because a longer DATA packet increases the probability of collisions. However, as shown in Figure 12, the packet delivery ratio of GOAL is almost constant. We believe this is because the reservation scheme in GOAL well prevents the collisions among DATA packets.

From Figure 13, we can see that both GOAL and VBF consume less energy as the size of the DATA packet becomes larger. This is because the longer DATA packet shortens the time of the idle state. We can still see that GOAL is more energy-efficient than VBF. The reason is the same as mentioned before. The reservation scheme in GOAL can largely reduce the collisions among DATA packets and therefore GOAL wastes less energy.

From Figure 14, we can observe that GOAL introduces more delivery delay than VBF. The reason is same as the explanation for Figure 11. We can also observe that the delivery delays of both GOAL and VBF become longer while the DATA packet size increases slightly. This is reasonable. A longer DATA packet implies longer transmission delay. Since transmission delay is a part of delivery delay, the delivery delay becomes longer.

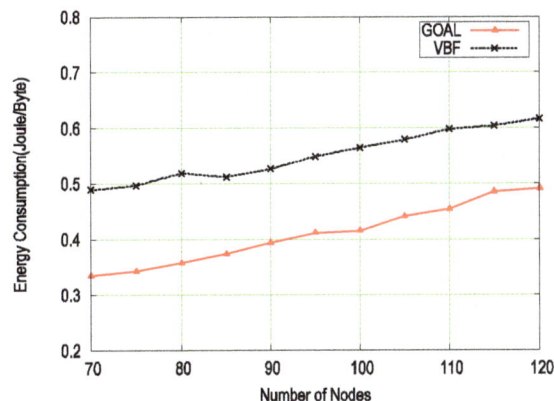

Figure 10. Energy consumption with varying number of nodes.

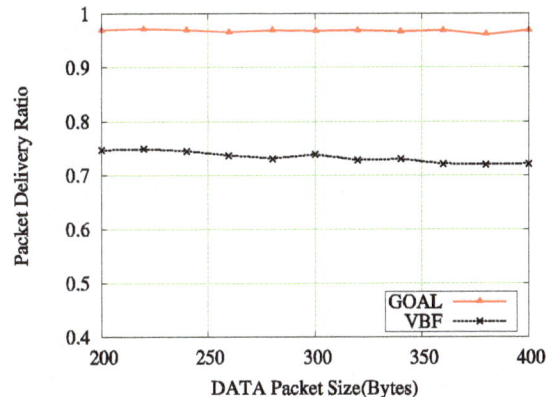

Figure 12. Packet delivery ratio with varying DATA packet size.

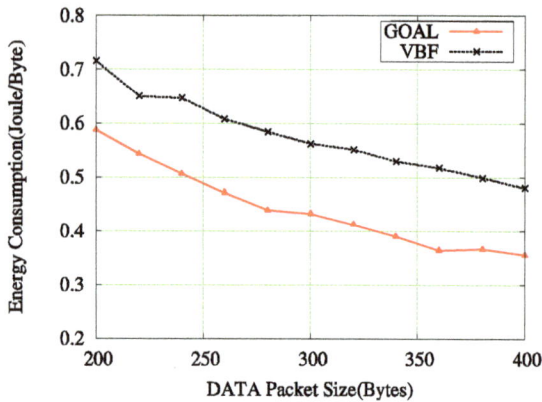

Figure 13. Energy consumption with varying DATA packet size.

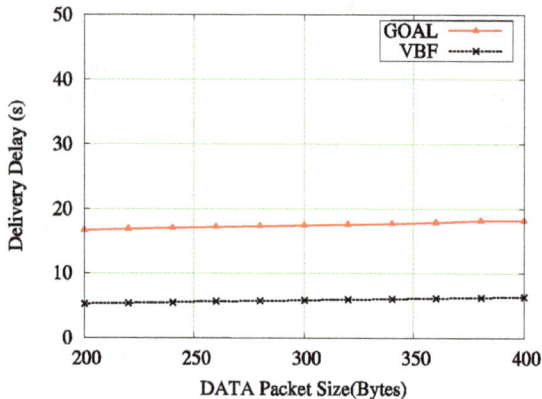

Figure 14. Delivery delay with varying DATA packet size.

5.3. Theoretical model verification

In this section, we perform two sets of simulation to verify the theoretical model of $\tilde{P}_{H}$. In the first set, GOAL is revised as mentioned in Section 4 and the relay traffic is disabled. With the results of this set, we can check the model accuracy. After all, it is not a general scenario if nodes do not relay packet. Therefore, we let nodes still forward DATA packets in the other simulation set. By comparing with the results of this set, we can find the gap between our model and the simulation in practical scenario. There are 200 stationary nodes with a transmission range of 100 m uniformly distributed within the network of size 500 m × 500 m × 500 m, and each node randomly selects a destination node. Then, we vary λ from 0.01 to 0.1 packets s^{-1} with step 0.01 to evaluate $\tilde{P}_{H}$.

The comparison between simulation results and our theoretical model is shown in Figure 15. From Figure 15, we can observe that the $\tilde{P}_{H}$ evaluated by our theoretical model is always tightly around the simulation results of GOAL without forwarding. This justifies that our theoretical model matches the simulation results, which also

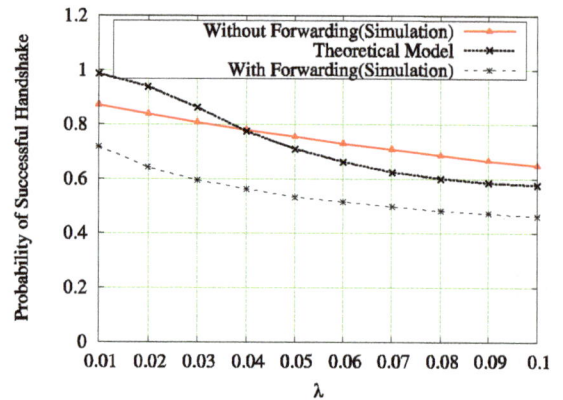

Figure 15. Comparison between theoretical model of $\tilde{P}_{H}$ and simulation results.

indicate that our model can well describe the probability of successful handshake.

At the same time, we can see the results of both the theoretical model of GOAL without forwarding are higher than that of GOAL with forwarding. This is reasonable. In GOAL with forwarding, nodes also send relay traffic to the channel and therefore the traffic in the channel is apparently much heavier than that in GOAL without forwarding. For that reason, there must be more collisions, which reduce the probability of successful handshake.

6. Conclusion

In this paper, GOAL, an efficient geo-routing aware MAC protocol, is proposed for underwater sensor networks. It is a reservation-based MAC protocol which can smoothly integrate with any existing geo-routing protocols with self-adaptation capability. Self-adaptation based REQ/REP handshake, geographic cyber carrier sensing, and implicit acknowledgement are used in GOAL to improve system performance. Although the end-to-end delivery delay increases because of the hop-by-hop retransmission mechanism in GOAL, it can achieve high end-to-end delivery ratio with low energy consumption. Plentiful simulation results show that GOAL outperforms existing VBF with broadcast MAC in both end-to-end delivery ratio and energy efficiency. Moreover, the simulation results demonstrate that our theoretical model can well describe the probability of a successful handshake.

References

[1] CUI, J., KONG, J., GERLA, M. and ZHOU, S. (2006) Challenges: building scalable mobile underwater wireless sensor networks for aquatic applications. *IEEE Network, Spec. Issue Wireless Sens. Networking* **20**(3): 12–18.

[2] HEIDEMANN, J., YE, W., WILLS, J., SYED, A. and LI, Y. (2006) Research challenges and applications for underwater sensor networking. In *Proceedings of IEEE Wireless*

Communications and Networking Conference (IEEE), 228–235.

[3] PARTAN, J., KUROSE, J. and LEVINE, B.N. (2006) A survey of practical issues in underwater networks. In *Proceedings of ACM WUWNet* (ACM), 1–8.

[4] LIU, L., ZHOU, S. and CUI, J. (2008) Prospects and problems of wireless communication for underwater sensor networks. *Wireless Commun. Mobile Comput.* **8**(8): 977–944.

[5] CHITRE, M., SHAHABUDDEN, S. and STOJANOVIC, M. (2008) Underwater acoustic communication and networks: recent advances and future challenges. *Mar. Technol. Soc. J.* **42**(1): 103–116.

[6] PERKINS, C.E. and ROYER, E.M. (1999) Ad hoc on-demand distance vector routing. In *Proceedings of IEEE Workshop on Mobile Computing Systems and Applications* (IEEE), 90–100.

[7] XIE, P., CUI, J. and LAO, L. (2006) VBF: vector-based forwarding protocol for underwater sensor networks. In *Proceedings of IFIP Networking* (Springer), 228–235.

[8] XIE, P., ZHOU, Z., PENG, Z., CUI, J. and SHI, J. (2009) Void avoidance in three-dimensional mobile underwater sensor networks. In *Proceedings of International Conference on Wireless Algorithms, Systems, and Applications (WASA)* (Springer), 305–314.

[9] YAN, H., SHI, Z. and CUI, J. (2008) DBR: depth-based routing for underwater sensor networks. In *Proceedings of IFIP Networking* (Springer), 72–86.

[10] CHENG, X., SHU, H., LIANG, Q. and DU, H. (2008) Silent positioning in underwater acoustic sensor networks. *IEEE Trans. Veh. Technol.* **57**(3): 1756–1766.

[11] EROL, M., VIERIRA, L.F.M. and GERLA, M. (2007) AUV-aided localization for underwater sensor networks. In *Proceedings of International Conference on Wireless Algorithms, Systems and Applications (WASA)* (Springer), 44–51.

[12] FRAMPTON, K.D. (2006) Acoustic self-localization in a distributed sensor network. *IEEE Sens. J.* **6**(1): 166–172.

[13] ZHOU, Z., CUI, J. and ZHOU, S. (2007). Localization for large scale underwater sensor networks. In *Proceedings of IFIP Networking* (Springer), 108–119.

[14] ZHOU, Z., CUI, J. and BAGTZOGLOU, A. (2008) Scalable localization scheme with mobility prediction for under-

water sensor networks. In *Proceedings of IEEE INFOCOM* (IEEE), 2198–2206.

[15] XIE, P., ZHOU, Z., CUI, J. and SHI, Z. (2007) R-MAC: an energy-efficient MAC protocol for underwater sensor networks. In *Proceedings of International Conference on Wireless Algorithms, Systems and Applications (WASA)* (Springer), 187–198.

[16] PARK, M.K. and RODOPLU, V. (2007) UWAN-MAC: an energy-efficient MAC protocol for underwater acoustic wireless sensor networks. *IEEE J. Oceanic Eng.* **32**(3): 710–720.

[17] SYED, A., YE, W. and HEIDEMANN, J. (2008) T-Lohi: a new class of MAC protocols for underwater acoustic sensor networks. In *Proceedings of IEEE INFOCOM* (IEEE), 231–235.

[18] FULLMER, C.L. and GARCIA-LUNA-ACEVES, J. (1995) Floor acquisition multiple access (FAMA) for packet-radio networks. In *Proceedings of ACM SIGCOMM* (ACM), 262–273.

[19] MOLINS, M. and STOJANOVIC, M. (2006) Slotted FAMA: a MAC protocol for underwater acoustic networks. In *Proceedings of IEEE OCEANS* (IEEE), 16–19.

[20] PENG, Z., ZHU, Y., ZHOU, Z. and CUI, J. (2010) COPE-MAC: a contention-based medium access control protocol with parallel reservation for underwater acoustic networks. In *Proceedings of IEEE OCEANS* (IEEE), 1–10.

[21] ZHOU, Z., PENG, Z., CUI, J. and SHI, Z. (2009) Analyzing multi-channel MAC protocols for underwater acoustic sensor networks. *UCONN CSE Technical Report: UbiNet-TR08-02.*

[22] XIAO, Y., ZHANG, Y., GIBSON, J.H. and XIE, G.G. (2009) Performance analysis of p-persistent aloha for multi-hop underwater acoustic sensor networks. In *Proceedings of ICESS* (IEEE), 305–311.

[23] BIANCHI, G. (2000) Performance analysis of the IEEE 802.11 distributed coordination function. *IEEE J. Sel. Areas Commun.* **18**(3): 535–547.

[24] XIE, P., ZHOU, Z., PENG, Z., YAN, H., HU, T., CUI, J., SHI, Z. *et al.* (2009) Aqua-Sim: an NS-2 based simulator for underwater sensor networks. In *Proceedings of IEEE/MTS Oceans* (IEEE), 1–7.

[25] LinkQuest, http://www.link-quest.com.

17

A Trace-Driven Analysis of Wireless Group Communication Mechanisms

Surendar Chandra[1,*], Xuwen Yu[2]

[1]FX Palo Alto Laboratory, Palo Alto, CA 94304, USA, surendar@acm.org
[2]VMware Inc., Palo Alto, CA 94304, USA, xyu@vmware.com

Abstract

Wireless access is increasingly ubiquitous while mobile devices that use them are resource rich. These trends allow wireless users to collaborate with each other. We investigate various group communication paradigms that underly collaboration applications. We synthesize durations when members collaborate using wireless device availability traces. Wireless users operate from a variety of locations. Hence, we analyzed the behavior of wireless users in universities, corporations, conference venues, and city-wide hotspots. We show that the availability durations are longer in corporations followed by university and then in hotspots. The number of simultaneously available wireless users is small in all the scenarios. The session lengths are becoming smaller while the durations between sessions are becoming larger. We observed user churn in all the scenarios. We show that synchronous mechanisms require less effort to maintain update synchronicity among the group members. However, distributed mechanisms require a large number of replicas in order to propagate updates among the users. For asynchronous mechanisms, we show that pull-based mechanisms naturally randomize the times when updates are propagated and thus achieve better performance than push based mechanisms. We develop an adaptive approach that customizes the update frequency using the last session duration and show that this mechanism exhibits good performance when the required update frequency intervals are large. We also show that for a given number of gossips, it is preferable to propagate updates to all available nodes rather than increasing the frequency while correspondingly reducing the number of nodes to propagate updates. We develop a middleware to illustrate the practicality of our approach.

Keywords: Wireless LAN, group communication mechanisms

1. Introduction

Wireless laptops are replacing desktops as the primary computing platform for many users. In an article published in 2006, USA today [1] described the emergence of about 30 million nomadic users in the US. Gartner Dataquest predicted a yearly growth of 10% of these users. Recently, global sales of laptops exceeded that of desktops [2]. On our university campus, between 2005 and 2008, the number of active wireless devices rose from 5,027 to 13,051 while the number of wired student desktops dropped from 7,035 to 1,323. Similar trends were observed at other universities [3].

Contemporary laptops are resource rich while high speed cellular and wireless LAN networks are ubiquitous. BuddeComm [4] estimated a global deployment of over 200,000 wireless hotspots. These hardware and network trends allow laptops to support collaborations in which groups of users modify shared objects. Synchronous approaches support simultaneous modifications by the group members, while asynchronous approaches eventually propagate updates to all participants. Each of these approaches can be implemented using a centralized or a distributed approach.

The performance of these systems depends on the availability of the users' devices. Prior systems were designed when users predominantly used wired desktops. We investigate the behavior of prior systems for wireless users. A significant deviation in the

*Corresponding author. Email: surendar@acm.org

behavior of wireless users from wired users could necessitate a redesign of existing collaboration systems.

Capturing such an update trace is difficult; collaboration applications may need to be instrumented in order to collect their usage statistics. A large scale deployment of modified applications is hard. Hence, we synthesize the duration when group members collaborate using information about when the users were online.

There are further complications in deciding when group members were online. Typical wireless LAN users exchange security credentials and associate with an access point (AP), acquire a DHCP address and other network resources and finally authenticate themselves with the system. Also, the collaboration system might be implemented as an application. Hence, users cannot be considered to be available until the user explicitly starts the necessary programs. Finally, these applications might allow the user to control when collaboration functionality was desirable. The trace collection mechanism should reflect the durations when the user is actually able to run and interested in the collaboration mechanisms.

Also, unlike desktop users [5], wireless users are mobile and work from many different locations [1, 6]: home, office and public venues. Ideally, we need to simultaneously monitor all wireless locations (schools, offices, hotspots etc.) in a geographic locale (city) in order to evaluate the system from the perspective of a mobile user who operates in these locales. However, such an analysis requires monitoring multiple administrative domains; currently unavailable to us and many researchers. An intrusive alternative is to install loggers in each monitored laptop.

In this paper, we analyzed the behavior of wireless users in different communities: university, corporate lab, conference venue and in a city-wide hotspot federation at different times and locations (spanning 2001 through 2008 at several locations in North America). We used publicly available traces from the CRAWDAD [7] archive. We also collected application level traces at a university. We compare the different traces from the perspective of group communication systems that might be utilized by these wireless users.

In all these traces, wireless LAN access was *free*. Hence, we assume that users exclusively used wireless LAN networks for collaboration (and did not also use a cellular network). Prior work [3, 8, 9] observed significant local wireless traffic within an university. An article in the Economist [10] highlighted attempts to encourage more interaction among hotspot users and suggests that the nomadic users themselves are supportive of such efforts because of common purpose, lending further credence to our assumption.

We show that the durations that users were online decreased while the durations between sessions increased. The unavailable durations were longer for hotspots as compared to the university users. The system was dynamic with constant churn.

The observed availability behavior leads to poor performance regardless of the update propagation policy. Synchronous mechanisms encounter few group members who will simultaneously modify the shared contents. Distributed approaches require as many as 150 replicas in order to achieve acceptable update availability in the hotspot scenario. In general, asynchronous approaches are preferable for disconnected access. For small group sizes, server mediated approaches achieve better performance although for large groups, a distributed approach out-performed a server based approach. In general, a pull based approach leverages the randomness of group availability to achieve better performance than push based schemes. We showed that the immediate past session duration of an user was sufficient to predict the current session duration. However, systems that required frequent update propagations did not benefit from history based prediction as much as systems that propagated updates less frequently. Our approach is practical; we used the policies developed to propagate updates in our moderated collaboration system called *flockfs* [11, 12]. We incorporate the lessons learnt into our Yenta middleware.

Section 2 analyzes the various availability traces. We investigate several synchronous and asynchronous group communications using these traces in Sections 3 and 4, respectively. We describe related work in Section 5 with concluding remarks in Section 6.

2. Wireless user availability analysis

Analysis of group communications depends on knowing when users create updates. One way to collect this information is to instrument and widely deploy several applications that follow the different group communication paradigms. Instead, we synthesize the durations when users update contents by extrapolating from durations when users were available.

2.1. Wireless traces analyzed

We used publicly available access traces collected from a corporate research lab, a conference and a city-wide hotspot federation. We also collected application level wireless user availability traces from an university.

University campus (academia). Some of the large scale wireless user availability traces were collected at Dartmouth College. These traces include SNMP logs from Fall 2003 and Winter 2004 [13] as well as syslog traces from Sep. 2005 through Oct. 2006 [14]. The syslog traces record the *association* and *disassociation* events from each wireless device to an access point; the time between these records show the duration when a particular user was online. However, the syslog traces were recorded using unreliable UDP datagrams; many of the trace datagrams were apparently lost (confirmed through private email correspondence). Our analysis of the syslog entries from the Dartmouth Aruba routers [14] (after removing extraneous association messages - we modified the syslog_parser script [15] to consider *'update station bssid to'* records for AP associations) showed that there were 428,703 events with matching *association* and *disassociation* records. We also observed 5,966,261 *association* events without the corresponding *disassociation* records (from the same access point) as well as *428,703 disassociation* records without the corresponding *association* records. Hence, these syslog traces were not suitable for our study. Older SNMP logs were analyzed in [3, 16]. We collected newer application level user availability information from our university.

On our campus, wireless networks are ubiquitous and deployed throughout the campus and the dormitories. The system used over 1,300 access points. We collected user availability traces using the Zeroconf protocol [17]. Users running Apple Mac OSX and Linux (with Avahi) reported when they became online (after successful association with an AP, acquiring an IP address and user authentication) and offline using the Zeroconf _workstation._tcp service. However, Zeroconf does not capture handover events across multiple access points within the same subnet. Users running the Apple iTunes client (on Microsoft Windows and Apple Mac OSX) report their availability using the _dacp._tcp service. Also, users who ran iTunes used the _daap._tcp service to advertise when the user explicitly consented to sharing their contents with other iTunes users. iTunes did not impose a tit-for-tat mechanism to fairly share contents; users are allowed to consume shared objects from other users while not sharing any objects on their own. Voida et al. [18] suggested that users share contents to express their individuality. Overall, the *workstation* reflects the user availability for asynchronous group communication systems, *dacp* highlights the potential availability for collaboration applications and *daap* shows the time that the user actually participated in iTunes like collaboration. iTunes only allows read-only sharing of objects and does not represent our target group collaboration systems. Regardless, these traces allow us to investigate the complexity of user availability.

Logistically, Zeroconf pushed the service availability information to the monitoring client using link local multicast. Since these packets are not routed, we required the service collection station to be co-located inside the monitored VLAN. We reconfigured the entire campus wireless network to use a single VLAN and placed our monitoring station on this wired distribution network of the wireless infrastructure. We used appropriate firewalls on the APs to prevent other types of network traffic from overwhelming the wireless networks. Without these filters, all broadcast packets will be sent to all the 1,300 APs. We collected the data for eleven days from Sep. 19, 2006 through Sep. 29, 2006 (our campus had 800 APs in 2006). We also collected the data from Dec. 3, 2007 through Aug. 25, 2008. This period included the end of Fall '07 semester, winter break, Spring '08 semester, spring break and Summer '08 sessions. We place particular importance on the two weeks starting at 12/04/2007 (when users were likely to be collaboratively working on course projects, though not necessarily using any of the mechanisms explored in this paper) as well as on 60 days starting from 12/04/2007 (includes the busy end of semester season, calm winter break and the beginning of a new semester). Throughout this paper, we refer to these traces as *workstation-2006, itunes-2006, share-2006, workstation-2008, itunes-2008* and *share-2008.* Note that the network administrators progressively removed groups of access points into their own (unmonitored) VLANs during the beginning of Spring '08 semester, spring break '08 and Summer '08, reducing the number of simultaneously available users. Our analysis of each segment showed that the user behavior was similar across all the time periods; albeit with fewer monitored users. Overall, the total number of users monitored in *workstation-2006, itunes-2006, share-2006, workstation-2008, itunes-2008* and *share-2008* was 2,036, 4,893, 1,702, 4,063, 3,745 and 2,391, respectively. The number of wireless devices was greater in 2008, although the number of iTunes users decreased from their 2006 levels. On the other hand, as compared to 34.78% of the iTunes users sharing their song collection in 2006, over 63.85% of iTunes users shared their songs in 2008.

Corporation. Balazinska et al. [19] collected user availability traces at IBM Research from Jul. 22, 2002 through Aug. 17, 2002. Although these traces were collected before wireless networks were ubiquitous to the broader community, wireless laptops were already in widespread use among corporate users. These traces also remain the only publicly available trace of wireless users in a corporate setting. During the data collection period, IBM used 177 APs spread across three buildings (one of which was reportedly ten miles from the other buildings). For their traces, they issued SNMP probes ranging from every 5 (55% of the traces) to up to 15 minute intervals. The probe duration affects the resolution of the user availability intervals. During the trace collection, they observed 1,366 wireless devices. They published their analysis of user and access point network load distribution in Mobisys 2003 [20]. We refer to these access traces as *corporate-2002.*

Conference. Conferences and other public congregations offer wireless access to an ephemeral collection of participants who likely share similar content interests.

Many computer science conferences install their own wireless APs to provide wireless acces in the conference halls. Lately, the trend has been to utilize the wireless infrastructure available at the conference venue itself and provide free coverage to the participants throughout the conference venue: halls as well as the guest rooms. We have been unsuccessful in convincing these hotels to allow us to monitor the behavior of conference participants. One publicly available trace that contains user availability information was collected by Balachandran et al. [21] at the SIGCOMM 2001 conference (hosted at the UCSD campus). The traces were collected over two and a half days. Wireless coverage was provided using four access points in a single hall. The traces contain 195 distinct devices. The authors focused on the wireless traffic created by these users and presented their analysis in Sigmetrics 2002 [22]. We refer to these traces as *sigcomm-2001*.

Note that the wireless trace from the SIGCOMM 2008 conference [23] is not useful for our study due to data corruption, as much of the syslog traces except for 82 minutes on the last day were lost. Those traces did not reliably capture a disassociation event, making it hard to know the session durations. Additionally, the traces only captured users who voluntarily connected to a special IEEE 802.11a AP.

Hotspot. Finally, we analyzed the availability traces from a hotspot federation. *île Sans Fil* (http://www.ilesansfil.org/welcome/) provides free coverage throughout Montreal, Canada using 206 hotspots; owners of public venues share their wireless networks that are then administered by *île Sans Fil*. Lenczner et al. [24] released access traces for the three years from Aug. 28, 2004 through Aug. 28, 2007. These traces were collected when users authenticate themselves to any of the hotspots controlled by *île Sans Fil*. The trace contained records of 69,689 wireless users. We place particular importance on the most recent two weeks (starting from 08/01/2007) for our analysis. We call these access traces *hotspot-2007*.

2.2. Number of simultaneously available users

First, we plot the number of simultaneously available users as a function of the time of day for the various traces in Figs. 1 and 2. When the number of simultaneously available users is high, synchronous mechanisms incur high overhead in propagating and maintaining updates from all the online members.

We note that users in all the traces exhibit a diurnal variation. Even corporate desktops which are stationary [5] exhibited some diurnal variations.

First we consider the behavior of users in academia. Fig. 1(d) and 1(a) plots the analysis of *workstation-2008* and *workstation-2006* traces, respectively. The first two weeks starting on 12/04/2007 of *workstation-2008* were

the weeks before final exams of the Fall 2007 semester. Many users were likely to be available working on their final projects. Hence, we pay particular attention to the first fifteen days of *workstation-2008*. Overall, we observed 2,036 new devices in *workstation-2006*, while in the first fifteen days of *workstation-2008*, we observed 2,729 new devices. In December 2007, we observed that the number of simultaneously available users reached a high of 410 users (15% of users) with a diurnal low of two users late on a Sunday night. Note that wireless access is ubiquitous on our campus and many of our students live in the dormitories that were monitored in this study. By comparison, from *workstation-2006* we detected as many as 775 users during the day times (38% of the population). Even during late night hours, we observed over 200 simultaneously available users. Note that we do not know whether a wireless device was a laptop or a desktop (desktops are likely to be left online even when the user is not actively using them).

Similarly, we analyzed the application level availability of *itunes-2008* and *itunes-2006* in Figs. 1(b) and 1(e), respectively. Overall, we observed 2,617 users in the first fifteen days of *itunes-2006* and 4,893 users in *itunes-2008*. The number of simultaneously available iTunes users reached 224 (8.5%) in *itunes-2008* and 989 (20.1%) in *itunes-2006*. This drop in popularity is attributable to the drop in the session durations (Section 2.3). We also analyzed the scenario where users control the sharing duration for *share-2008* and *share-2006* in Figs. 1(c) and 1(f), respectively. Of the 1,702 unique users from *share-2006*, we observed between 100 and 250 users (6%-15%) were simultaneously available. For the first fifteen days of *share-2008*, we observed between 10 and 54 users (from a total of 2,391 users or 0.4%-2.3%). The number of users who were willing to simultaneously share their iTunes collection dropped significantly between 2006 and 2008 even though the number of unique users sharing their contents had increased. User availability is complex and depends on the mechanism used for collecting the traces. Lower level mechanisms over-estimate the number of users that were willing to share their contents through iTunes. We expand on this observation in Section 2.3.

Next we analyzed the behavior of *corporate-2002* traces in Fig. 2(a). We observed 1,366 unique devices. During the weekdays, the number of simultaneously available users reached as high as 529 (38.7%). Even during late nights, we observed over 200 users who were available; but these were likely devices without any users actively using them. Over the weekend, the number of devices drops further to about 160; its likely that about 40 users (who left their wireless devices at work during the weekday nights) took their wireless devices home for the weekend. Similar analysis of *sigcomm-2001* data in Fig. 2(b) showed that 119 of the 195 devices were available during the day time. Even

(a) workstation-2006   (b) itunes-2006   (c) share-2006

(d) workstation-2008   (e) itunes-2008   (f) share-2008

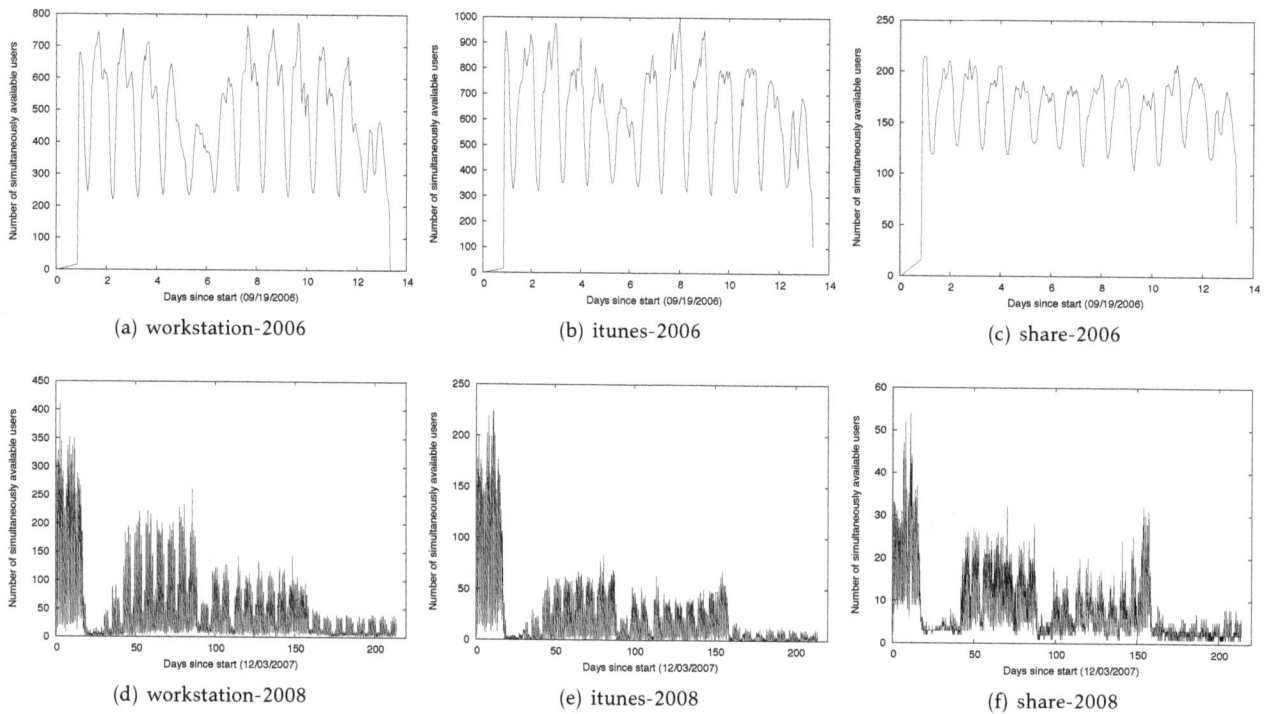

Figure 1. Number of simultaneously available campus users (academia)

during night times when the conference participants were likely to be unavailable, we still observed over 20 devices that were associated with the AP. Perhaps, these conference attendees were collaborating with colleagues in their home institutions located in a different time zone.

Hotspot-2007 traces span over three years with 70,000 users observed during the entire duration. In Fig. 2(c), we observed a steady increase in the number of simultaneously available users. During a two week duration starting in 8/1/2007 (Fig. 2(d)), we observed 2,725 unique devices; about 120 of them (4.4%) were simultaneously available during the day.

2.3. Session duration

We analyzed the session characteristics and plot the amount of time that a user was available and the time between sessions for the various traces in Figs. 3 and 4.

First, we plot the session behavior for the *workstation-2008* and *workstation-2006* traces in Fig. 3(a). We also plot the values for the two weeks starting in 12/04/2007. We note that the session durations in *workstation-2008* traces were short; 50% of the sessions were under 20 minutes and 95% of the sessions were less than 75 minutes. Focusing our attention on the two weeks after 12/04/2007, we note that 50% of the sessions were under 20 minutes and 95% of the sessions were less than 70 minutes. Sessions from *workstation-2006* were longer; 50% of the sessions were under

one hour with 95% of the sessions were under 6.7 hours. The durations between sessions had increased from *workstation-2006* to *workstation-2008*. For the *workstation-2008* trace, 50% of the durations between user sessions were less than 1.4 hours while 29% were longer than ten hours. Specifically focusing on the first two weeks in December 2007, these values were 1.2 hours for the median and 15% of the durations were longer than ten hours. On the other hand, *workstation-2006* showed that 50% of the durations between sessions was less than 47 minutes while 10.5% of the durations between sessions was larger than ten hours.

Even though the number of devices had increased (from 2,036 in *workstation-2006* to 2,730 devices in the first two weeks of 2007), the session durations had decreased. Kotz et al. observed a median session length of 16.6 minutes in 2002 [16] and a reduction to under ten minutes in 2004 [3]. Chinchilla et al. [25] also observed that only 16.2% of sessions were less than one minute. Note that our trace collected using Zeroconf does not include IP resource acquisition and authentication durations included when analyzing AP SNMP logs. This change might be either because users were using less computing time or that the newer laptops offer a more reliable energy saving mode allowing the users to sleep (and hence become unavailable) longer. To understand this behavior, we analyzed the duration between successive arrivals of a particular user in order to understand whether the

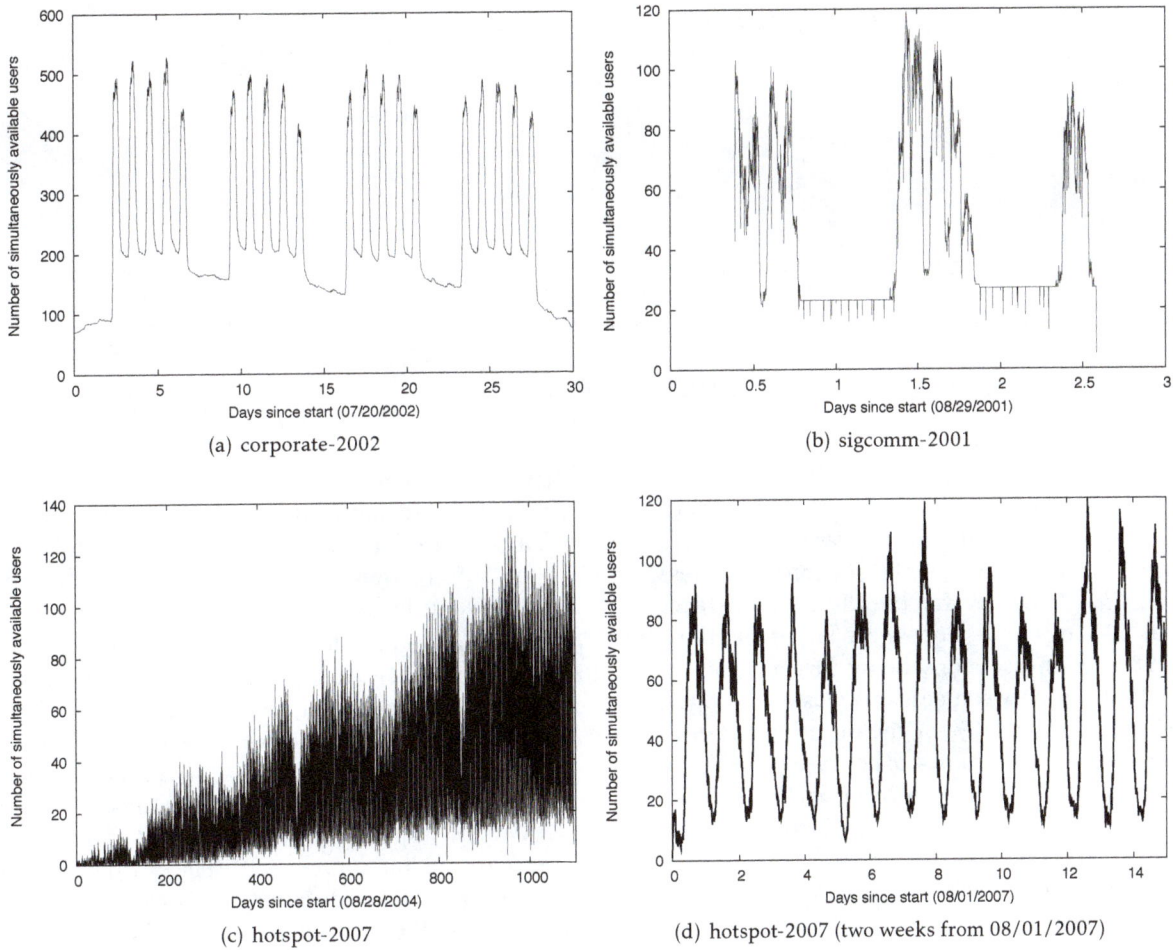

(a) corporate-2002

(b) sigcomm-2001

(c) hotspot-2007

(d) hotspot-2007 (two weeks from 08/01/2007)

Figure 2. Number of simultaneously available users

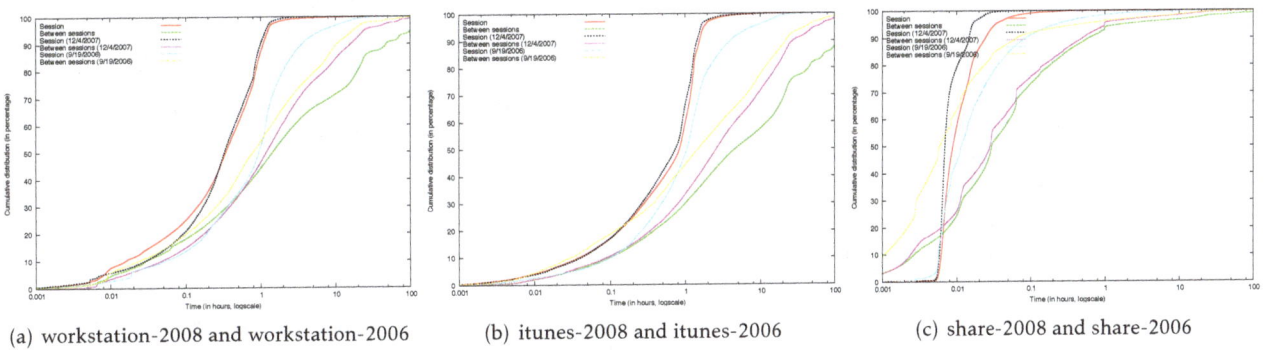

(a) workstation-2008 and workstation-2006

(b) itunes-2008 and itunes-2006

(c) share-2008 and share-2006

Figure 3. Online availability behavior of campus users (academia)

shortening session durations equaled the increase in duration between sessions. For the first two weeks in December 2007, median values were 1.78 hours while 75% of users online every 5.5 hours. In 2006, the median values were 2.52 hours with 75% of users online every 6.9 hours. The users whom came online often remained online for shorter durations. We will later

show the performance reduction caused by this session duration change.

Next, we analyzed the session behavior for the *itunes-2008* and *itunes-2006* traces (Fig. 3(b)). Note that prior trace collection efforts had not measured application level session availability durations. We note that 50% of the iTunes sessions from *itunes-2008* were less than 48

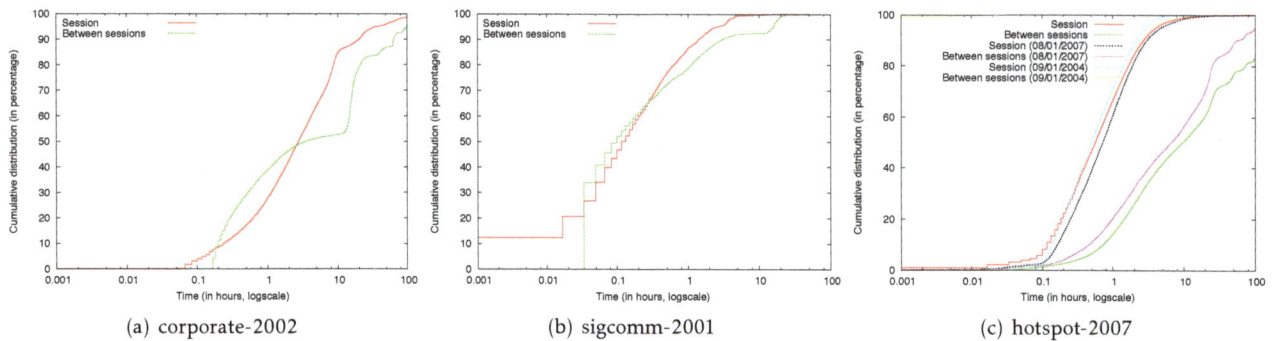

(a) corporate-2002   (b) sigcomm-2001   (c) hotspot-2007

Figure 4. Online availability behavior

minutes, while 95% of the sessions were less than 1.75 hours. Specifically focusing on the two periods since 12/04/2007, we note that 50% of the sessions were less than 45 minutes, while 95% of the sessions were less than 100 minutes. However, for the *itunes-2006* traces we note that 50% of the sessions were larger than one hour with 95% of the sessions being larger than 7.7 hours. The duration between sessions was even larger; *itunes-2008* showed that 50% of the duration between iTunes sessions was less than 4.5 hours with 95% of the durations over ten hours. However, analysis of the *itunes-2006* traces showed that 50% of the duration between iTunes sessions lasted for less than 1.5 hours with 22% over ten hours. These trends are similar to the *workstation* traces.

We plot the session characteristics for the *share-2008* and *share-2006* traces in Fig. 3(c). Session durations when users shared their iTunes collection were small. For the *share-2008* traces, over 96% of the sessions were under 0.01 hours while 50% of the sessions were under 0.01 hours for the *share-2006* traces. Over 90% of the durations between sessions for the *share-2008* traces were under an hour while the corresponding measures for the *share-2006* traces were under 0.1 hours. The users exert control over when they shared their songs.

For the *corporate-2002* traces (Fig. 4(a)), we note that 50% of the sessions were longer than 2.8 hours with 95% of the sessions less than 36 hours. Analyzing the duration between sessions, we note that 50% of them were over 3.5 hours with 48% over 10 hours. Similarly, for the *sigcomm-2001* traces (Fig. 4(b)), we note that most sessions and duration between sessions were small; 86% of the sessions were less than one hour long with only 22% of the duration between sessions longer than one hour. It is likely that conference attendees were paying attention to the conference and were using wireless networks for brief intervals. On the other hand, analyzing the *hotspot-2007* traces (Fig. 4(c)), we note that 50% of the sessions were smaller than 35 minutes with 95% of the sessions less than four hours. Similarly, analyzing the Verizon hot spot network in

Manhattan, Blinn et al. [26] observed that 45.74% of the sessions lasted more than one hour. The duration between sessions for our hotspot traces was over 9.6 hours for 50% of the cases with 49% of the duration between sessions longer than ten hours.

We also analyzed the long term daily average session duration for *workstation-2008*, *itunes-2008*, *share-2008* and *hotspot-2007* and plot the results in Fig. 5. For *workstation-2008* (Fig. 5(a)), we note that the session lengths are longer during breaks (weekend, winter, spring and summer breaks) than when classes were in session. At the application level, these variations were less pronounced for *itunes-2008* (Fig. 5(b)) where we saw a slight trend towards longer session durations. The session lengths for *share-2008* remained small during the semester and increased during the breaks. For the *hotspot-2007* traces (Fig. 5(d)), we note a slight increase in the session duration with less variability, especially when compared to the first six months.

By contrast, the Farsite study [5] used ping messages on a corporate desktop environment and reported that most machines were always available with unavailable intervals occurring during weekends. They reported that the median number of machines were available for more than 95% of the time. The Farsite study observed that the unavailable durations occurred during late nights and on the weekends.

Variations based on the capture mechanism. The session analysis (Section 2.3) showed the differences in session lengths depend on whether the data was captured at the workstation or by a sharing application (e.g., iTunes). In order to understand the availability, we breakdown the session analysis for the 2006 traces into different time ranges and report our observations in Fig. 6. For these experiments, we divided the day into three time ranges: 12 AM to 9 AM (late night to early morning), 9 AM to 5 PM (typical corporate work time) and 5 PM to 12 PM (evening). We analyzed the session durations and the period of unavailability between sessions for the three traces studied: *workstation-2006*, *itunes-2006* and

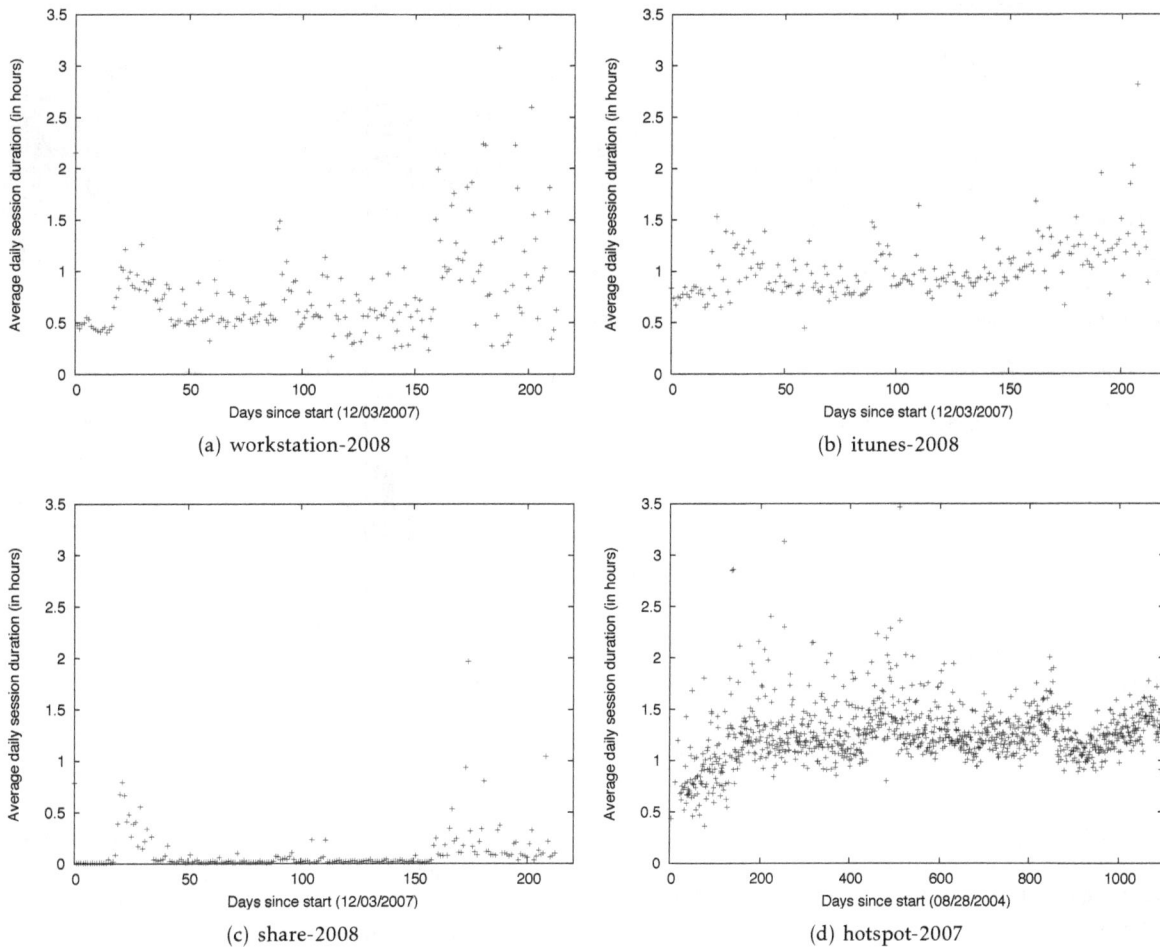

Figure 5. Evolution of session duration

share-2006 and plotted the cumulative distributions. For sessions that spanned multiple time ranges, we attributed the durations to the starting time range. Most sessions were short and so this choice did not have an significant impact.

Analyzing the session durations for the *workstation-2006* (Fig. 6(a)), we noted that the system behavior was similar for the duration of 9 AM - 5 PM and 5 PM - 12 PM. The session durations were shorter during late night hours. For example, the median durations were about twenty minutes as compared to an hour for the rest of the day. 20% of the day time and evening users were available for more than 1.5 hours. Also, an analysis of unavailable durations from Fig. 6(d) showed that late night users were either available for two hours or available for long durations; 70% were available for less than two hours while the remaining 30% were available for over 6 hours. This behavior was consistent with users who either turn off their laptops before they go to sleep or leave them ON throughout the night. We observed no such behavior during the other durations. Over 90% of the evening users returned after less than three hours, while 10% of the morning users were unavailable for over eighteen hours. One likely explanation was that the work day users included staff who returned the following day while evening users included dormitory students who became available after dinner.

itunes-2006 traces were similar to the *workstation-2006* traces (Figs. 6(b) and 6(e)). 60% of the users were unavailable for less than an hour or for the entire eight hours. The session duration did not show significant difference across the various time ranges. However, *share-2006* traces exhibited oscillatory behavior where users became available and unavailable for short periods of time (both for the session durations (Fig. 6(c)) as well for the unavailable durations (Fig. 6(f))).

Our analysis shows the complexity of user availability. Further work is required to capture availability traces from deployed applications. We are collecting such traces for our collaborative system [11, 12] designed for wireless users.

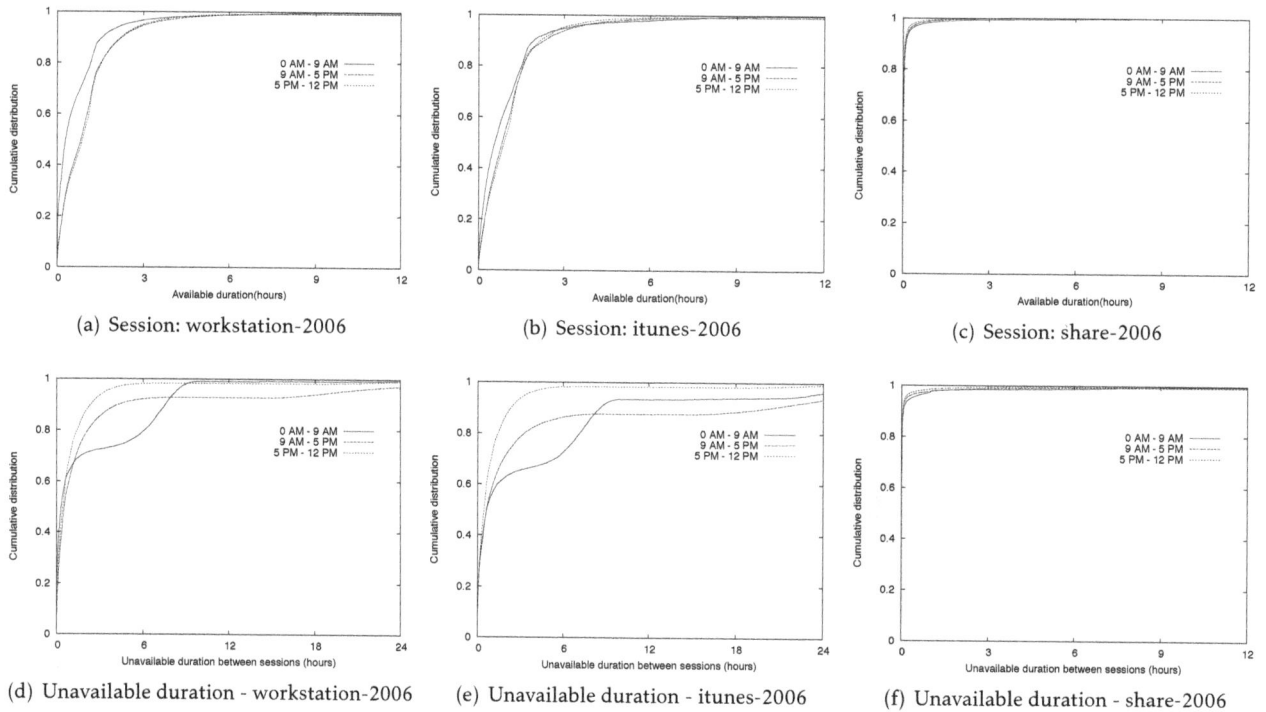

(a) Session: workstation-2006   (b) Session: itunes-2006   (c) Session: share-2006

(d) Unavailable duration - workstation-2006   (e) Unavailable duration - itunes-2006   (f) Unavailable duration - share-2006

Figure 6. Session durations in different time ranges

2.4. User churn behavior

Next, we investigate the introduction and attrition of nodes for the various traces in Fig. 8. A new node will need to receive all the updates from prior users. In the academic scenario, we expect new users to arrive at the beginning of a semester while generally remaining stable throughout the semester. Analyzing the *workstation-2008* traces in Fig. 7(d), we note that over 2,750 users (of the total 4,000 users) were seen within the first few weeks. However, after the winter break (about 50 days since the start of the trace), new users were steadily added even though the overall number of monitored APs was continuously reduced. On the other hand, about 500 users were never seen after the Fall 2007 semester (first few weeks). During the Spring 2008 semester, users constantly left the system until about 150 days into the trace. Note that APs were removed from our monitoring abruptly in the middle of the spring break (and not gradually as indicated by Fig. 7(d)). Similar results were observed for analyzing the *itunes-2008* trace (Fig. 7(e)) and the *corporate-2002* traces (Fig. 8(a)) where about 50% of the users appeared at the beginning of the traces (delayed because some users were likely offline during the start of the trace collection). However, the rest of the users continued to trickle in during the remainder of the trace duration (with a two day jump, likely because of the weekend). We even observed this effect in the *sigcomm-2001* traces (Fig. 8(b)) where the short duration of the

traces made such a behavior seemingly unlikely; we observed that about 35 users joined the system after the first day. Finally, we observed that new users were constantly entering the system and existing users were similarly leaving the system by analyzing the *hotspot-2007* traces (Fig. 8(c)). This is likely typical in hotspots.

2.5. Summary of results

We analyzed the availability traces from academia, a corporate lab, a conference and a hotspot federation. An analysis of users in academia showed that session durations are becoming shorter and period between sessions are getting larger; both at the machine level as well as for the iTunes sessions. Students in academia are inherently mobile; roaming between their dormitory, classes, cafeterias etc. Anecdotally, newer wireless devices are more reliable for power saving sleep cycles; perhaps users are only using their wireless devices when they need to. In the corporate setting traces from 2002, the session durations were longer. In a corporate setting, most users have an office which will allow them to remain online for prolonged durations. The users in the hotspot remain online for short periods and then stay offline for long durations. These observations manifest themselves in fewer users (e.g., a smaller percentage of users) being simultaneously available in the recent traces from academia and in the wireless hotspots. In the next sections, we

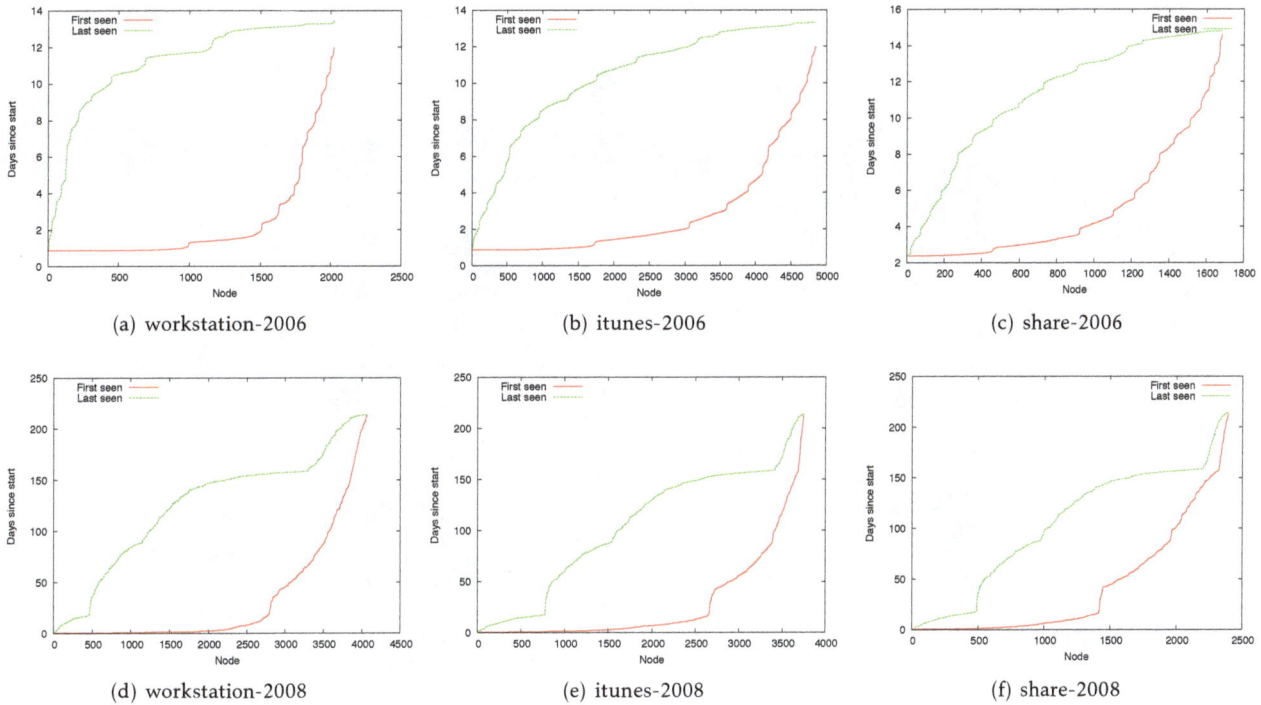

(a) workstation-2006

(b) itunes-2006

(c) share-2006

(d) workstation-2008

(e) itunes-2008

(f) share-2008

Figure 7. User churn behavior (academia)

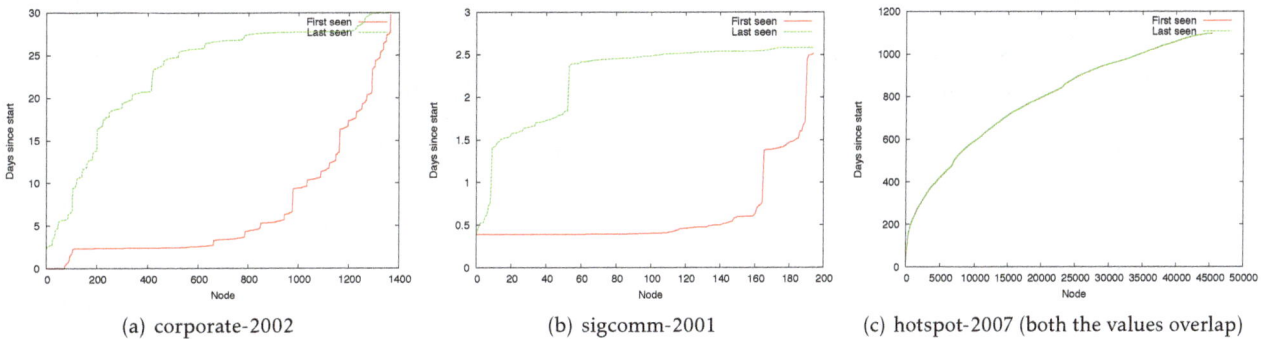

(a) corporate-2002

(b) sigcomm-2001

(c) hotspot-2007 (both the values overlap)

Figure 8. User churn behavior

investigate the implications of these observations for group communication mechanisms.

Next, we investigate various group communications mechanisms. Collaboration applications use group communication mechanisms to implement their collaboration functionality. We consider applications which allow any group member to modify shared contents. The system attempts to reflect each update to all the other group members. The group communication mechanism itself does not guarantee any message delivery order; although group collaboration applications may impose application specific ordering constraints. Also, we investigate the average behavior among a random group of wireless users. This random group should represent a lower bound; actual collaborating group members will exhibit higher correlated availability. We choose representative traces for ease of illustration.

3. Synchronous mechanism

Synchronous mechanisms support concurrent modifications of shared objects by all group members. Conflicting updates by different group members are addressed by either exclusively locking the shared object or using optimistic mechanisms [27]. Without conflicting updates, centralized approaches spend little effort in maintaining consistency. However, distributed approaches still need to propagate updates from the group member who updated the contents to other group members. Some examples of centralized synchronous systems include storage systems

(a) workstation-2008 (max values in a day)

(b) itunes-2008 (max values in a day)

(c) workstation-2006

(d) itunes-2006

(e) corporate-2002

(f) sigcomm-2001

(g) hotspot-2007

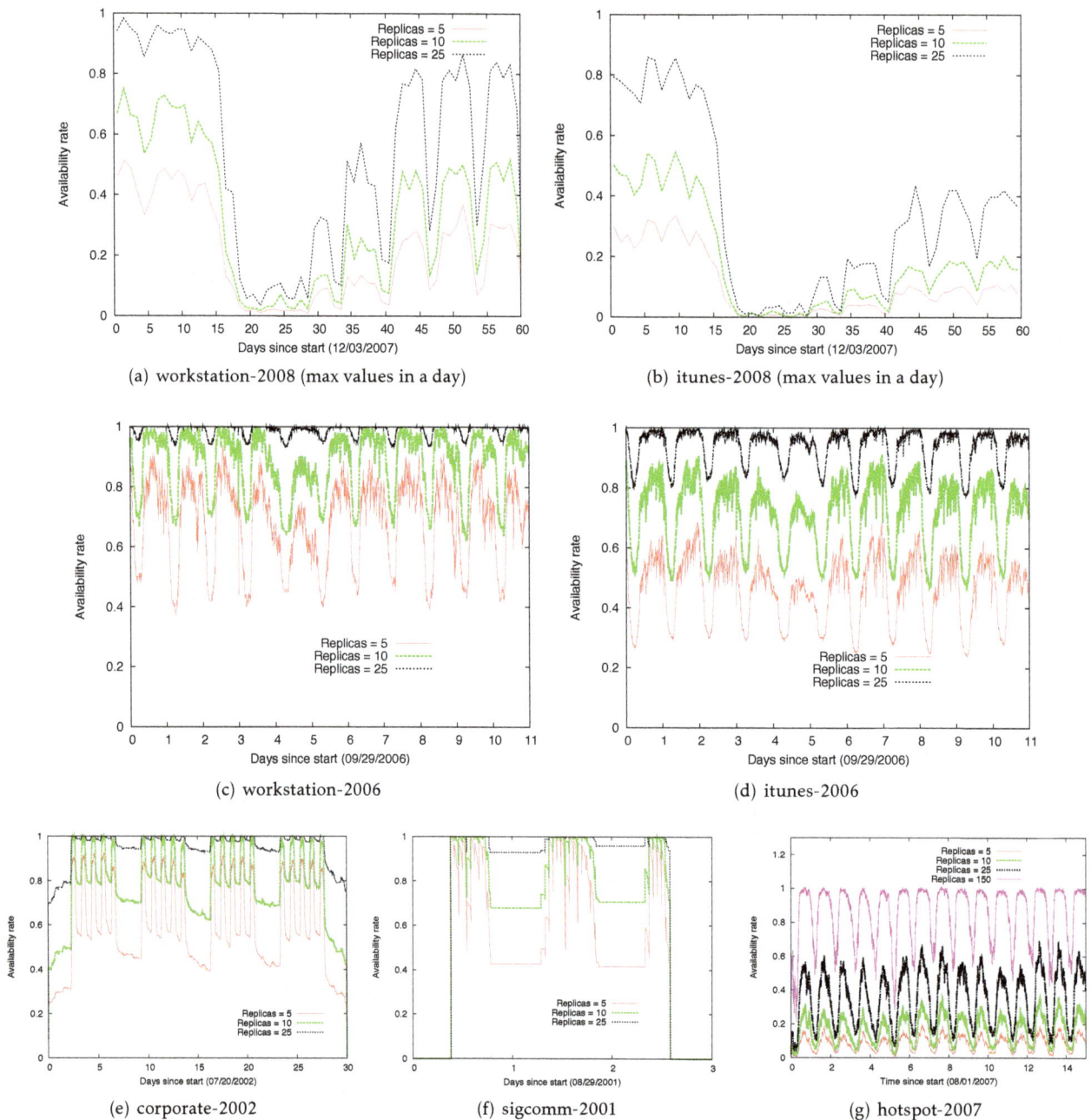

Figure 9. Improving availability of local updates for other group members by replicating objects among wireless devices

such as AFS [28] and NFS [29] as well as applications such as MoonEdit (moonedit.com) and Google-Docs (docs.google.com). Applications such as UNA (n-brain.net) and SubEthaEdit (codingmonkeys.de/subethaedit) use a distributed approach.

In Section 2.2, we showed that few users were simultaneously available. For the more recent wireless users in academia, only 15% of the users were simultaneously available during the daytimes. These figures were even lower for the hotspot users where only 4% of the users were simultaneously available.

On the other hand, 38.7% of corporate users were simultaneously available during the daytimes. These figures were far lower during night times. By comparison, Farsite [5] observed high availability among wired corporate desktops. On average, the number of conflicting updates among wireless group members who were simultaneously available is likely to be small. Conflict resolution techniques designed for wired users should be adequate for wireless users.

Since not all group members are simultaneously available, the challenge is to propagate each update to other group members as they become available. One way to increase availability is to use a wired server for storage. Another alternative is to use the wireless devices themselves. Since contemporary wireless devices are resource rich, we evaluate the amount of replication required to achieve good availability of specific contents. Such evaluations had been performed using desktop storage [30, 31]. Using corporate desktops, Farsite [32] achieved four nines of availability using just three replicas.

We randomly select five, ten and twenty five replicas. We assume that these nodes were simultaneously available with the originating node and so the updates were directly replicated in all these nodes. Using asynchronous mechanisms [33] to propagate updates to these replicas can itself introduce long delays. We calculate the availability of this particular content to others throughout our trace interval. In effect, this is the best case scenario where the group member that requires the update can be any node from our trace other than the replica nodes. Note that we only need one of these replicas to be online in order to make the contents available. We repeated each experiment 1,000 times and plot the average availability by the time of the day for the various traces in Fig. 9.

First we analyzed the system in academia. For the *workstation-2008* and *itunes-2008* traces, we improve readability by plotting the maximum availability values in a particular day. In general, the availability follows a diurnal as well as a seasonal pattern (end of semester, winter break, beginning of new semester). Analyzing the *workstation-2008* traces (Fig. 9(a)), note that five replicas only achieved about 50% availability during December 2007. The availability reduced to almost 0% during the winter break before recovering to about 20%-30% in the Spring 2008 semester. Increasing the replication rates to ten replicas improved the availability in December 2007 to 70% while 25 replicas are required to achieve 100% availability (albeit briefly). Comparing these results with the *workstation-2006* traces (Fig. 9(c)), we observed that five replicas achieved up to 90% availability while 25 replicas achieved 100% availability consistently during daytimes. Reduced session durations is worsening the prospect for distributed update propagation.

An analysis of *itunes-2008* showed that the availability was worse for iTunes users. Fig. 9(b) shows that *itunes-2008* users only achieve about 40% availability for using five replicas in December 2007. Even when we increase this amount to 25 replicas, the availability only reaches 90%. On the other hand, by the beginning of Spring 2008, even with 25 replicas, the maximum availability only reaches around 40%. By contrast, an analysis of *itunes-2006* in Fig. 9(d), shows that 25 replicas was enough to provide 100% availability during

the daytimes. Even five replicas achieved about 60% availability for a targeted contents during the day times.

Next we analyzed the behavior in a corporate setting using *corporate-2002*. We note that (Fig. 9(e)) five replicas provide about 90% availability during the weekday business hours. During the evenings, we achieve availability of about 60% which reduces to about 45% by the weekend. Replications of about 25 copies achieve about 95% availability even during the weekend. Note that Farsite [32] achieved excellent availability using only three desktop replicas.

The *sigcomm-2001* traces (Fig. 9(f)) show that even five replicas achieved good availability during the conference durations. However, 25 replicas are required to achieve good availability during the night hours when the conference was not in session. Finally, we analyzed the behavior for the *hotspot-2007* traces. In Fig. 9(g), we note that five replicas only achieve about 10% availability. Even 25 replicas only achieves about 60% availability during daytimes. We required about 150 replicas to achieve 100% availability during the weekdays. Distributed update propagation is not feasible among wireless devices in a hotspot; the hotspot provider should provide the storage infrastructure necessary to make shared contents available to other group members.

To summarize, the number of simultaneously available users and the duration that each user was available is decreasing. Unlike wired scenarios, this trend reduces the effort required to maintain consistency among updates. However, this exacerbates the problem of distributing updates from one group member to all others. Distributed approaches require a large number of replicas to forward updates to other members. Even though laptops are resource rich and capable of providing service to others, the replica requirement makes a distributed approach impractical, especially in hotspots where we require over 150 replicas. Hotspots are also unlikely to provide the necessary storage servers. The performance achieved in a university and corporate scenarios were tolerable, especially during durations when all the users were active (e.g., daytimes). Next, we investigate asynchronous mechanisms.

4. Asynchronous mechanism

Asynchronous mechanisms modify local copies of shared contents. Updates are eventually propagated to other group members. Because of the update propagation delay, collaboration applications explicitly reconcile updates from other group members. Propagation and reconciliation can be mediated by servers (e.g., Coda [34], Apple iDisk (apple.com/mobileme) and Windows SkyDrive (skydrive.live.com)) or through distributed mechanisms (e.g., Ficus [35], Bayou [36] and Windows Live Sync (sync.live.com)).

Update propagation can be implemented as a daemon process that starts as soon as the users authenticate themselves with the system. Since *workstation-2008* traces capture this duration, we illustrate the system performance using this trace.

4.1. Propagation policies

A distributed approach propagates the updates to simultaneously online group members, likely delayed from when the update was created. Each update is subsequently propagated to other nodes through successive gossips in order to eventually reach all group members. Some of these gossip sessions are unnecessary because the corresponding pair of nodes already have all the updates available. Frequent propagation attempts will improve system performance while incurring a high number of unnecessary gossips. For example, Bayou [36] used version vectors to identify updates that needed to be propagated during a pairwise anti-entropy session. We investigate the impact of various policy parameters for update propagation.

Updates can either be pushed by the originating node or pulled by other nodes. We refer to these policies as *P2P-push* and *P2P-pull*, respectively. Similarly, a server mediated approach uses a (always online) server. This mediation can either be initiated by the node or the server. In server initiated mechanisms, the server periodically pulls updates from nodes that are online and then pushes them to other online nodes. We refer to this policy as *Svr-ServInit*. On the other hand, in a node initiated policy, the node that created updates periodically pushes the updates to the server while also retrieving updates from other nodes from the server. We refer to this policy as *Svr-NodeInit*. When they come online, nodes always initiate a push or pull operation.

The time to perform each push and push operation depends on the size of the update as well as on the available network bandwidth. Given the bandwidth availability on wireless LANs, we do not model this duration (similar assumptions as Demers et al. [37] and Birman et al. [38]). We assume that nodes go offline without explicitly pushing its updates to other nodes.

Depending on the group size, we also need to choose several other propagation parameters. During the initial phases, few nodes contain the updates and hence the system has to aggressively propagate the updates before the nodes that contain them go offline. This is particularly important while the update was only available in the node that created the update.

4.2. Performance metrics

Unlike metrics used by Vahdat et al. [39] that measured the time required to propagate a single update from a random node to every other group member, we use metrics that capture a dynamic system where

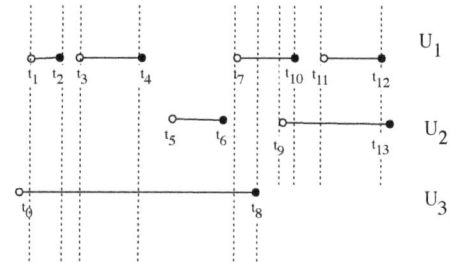

Figure 10. illustration of availability behavior of three users

new updates are continuously created by each group member. Consider a group of three nodes (U_1, U_2 and U_3) and their availability durations as illustrated in Fig. 10. U_1 is available from $t_1 \ldots t_2$, $t_3 \ldots t_4$, $t_7 \ldots t_{10}$ and $t_{11} \ldots t_{12}$; U_2 is available from $t_5 \ldots t_6$ and $t_9 \ldots t_{13}$ and U_3 is available from $t_0 \ldots t_8$. Say at t_8, our metrics quantify the amount of updates from U_1, U_2 and U_3 that are unavailable in the other nodes. Rather than choosing arbitrary values for update creation rates, we assume that nodes create updates at a constant rate. Hence, the amount of time that a node was available measures the updates created by the node. Consider a node that was missing updates created by another node that was online for two hours. A target application might create updates at the rate of one update per hour. For this application, this system performance would translate to missing two updates.

We quantify the number of updates using a lag metric and the network overhead by the number of gossips.

- *lag metric:* We measure the average number of updates that are unavailable at a node using *lagAmount*. Consider the illustration in Fig. 10 (ignoring U_3 for now). At t_5, U_2 does not have updates ($t_1 \ldots t_2$) and ($t_3 \ldots t_4$) from U_1; the *lagAmount* at U_2 is $(t_2 - t_1) + (t_4 - t_3)$. At t_6, the *lagAmount* of U_1 is $t_6 - t_5$ and for U_2 is $(t_2 - t_1) + (t_4 - t_3)$ for an average *lagAmount* of $\frac{(t_2-t_1)+(t_4-t_3)+(t_6-t_5)}{2}$. The lag values depend on the propagation policies (further explored in Section 4.4). For example, U_3 could help in ferrying updates between U_1 and U_2.

- *Gossips: numGossips* measures the number of pairwise anti-entropy operations. If a particular gossip only considered updates from the corresponding node (without propagating updates from other peers), we consider these to be wasted gossips (*numWGossips*). Note that these metrics do not account for the size of updates (for example, measured in kilobytes) that were actually propagated during a successful gossip operation; this assumption was also used by prior work [37, 38].

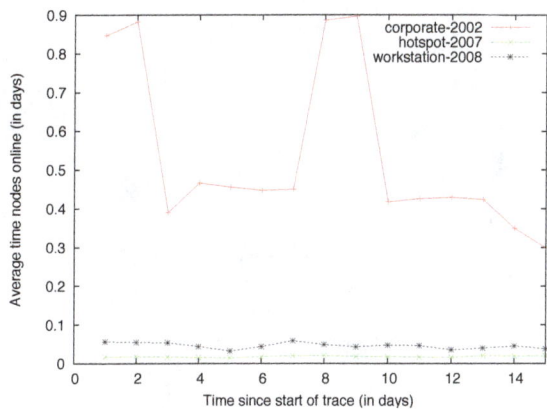

(a) Average time nodes online

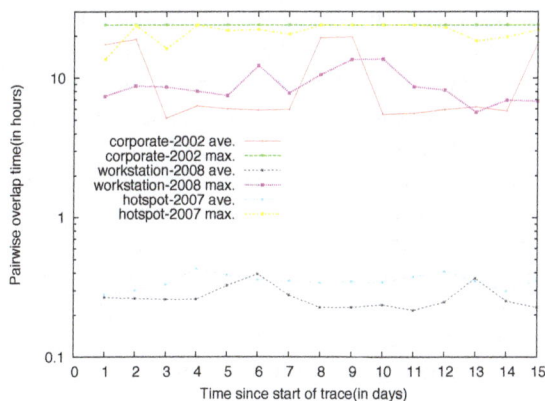

(b) Overlap time among user pairs

Figure 11. Availability parameters that affect the lag metrics

The number of node pairs is $O(n^2)$ of the group size, the lag metrics depend on the group size ($O(n)$). The lag metrics depend on the number of updates created; hence given our assumptions, a locale where users are available for long durations will create large numbers of updates which can potentially lead to higher *lagAmount*s. We plot the average number of updates created per day of our traces in Fig. 11(a). *Corporate-2002* users were available for long durations. Hence they were assumed to be creating large number of updates. Users were online for over 0.4 and 0.9 days on weekdays and weekends, respectively. However, the number of users was smaller during the weekends. These values were smaller for the other traces, about 0.02 days for the *hotspot-2007* trace and 0.04 days for the *workstation-2008* traces.

For distributed scenarios, the lag metrics also depend on the periods when nodes overlap. We plot the amount of pair-wise node overlap for the various traces in Fig. 11(b). The average overlap between pairs of nodes was 0.35 hours for both the *hotspot-2007* and *workstation-2008* traces. However, the duration was about six hours during the weekdays and over 11 hours over

the weekend for the *corporate-2002* users. Unlike the *corporate-2002* trace where some nodes overlapped for 24 hours, the pairs of *university* users never overlapped for more than 10.5 hours. Thus, the number of updates as well as the rate of update propagation are likely to be high in the *corporate-2002* scenario.

We randomly created groups of three, fifteen and thirty users and investigate the effects of the update frequency using the lag and gossip metrics. Three is the smallest meaningful group size for asynchronous propagation while thirty is a large group. We repeat the simulations for fifty different groups.

4.3. Limitations of asynchronous propagation

First we analyzed the performance limit. The best performance is achieved in a *Svr-ServInit* scenario where the update duration is zero seconds; i.e., the server constantly polls all nodes; any updates are immediately pulled by the server and sent to all the other nodes that are also online. Still, updates cannot be sent to nodes that are not online. Thus, the best performance is dependent on the user availability. Also, this policy is not practical because it will incur tremendous network overhead for constantly pulling updates. Practical systems delay this operation and propagate updates at less frequent intervals; we investigate these intervals in Section 4.4.

We plot the *lagAmount* for groups of size three, fifteen and fifty in Fig. 12. We note that the lag metrics continue to rise throughout the trace duration. This phenomenon is caused by node churn. When a node leaves the system, we continue to compute its lag values for updates created by other nodes. Our propagation policies described in Section 4.4 counteract this effect by reporting the relative difference from the best policy.

The lag amounts were high and depend on the group size. From Figs. 12(a), 12(b) and 12(c), we note that the number of updates that are yet to propagate to a node is as high as forty four days by the end of the trace for the *corporate-2002* trace. The session durations for the *hotspot-2007* and *workstation-2008* traces were smaller (Section 2), producing smaller lag amounts.

Each day, nodes did not receive updates that were created in about $\frac{1}{5}$ days from other nodes. The number of updates accumulated depends on the node session duration and churn; *corporate-2002* users exhibit large amounts of pending updates. Unless efforts to improve the wireless user availability are successful, asynchronous mechanisms are not suitable for applications which require tighter propagation.

4.4. Performance of propagation policies

Next, we investigate the extra overhead imposed by various update propagation policies as compared to the best policy (Section 4.3). For example, if the *lagAmount*

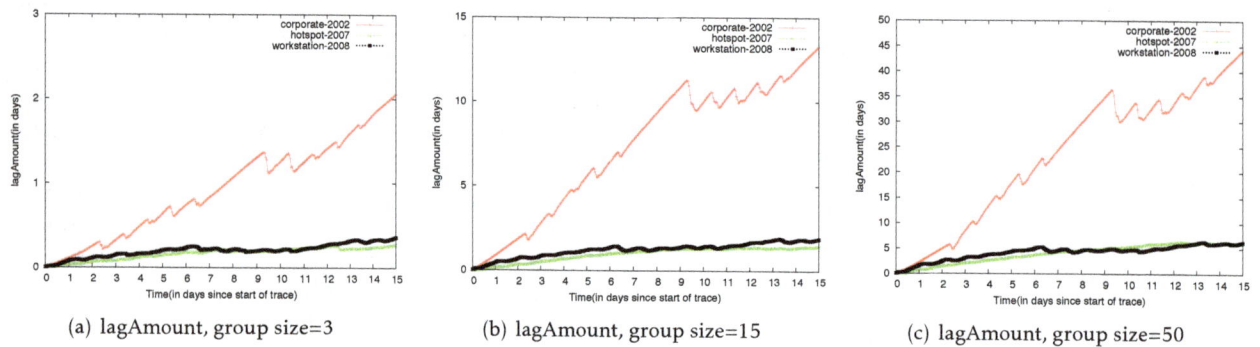

(a) lagAmount, group size=3

(b) lagAmount, group size=15

(c) lagAmount, group size=50

Figure 12. Best performance achievable for the various traces

on a particular day for the *P2P-pull* policy was four while the corresponding values for the best policy was three, we report a relative *lagAmount* of one day. Analyzing the relative costs has the added benefit of eliminating the lag values accumulated due to not propagating updates to nodes that have left the system. However, updates which were created on the node that will leave the system and that was not propagated to other nodes will still affect the relative costs. We prefer a policy that minimizes the relative performance.

Investigation of propagation frequency. Nodes propagate updates when they become online with subsequent propagations at regular intervals. This frequency is application driven and depends on the user availability. If the node becomes unavailable before update propagation, then the updates created during this session will be delayed. It is impractical to require nodes to propagate updates before going offline. Hence, we investigate whether we can predict when a node will go offline and correspondingly adapt the update frequency. We investigate history based approaches that use the past session durations to predict the expected session duration. For this analysis, we use various history values of one, two and three. For each of these values, we choose base frequencies of five min., fifteen min., thirty min. or an hour. For each base frequency, if the predicted session duration was less than the predicted duration, we reduce the update frequency to the predicted value. We tabulate the cumulative percentage of times when a node goes offline before subsequent update propagation for our traces in Table 1. The table is read as follows: for the *workstation-2008* trace, using a base frequency of an hour, 90.1% of the time, the node will go offline before subsequent propagation. Given the average session duration of 20 mins. (Section 2), most of these updates are unduly delayed until the node came back online; average time between sessions was 1.78 hours for this trace (Section 2). The lag values are reduced when the times when no

further updates were propagated was low; using the last session duration can reduce this value to 48.9%.

From Table 1, we note that the percentage of time when nodes did not subsequently propagate updates was higher for *workstation-2008*. In all the traces, the best performance was achieved when the base update frequency was high; this observation needs to be reconciled with the number of unnecessary gossips. An adaptive policy that used the recent past session duration was also effective. We investigate our propagation policies using the adaptive as well as application specified propagation durations.

P2P-pull. We plot the relative *lagAmount* and the number of gossips and unnecessary gossips in Fig. 13. As nodes come online, they pull updates from other group members that are also online at frequent intervals that are either fixed or adaptive (Section 4.4). The times that nodes come online are unpredictable. Hence the pull operations to propagate updates created at any node to other nodes is also random. Note that the adaptive duration allows the local node to pull updates from other nodes before going offline; it does not affect when the local updates are sent to other nodes. Adaptive policies only made significant improvements in the *workstation-2008* scenario.

Consider the relative *lagAmount* (Fig. 13(a)). We note that the *lagAmounts* continue to increase because of the residual updates left in a node that were not pulled by other nodes before it left the system; the best policy (Section 4.3) would have propagated these updates. This effect was pronounced in the *hotspot-2007* scenarios. For a group of size 15 in *hotspot-2007* scenario with an adaptive policy, base frequency of one hour and history depth of one, the relative *lagAmount* on the fifteenth day was over 1.2 days; i.e., each pair of nodes did not have about $\frac{1.2}{14} = 0.086$ days worth of updates. These values were slightly bigger for smaller groups which had fewer opportunities for update propagation (pairwise *lagAmount* of $\frac{0.2}{2} = 0.1$). We also observed that the adaptive policy for the *workstation-2008* scenario

trace	base freq = 5 min				base freq = 15 min				base freq = 30 min				base freq = 60 min			
	0	1	2	3	0	1	2	3	0	1	2	3	0	1	2	3
corporate-2002	1.4	1.3	1.4	1.4	10.0	8.9	9.7	9.7	17.5	14.6	16.0	16.3	28.5	22.3	24.7	25.2
hotspot-2007	2.5	2.4	2.5	2.5	20.1	16.7	17.4	17.2	38.8	28.7	30.2	30.1	60.6	39.4	41.4	41.1
workstation-2008	18.9	15.7	17.3	17.8	40.0	30.1	33.9	35.5	63.4	41.6	47.1	49.2	90.1	48.9	52.4	53.7

Table 1. Cumulative percentage of times when node went offline before subsequent update propagation

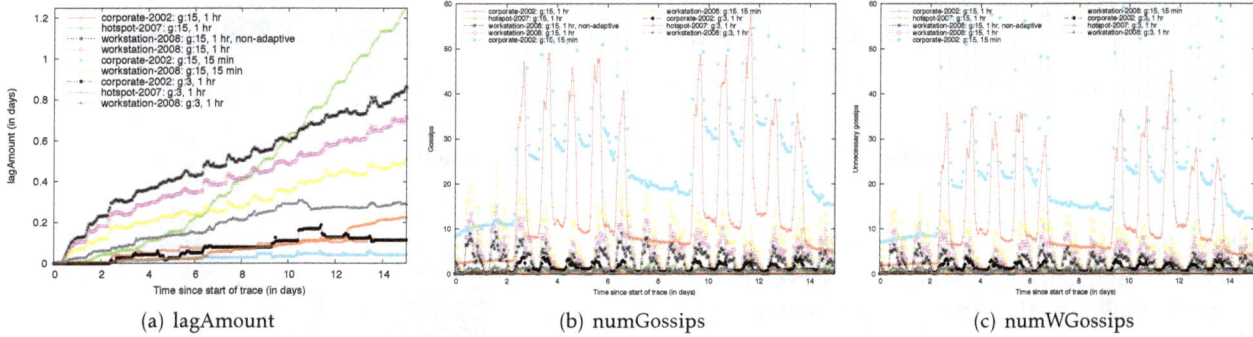

(a) lagAmount (b) numGossips (c) numWGossips

Figure 13. P2P-pull (adaptive duration by default)

showed a residual relative *lagAmount* of 0.7 days vs 0.88 days for a non-adaptive policy.

These updates were spread over long durations, especially in the *hotspot-2007* scenario. The improvements of the adaptive policy was more modest even for the *workstation-2008* setting. Investigating the number of gossips and the number of unnecessary gossips in Figs. 13(b) and 13(c), respectively, we note the high number of gossips as well as unnecessary gossips in the corporate setting. For other scenarios which exhibit poorer user availability, the number of gossips were low; about ten for groups of size fifteen. Later, we show that these parameters are competitive with server mediated approaches.

P2P-push. For the *P2P-push* policy (Fig. 14), the system is far more sensitive to the propagation rate of the local node. Note that for *P2P-pull*, it is the responsibility of other nodes to pull local updates. For a group of size 15, there are 14 other nodes which are attempting to perform this propagation vs one for the *P2P-push* policy. An adaptive policy may push its updates to another node before going offline. However, any local update that was not pushed by the source to another node will not propagate to any other nodes; increasing the overall lag metrics. Overall, the system performance is slightly worse than the performance of the *P2P-pull* policy. An adaptive policy for *workstation-2008* setting for a group of size 15 with base propagation frequency of one hour experiences an *lagAmount* of 0.75 days (as compared to 0.7 days for the *P2P-pull* policy). Hence, *P2P-pull* is preferable to *P2P-push*.

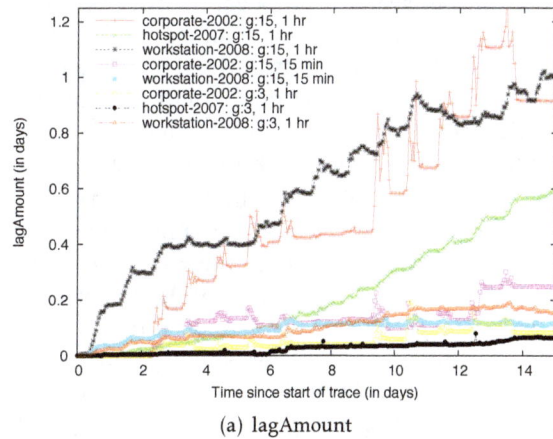

(a) lagAmount

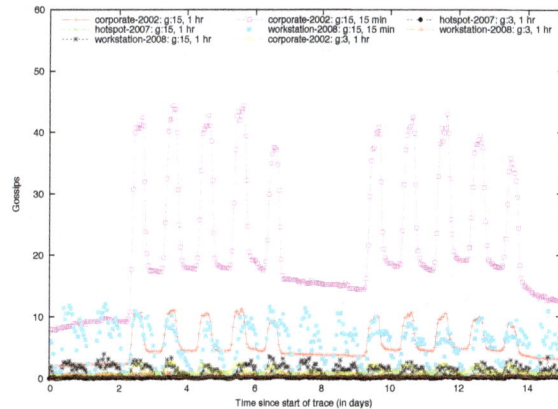

(b) numGossips

Figure 15. Svr-SrvInit

(a) lagAmount

(b) numGossips

(c) numWGossips

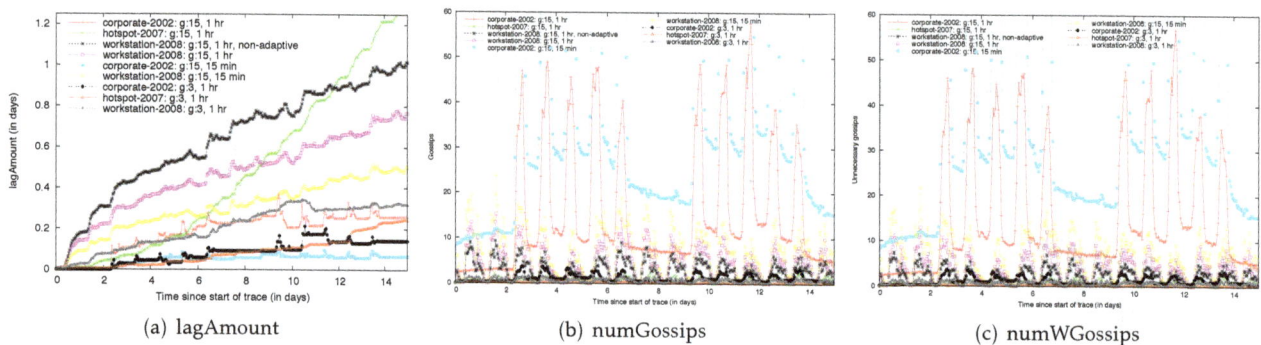

Figure 14. P2P-push (adaptive duration by default)

Svr-ServInit. Next, we investigate the relative performance of server assisted approaches. The server periodically pulls new updates from online nodes and sends them to other online nodes. We plot the relative *lagAmount* and the number of gossips in Fig. 15. Note that the server cannot adapt to the session durations of individual nodes; if a particular node came online and went offline between the propagation interval, then its updates are not propagated to other nodes even though the server is continuously available. Each gossip as compromised of two operations; a pull of contents form a particular node and push these contents to other nodes. All the gossips in this scenario are useful.

From Fig. 15(a), we observed that the *lagAmount* progressively increases with time. Even if the server was continuously available, it is possible for a node to come online and go offline without propagating its updates to other nodes (especially if the session duration was below the propagation rate). For small group sizes, the *Svr-ServInit* policy exhibits better performance than distributed approaches. For example, for groups of size three in a *workstation-2008* with base updates every hour, the adaptive *P2P-pull* policy experienced a relative *lagAmount* of 0.3 days at the end of the 15th day. However, the *Svr-ServInit* policy for similar settings experienced a *lagAmount* of 0.15 days. However, the distributed approaches are more competitive for larger groups. For the *Svr-ServInit* policy in the *workstation-2008* scenario with a group size of 15 and a propagation frequency of an hour, the relative *lagAmount* is about a day. The corresponding values for an adaptive *P2P-pull* was only 0.7 days. The improvements were even more pronounced for groups of size 50. Since the distributed approaches allow all the nodes to initiate the propagation operation, P2P schemes achieve better randomization of update propagation.

Svr-NodeInit. *Svr-NodeInit* policy (Fig. 16) is similar to the *P2P-push* policy in that nodes periodically sends its updates to the server while simultaneously downloading new updates from other nodes. Thus, this

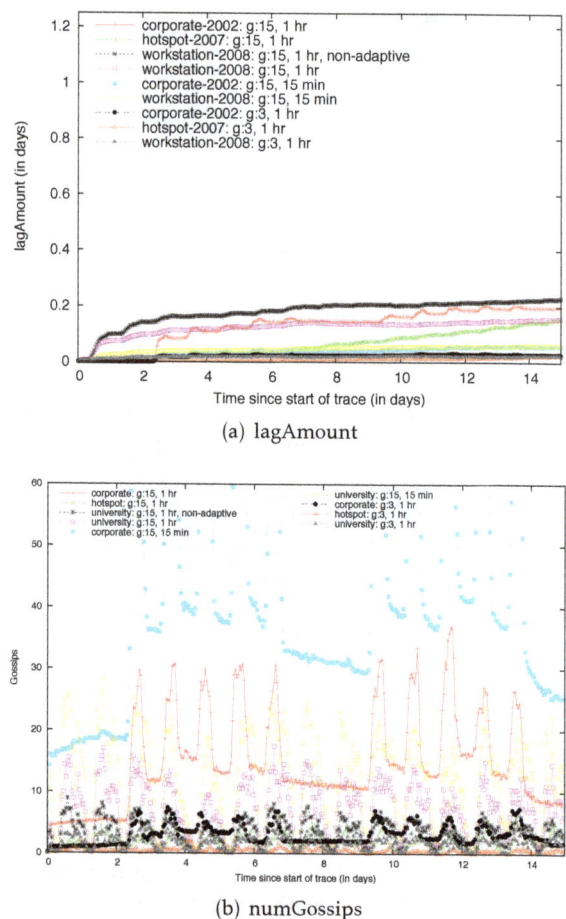

(a) lagAmount

(b) numGossips

Figure 16. Svr-NodeInit

policy leverages the availability of server nodes and propagates its updates quickly. Fig. 16(a) shows that even after fifteen days, the *lagAmount* was a modest 0.2 days (this value can be as high as 1.2 days for the other propagation mechanisms).

Summary. We investigated the performance bounds of a best policy that instantaneously propagated the updates created at any node to other nodes. We then compared the performance of several distributed and server mediated approaches with this policy. The fundamental limitation of these practical policies was the mismatch between the propagation rates and the times when nodes go offline before propagating their updates to other nodes. Frequent propagations correspondingly increase the number of unnecessary gossips; an adaptive policy that tuned the frequencies to the node session duration improved the system performance for distributed approaches. These adaptive policies only required the last session duration of any node. We showed that the *Svr-NodeInit* policy achieved the best performance with the *P2P-pull* policy exhibiting competitive performance. Overall, distributed policies are competitive, especially for larger groups.

4.5. Peer selection mechanisms

Increasing the propagation rate improves system performance while also requiring a higher number of gossips. One approach to address this concern is to choose higher propagation rates while proportionately reducing the number of nodes to propagate updates. We investigate this tradeoff for our wireless users.

We plot the relative performance for varying the number of peers used for update propagation using the non-adaptive *P2P-pull* policy for groups of size 15 in Fig. 17. We reduce the percentage of users from 100% to 50% and 25% while increasing the propagation rates from 1 hour to 30 min. and 15 min. For the *corporate-2002* trace, choosing fewer nodes has relatively minor effect on the *lagAmounts*. At the end of fifteen days, the *lagAmount* was about 0.1 days for propagating to 100% as well as 50% of the nodes (while doubling the propagation rates). Propagating to 25% of the nodes (while quadrupling the propagation rate) increased the *lagAmount* to about 0.15 days. The corresponding reduction in unnecessary gossips was from 24 to 16 and 14, respectively. Reducing the percentage of nodes to propagate updates while increasing the update frequency is a viable approach for large *corporate-2002* groups. For the *hotspot-2007* trace, reducing the frequency increased the *lagAmount* from 1.3 days to 1.5 days with small differences in the number of unnecessary gossips. On the other hand, for the *workstation-2008* trace, reducing the percentage of online nodes to propagate updates from 100% to 50% and 25% (while exponentially increasing the propagation rates) drastically worsened the *lagAmount* from about 0.8 days to 1.3 days and 1.6 days, respectively. The number of unnecessary gossips showed a small improvement.

A similar analysis for groups of size fifty (Fig. 18) showed a worse performance for both the *hotspot-2007* and the *workstation-2008* traces. For the *hotspot-2007* traces, the *lagAmounts* increased from 2 days to 2.6

and 3.2 days, respectively. For the *workstation-2008* traces, the *lagAmounts* increased from 0.6 days to 1.2 and 1.55 days, respectively. The number of unnecessary gossips also worsened for the *corporate-2002* trace; from 50 gossips to 225 and 125 gossips, respectively. For scenarios which exhibit poorer availability, reducing the percentage of nodes while increasing the update propagation rate is not a viable option.

5. Related work

Bolosky et al. [5] analyzed the availability of wired corporate desktops. They [32] showed that the availability was sufficient to support up to four nines availability using three replicas in the Farsite [30] storage system.

Prior studies analyzed user mobility, the network traffic and load characteristics of wireless users in various deployment scenarios. We used some of these traces to analyze the behavior of group communications amongst these wireless users. Tang et al. [40] monitored the WLAN network in a university building using *tcpdump* traces, AP SNMP and authentication logs in order to understand the network traffic and load characteristics. Analyzing the authentication logs, they observed that session lengths were longer than twelve hours. Kotz et al. [16] investigated a campus wide WLAN network using packet dumps (*tcpdump*) as well as access point SNMP and association logs. They note that HTTP protocol and file backups dominated the wireless traffic. They also conducted a follow-up analysis [3] on the evolution of their campus WLAN deployment which showed that the traffic had evolved to include significant amounts of traffic for streaming multimedia, VOIP and P2P traffic. They also observed that on-campus traffic outstripped off-campus traffic, an indication of the importance of group communications among wireless users. Chandra et al. [9] analyzed the popularity and nature of objects shared by iTunes users in a campus setting. Balachandran et al. [22] analyzed the wireless traffic from SIGCOMM 2001 participants and showed that Web and SSH dominated the traffic at over 64% of the traffic. Balazinska et al. [20] analyzed the access point load and user mobility behavior of corporate wireless LAN users by using SNMP probes of the access points. Papadopouli et al. [41] analyzed the wireless user mobility pattern at the UNC campus. On the Verizon hotspot network, Blinn et al. [26] observed that 45.74% of the user sessions lasted more than one hour. Hsu et al. [42] presented a comprehensive analysis of the user mobility behavior across four different university campuses using access point logs. McNett et al. [43] focused their mobility analysis to PDA users in a university setting.

Song et al. [44] used the AP records to synthesize contact patterns among wireless users. They note that asynchronous update propagation can be unacceptably

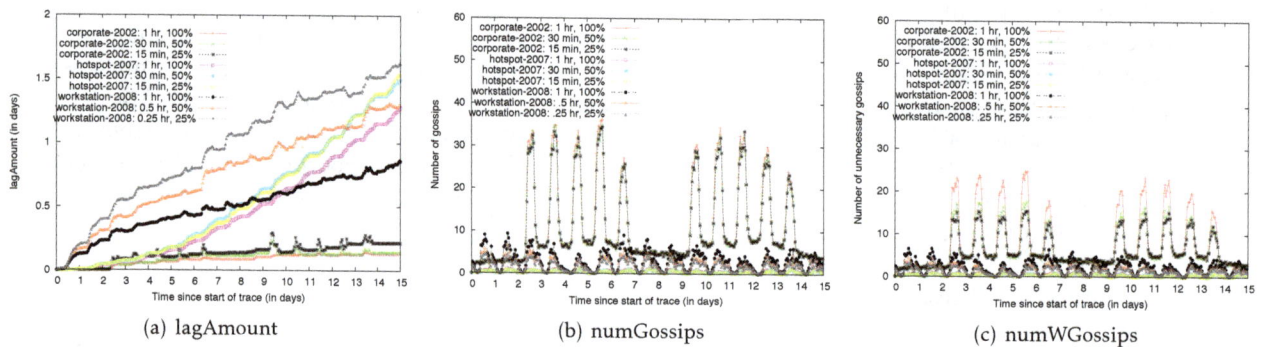

(a) lagAmount   (b) numGossips   (c) numWGossips

Figure 17. Tradeoff between update frequency and choosing fewer online nodes (group size: 15, *P2P-pull*)

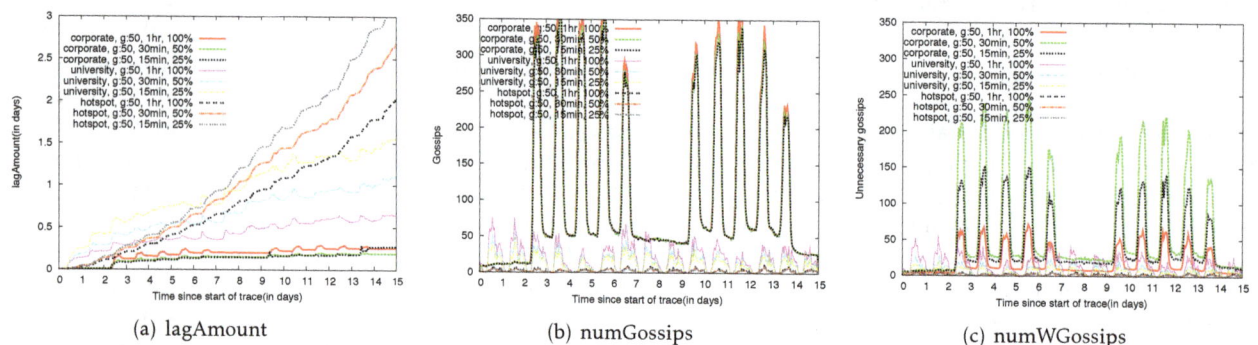

(a) lagAmount   (b) numGossips   (c) numWGossips

Figure 18. Tradeoff between update frequency and choosing fewer online nodes (group size: 50, *P2P-pull*)

long, especially among casual users (some users never again meet each other). We assume the availability of a wireless distribution infrastructure. This allowed us to ignore the spatial mobility patterns of the users; we consider any two nodes that were online anywhere on the wireless trace to be accessible. We expect our node availability to be far better than that was observed using user vicinity contact measurements. However, we observed limited system performance. This places serious doubts on the viability of mechanisms that require collaborating groups be co-located ([44]).

Demers et al. [37] describe epidemic algorithms and the parameters that affect system performance. Epidemic algorithms are designed to asynchronously propagate updates; particularly in scenarios with poor user availability. They operate without strict bounds on the time to propagate updates to all the participants. Birman et al. [38] surveyed the recent developments on the strengths and limitations of gossip protocols. They showed that gossip based protocols are not designed to expeditiously propagate updates. We use empirical wireless user availability data to investigate the impact of the propagation parameters.

Vahdat et al. [39] used epidemic routing to propagate updates in an ad hoc network. They simulated a random node mobility and showed that epidemic routing

achieved eventual delivery of 100% of messages. Similarly, Davis et al. [45] investigated propagation among wearable computers using simulated human mobility. They investigated the effects of message duplication and buffer overhead. Recently, delay tolerant network technologies (DTN) are used to asynchronously propagate updates among a set of clients. Fall [46] introduced a network architecture that operated without continuous network connectivity among the participating nodes. In their followup work, Jain et al. [47] investigated the routing behavior across a DTN. They used simulations and progressively increased the amounts of network topology information available to the routing mechanism. They showed that the systems performed better with the addition of more topology information. We validate propagation rates using empirical node availability.

Bakhshi et al. [48] surveyed analysis techniques for gossiping protocols. Our evaluation metrics were influenced by consistency count metric (Kuenning et al. [49]). Jelasity et al. [50] addressed the problem of selecting peers for gossiping. Kwiatkowska et al. [51] evaluated gossip protocols using probabilistic model checking. In contrast to simulation based studies, they provide both an exhaustive search of all possible behaviors of the system, including best and worst-case scenarios and exact quantitative results. They were concerned with identifying the set of gossip peers; each node maintains a relatively small local membership table providing a partial view of the network. Using empirical wireless users availability, we show that the system should gossip with all the available peers.

Saito et al. [52] surveyed a number of optimistic replication algorithms. Bayou [53] provides an user-level storage for asynchronous collaborative applications. Nodes exchange version vectors to identify updates that needed to be exchanged during a pair-wise anti-entropy protocol. Updates eventually reach all the participants. The system provides some bounds by using a primary commit protocol. Refdbms [54] used a similar anti-entropy protocol to disseminate bibliographic entries. Allavena et al. [55] described a scalable gossip-based for local view maintenance. Khelil et. al. [56] developed an epidemic model for an information diffusion algorithm. Motani et. al. [57] developed a wireless virtual social network (PeopleNet) which mimics the way people seek information via social networking. It used the infrastructure to propagate queries of a given type to users in specific geographic locations, called bazaars. Within each bazaar, the query was further propagated between neighboring nodes via a peer-to-peer connectivity until a query match. Rivière et al. [58] described an architecture to develop and reuse epidemic based systems. Zhuang et al. [59] described an Internet indirection infrastructure [60] among mobile users. Our empirical analysis of the availability pattern of wireless users showed that the system that exclusively used other peers was likely to enjoy similar performance to a system that used servers for assistance in propagation, especially for larger group sizes.

6. Conclusions

Modern wireless devices are resource rich and ubiquitous. In this work, we analyzed the behavior of various group communications mechanisms at a university, corporate lab, conference venue and a hotspot federation using empirical user availability traces. Our analysis is agnostic to the requirements of specific applications, both in terms of the frequency of updates to the shared contents as well as the size of these updates. We show that the availability behavior was better in corporate settings where the users were available for longer durations with a larger percentage of the wireless users available on weekdays. The session durations were smaller in academia and are becoming shorter. The systems exhibited constant node churn which places heavy load on asynchronous group communication systems that need to transmit prior updates to these newer nodes. We developed a lag metric to measure the performance of our system. We show that, regardless of the

propagation policy, factors such as node availability and node churn play an important role in the system performance. Distributed policies are competitive, especially for larger groups. We show that for wireless users, propagation mechanisms should propagate update to all the nodes. Our work highlights the need for robust expiration mechanisms for older updates. The amount of updates created is small enough to make distributed approaches viable. We built a middleware called Yenta to implement the lessons learn from our analysis. The algorithms also form the basis for our *flockfs* moderated collaboration group-ware. Stable versions of Yenta will be published at http://yenta.sourceforge.net/.

Acknowledgements.   We thank Kevin Smyth for his help in configuring the APs. This work was supported in part by the U.S. National Science Foundation (CNS-0447671).

References

[1] DELLA CAVA, M.R. (2006), Working out of a 'third place', USA Today. 10/5/2006.

[2] iSUPPLI CORP. (2008), Notebook pc shipments exceed desktops for first time in Q3, isuppli.com/NewsDetail.aspx?ID=19823.

[3] HENDERSON, T., KOTZ, D. and ABYZOV, I. (2004) The changing usage of a mature campus-wide wireless network. In *MobiCom '04*: 187–201.

[4] BUDDECOMM (2008), 2008 global wireless broadband - next generation mobility.

[5] BOLOSKY, W.J., DOUCEUR, J.R., ELY, D. and THEIMER, M. (2000) Feasibility of a serverless distributed file system deployed on an existing set of desktop pcs. In *ACM SIGMETRICS*: 34–43.

[6] ROSENWALD, M.S. (2009) Digital nomads choose their tribes. *Washington Post* .

[7] KOTZ, D. and HENDERSON, T. (2005) Crawdad: A community resource for archiving wireless data at dartmouth. *IEEE Pervasive Computing* 4(4): 12–14.

[8] CHEN, G. and KOTZ, D. (2005) *Structural Analysis of Social Networks with Wireless Users*. Tech. Rep. TR2005-549, Dept. of Computer Science, Dartmouth College.

[9] CHANDRA, S. and YU, X. (2007) Share with thy neighbors. In *ACM/SPIE Multimedia Computing and Networking (MMCN 2007)* (San Jose, CA).

[10] (2008), The new oases, The Economist.

[11] CHANDRA, S. and REGOLA, N. (2009) flockfs, a moderated group authoring system for wireless workgroups. In *Mobiquitous '09* (Toronto, Canada).

[12] CHANDRA, S. (2011) Moderated group authoring system for campus-wide workgroups. *IEEE Transactions on Mobile Computing* 99(PrePrints).

[13] KOTZ, D., HENDERSON, T. and ABYZOV, I. (2004), CRAWDAD trace set dartmouth/campus/snmp (v. 2004-11-09).

[14] HENDERSON, T. and KOTZ, D. (2007), CRAWDAD trace dartmouth/campus/syslog/05_06 (v. 2007-02-08), http://crawdad.cs.dartmouth.edu/dartmouth/campus/syslog/05_06.

[15] HENDERSON, T. (2006), CRAWDAD tool tools/process/syslog/syslog_parser (v. 2006-11-01), http://crawdad.cs.dartmouth.edu/tools/process/syslog /syslog_parser.

[16] KOTZ, D. and ESSIEN, K. (2002) Analysis of a campus-wide wireless network. In *ACM MobiCom '02*: 107–118.

[17] Zero configuration networking (zeroconf), http://www.zeroconf.org/.

[18] VOIDA, A., GRINTER, R.E., DUCHENEAUT, N., EDWARDS, W.K. and NEWMAN, M.W. Listening in: practices surrounding itunes music sharing. In *CHI '05*: 191–200.

[19] BALAZINSKA, M. and CASTRO, P. (2003), CRAWDAD data set ibm/watson (v. 2003-02-19), http://crawdad.cs. dartmouth.edu/ibm/watson.

[20] BALAZINSKA, M. and CASTRO, P. (2003) Characterizing mobility and network usage in a corporate wireless local-area network. In *MobiSys '03*: 303–316.

[21] BALACHANDRAN, A., VOELKER, G.M., BAHL, P. and RANGAN, P.V. (2002), CRAWDAD data set ucsd/sigcomm2001 (v. 2002-04-23), http://crawdad.cs.dartmouth.edu/ucsd/sigcomm2001.

[22] BALACHANDRAN, A., VOELKER, G.M., BAHL, P. and RANGAN, P.V. (2002) Characterizing user behavior and network performance in a public wireless lan. In *ACM SIGMETRICS '02*: 195–205.

[23] SCHULMAN, A., LEVIN, D. and SPRING, N. (2009), CRAWDAD data set umd/sigcomm2008 (v. 2009-03-02), Downloaded from http://crawdad.cs.dartmouth.edu/umd/sigcomm2008.

[24] LENCZNER, M., GREGOIRE, B. and PROULX, F. (2007), CRAWDAD trace ilesansfil/wifidog/session/04_07 (v. 2007-08-27), http://crawdad.cs.dartmouth.edu/ ilesansfil.

[25] CHINCHILLA, F., LINDSEY, M. and PAPADOPOULI, M. (2004) Analysis of wireless information locality and association patterns in a campus. In *IEEE INFOCOM '04*.

[26] BLINN, D.P., HENDERSON, T. and KOTZ, D. (2005) Analysis of a Wi-Fi hotspot network. In *International Workshop on Wireless Traffic Measurements and Modeling (WiTMeMo '05)*: 1–6. URL http://www.cs.dartmouth. edu/~dfk/papers/blinn:hotspot.pdf.

[27] ELLIS, C.A. and GIBBS, S.J. (1989) Concurrency control in groupware systems. *SIGMOD Rec.* 18(2): 399–407.

[28] MORRIS, J.H., SATYANARAYANAN, M., CONNER, M.H., HOWARD, J.H., ROSENTHAL, D.S. and SMITH, F.D. (1986) Andrew: a distributed personal computing environment. *Commun. ACM* 29(3): 184–201.

[29] SHEPLER, S., CALLAGHAN, B., ROBINSON, D., THURLOW, R., BEAME, C., EISLER, M. and NOVECK, D. (2003), Network file system (nfs) version 4 protocol, RFC 3530.

[30] ADYA, A., BOLOSKY, W.J., CASTRO, M., CERMAK, G., CHAIKEN, R., DOUCEUR, J.R., HOWELL, J. *et al.* (2002) Farsite: federated, available, and reliable storage for an incompletely trusted environment. *SIGOPS Oper. Syst. Rev.* 36(SI): 1–14.

[31] VAZHKUDAI, S.S., MA, X., FREEH, V.W., STRICKLAND, J.W., TAMMINEEDI, N., SIMON, T. and SCOTT, S.L. (2006) Constructing collaborative desktop storage caches for large scientific datasets. *Trans. Storage* 2(3): 221–254.

[32] DOUCEUR, J.R. and WATTENHOFER, R.P. (2001) Optimizing file availability in a secure serverless distributed file system. In *20th IEEE Symposium on Reliable Distributed Systems* (New Orleans, LA): 4–13.

[33] YU, X. and CHANDRA, S. (2008) Campus-wide asynchronous lecture distribution using wireless laptops. In *ACM/SPIE MMCN'08* (San Jose, CA), 6818.

[34] SATYANARAYANAN, M., KISTLER, J.J., KUMAR, P., OKASAKI, M.E., SIEGEL, E.H. and STEERE, D.C. (1990) Coda: A highly available file system for a distributed workstation environment. *IEEE Transactions on Computers* 39(4).

[35] PAGE, T.W., GUY, R.G., HEIDEMANN, J.S., RATNER, D., REIHER, P., GOEL, A., KUENNING, G.H. *et al.* (1998) Perspectives on optimistically replicated peer-to-peer filing. *Software—Practice and Experience* 28(2): 155–180. URL http://www.isi.edu/~johnh/PAPERS/ Page98a.html.

[36] DEMERS, A., PETERSEN, K., SPREITZER, M.J., TERRY, D., THEIMER, M. and WELCH, B. (1994) The bayou architecture: support for data sharing among mobile users. In *Workshop on Mobile Computing Systems and Applications* (Santa Cruz, CA): 2–7.

[37] DEMERS, A., GREENE, D., HAUSER, C., IRISH, W., LARSON, J., SHENKER, S., STURGIS, H. *et al.* (1987) Epidemic algorithms for replicated database maintenance. In *PODC*: 1–12.

[38] BIRMAN, K. (2007) The promise, and limitations, of gossip protocols. *SIGOPS Oper. Syst. Rev.* 41(5): 8–13.

[39] VAHDAT, A. and BECKER, D. (2000) *Epidemic Routing for Partially Connected Ad Hoc Networks*. Tech. Rep. CS-2000-06, Duke University.

[40] TANG, D. and BAKER, M. (2000) Analysis of a local-area wireless network. In *ACM Mobicom '00*: 1–10.

[41] PAPADOPOULI, M., SHEN, H. and SPANAKIS, M. (2005) Characterizing the duration and association patterns of wireless access in a campus. In *11th European Wireless Conference* (Nicosia, Cyprus).

[42] JEN HSU, W. and HELMY, A. (2005) *IMPACT: Investigation of Mobile-user Patterns Across University Campuses using Wlan Trace Analysis*. Tech. rep., USC.

[43] MCNETT, M. and VOELKER, G.M. (2005) Access and mobility of wireless pda users. *SIGMOBILE Mob. Comput. Commun. Rev.* 9(2): 40–55.

[44] SONG, L. and KOTZ, D. (2007) Evaluating opportunistic routing protocols with large realistic contact traces. In *ACM MobiCom workshop on Challenged Networks (CHANTS 2007)*.

[45] DAVIS, J.A., FAGG, A.H. and LEVINE, B.N. (2001) Wearable computers as packet transport mechanisms in highly-partitioned ad–hoc networks. In *IEEE Intl. Symp. on Wearable Computers*: 141–148.

[46] FALL, K. (2003) A delay-tolerant network architecture for challenged internets. In *SIGCOMM '03*: 27–34.

[47] JAIN, S., FALL, K. and PATRA, R. (2004) Routing in a delay tolerant network. In *Sigcomm '04*: 145–158.

[48] BAKHSHI, R., BONNET, F., FOKKINK, W. and HAVERKORT, B. (2007) Formal analysis techniques for gossiping protocols. *SIGOPS Oper. Syst. Rev.* 41(5): 28–36.

[49] KUENNING, G.H., BAGRODIA, R., GUY, R.G., POPEK, G.J., REIHER, P.L. and WANG, A.I. (1998) Measuring the quality of service of optimistic replication. In *ECOOP '98: Workshop ion on Object-Oriented Technology* (London, UK: Springer-Verlag): 319–320.

[50] JELASITY, M., GUERRAOUI, R., KERMARREC, A.M. and van STEEN, M. (2004) The peer sampling service: experimental evaluation of unstructured gossip-based implementations. In *Middleware '04* (New York, NY, USA: Springer-Verlag New York, Inc.): 79–98.

[51] KWIATKOWSKA, M., NORMAN, G. and PARKER, D. (2008) Analysis of a gossip protocol in prism. *SIGMETRICS Perform. Eval. Rev.* **36**(3): 17–22.

[52] SAITO, Y. and SHAPIRO, M. (2005) Optimistic replication. *ACM Comput. Surv.* **37**(1): 42–81.

[53] TERRY, D.B., THEIMER, M.M., PETERSEN, K., DEMERS, A.J., SPREITZER, M.J. and HAUSER, C.H. (1995) Managing update conflicts in bayou, a weakly connected replicated storage system. In *ACM symposium on Operating systems principles* (Copper Mountain, CO): 172–182.

[54] GOLDING, R.A., LONG, D.D.E. and WILKES, J. (1994) The refdbms distributed bibliographic database system. In *Winter Usenix Conference* (San Francisco, CA): 47–62.

[55] ALLAVENA, A., DEMERS, A. and HOPCROFT, J.E. (2005) Correctness of a gossip based membership protocol. In *PODC '05* (New York, NY, USA: ACM): 292–301.

[56] KHELIL, A., BECKER, C., TIAN, J. and ROTHERMEL, K. (2002) An epidemic model for information diffusion in manets. In *MSWiM '02*: 54–60.

[57] MOTANI, M., SRINIVASAN, V. and NUGGEHALLI, P.S. (2005) Peoplenet: engineering a wireless virtual social network. In *MobiCom '05*: 243–257.

[58] ÉTIENNE RIVIÈRE, BALDONI, R., LI, H. and PEREIRA, J. (2007) Compositional gossip: a conceptual architecture for designing gossip-based applications. *SIGOPS Oper. Syst. Rev.* **41**(5): 43–50.

[59] ZHUANG, S., LAI, K., STOICA, I., KATZ, R. and SHENKER, S. (2003) Host mobility using an internet indirection infrastructure. In *MobiSys '03* (New York, NY, USA: ACM): 129–144.

[60] STOICA, I., ADKINS, D., ZHUANG, S., SHENKER, S. and SURANA, S. (2004) Internet indirection infrastructure. *IEEE/ACM Trans. Netw.* **12**(2): 205–218.

Permissions

All chapters in this book were first published in MCA, by EAI European Alliance for Innovation; hereby published with permission under the Creative Commons Attribution License or equivalent. Every chapter published in this book has been scrutinized by our experts. Their significance has been extensively debated. The topics covered herein carry significant findings which will fuel the growth of the discipline. They may even be implemented as practical applications or may be referred to as a beginning point for another development.

The contributors of this book come from diverse backgrounds, making this book a truly international effort. This book will bring forth new frontiers with its revolutionizing research information and detailed analysis of the nascent developments around the world.

We would like to thank all the contributing authors for lending their expertise to make the book truly unique. They have played a crucial role in the development of this book. Without their invaluable contributions this book wouldn't have been possible. They have made vital efforts to compile up to date information on the varied aspects of this subject to make this book a valuable addition to the collection of many professionals and students.

This book was conceptualized with the vision of imparting up-to-date information and advanced data in this field. To ensure the same, a matchless editorial board was set up. Every individual on the board went through rigorous rounds of assessment to prove their worth. After which they invested a large part of their time researching and compiling the most relevant data for our readers.

The editorial board has been involved in producing this book since its inception. They have spent rigorous hours researching and exploring the diverse topics which have resulted in the successful publishing of this book. They have passed on their knowledge of decades through this book. To expedite this challenging task, the publisher supported the team at every step. A small team of assistant editors was also appointed to further simplify the editing procedure and attain best results for the readers.

Apart from the editorial board, the designing team has also invested a significant amount of their time in understanding the subject and creating the most relevant covers. They scrutinized every image to scout for the most suitable representation of the subject and create an appropriate cover for the book.

The publishing team has been an ardent support to the editorial, designing and production team. Their endless efforts to recruit the best for this project, has resulted in the accomplishment of this book. They are a veteran in the field of academics and their pool of knowledge is as vast as their experience in printing. Their expertise and guidance has proved useful at every step. Their uncompromising quality standards have made this book an exceptional effort. Their encouragement from time to time has been an inspiration for everyone.

The publisher and the editorial board hope that this book will prove to be a valuable piece of knowledge for researchers, students, practitioners and scholars across the globe.

List of Contributors

Stephan Olariu
Department of Computer Science, Old Dominion University, Norfolk, Virginia, USA

Mohamed Eltoweissy
Pacific Northwest National Laboratory, Richland, Washington DC, USA

Mohamed Younis
Department of Computer Science and Electrical Engineering, University of Maryland, College Park, MD 20742, USA

Andrey Garnaev
WINLAB, Rutgers University, North Brunswick, USA

Wade Trappe
WINLAB, Rutgers University, North Brunswick, USA

Nathalie Mitton and David Simplot-Ryl
Inria Lille - Nord Europe - firstname.lastname@inria.fr

Jun Zheng
National Mobile Communications Research Lab - Southeast University, China

Sylwia Romaszko
Institute for Networked Systems, RWTH Aachen University, Kackertstrasse 9, 52072 Aachen, Germany

Daniel Denkovski, Valentina Pavlovska and Liljana Gavrilovska
Faculty of Electrical Engineering and Information Technologies, Ss. Cyril and Methodius University in Skopje, Macedonia

Paolo Bellavista, Antonio Corradi and Andrea Reale
DISI - University of Bologna, Italy

Seyedali Hosseininezhad and Victor C. M. Leung
Department of Electrical and Computer Engineering, The University of British Columbia, Vancouver, BC, Canada

Donglin Hu and Shiwen Mao
Department of Electrical and Computer Engineering, Auburn University, Auburn, AL 36849-5201, USA

Wissam Chahin and Rachid El-Azouzi
CERI/LIA, University of Avignon, 339, chemin des Meinajaries, Avignon, France

Francesco De Pellegrini
CREATE-NET Via alla Cascata 56/D, Povo, Trento, Italy

Amar Prakash Azad
SOE, UCSC, USA

Jun Li, Mylène Toulgoat, Yifeng Zhou and Louise Lamont
Communications Research Centre Canada, 3701 Carling Avenue, Ottawa, ON. K2H 8S2 Canada

David Cairns and Marwan Fayed
Computing Science and Math, University of Stirling, Stirling, FK9 4LA, UK

Hussein T. Mouftah
SITE, University of Ottawa, Ottawa, ON, K1N 6N5, Canada

Ahmad Radaideh and John N. Daigle
University of Mississippi, University, MS 38677, USA

Jose Oscar Fajardo, Ianire Taboada and Fidel Liberal
NQaS Research Group, Department of Electronics and Telecommunications, University of the Basque Country (UPV/EHU), ETSI Bilbao, 48013 Bilbao, Bizkaia, Spain

Donglin Hu and Shiwen Mao
Department of Electrical and Computer Engineering, Auburn University, Auburn, AL 36849-5201, USA

Andrew Tinka
Electrical Engineering and Computer Sciences, University of California, Berkeley, CA, USA

Thomas Watteyne
Berkeley Sensor & Actuator Center, University of California, Berkeley, CA, USA

Currently with Dust Networks, Hayward, CA, USA

Kristofer S. J. Pister
Berkeley Sensor & Actuator Center, University of California, Berkeley, CA, USA

Alexandre M. Bayen
Systems Engineering, Department of Civil and Environmental Engineering, University of California, Berkeley, CA, USA

Alexandre Bouard and Dennis Burgkhardt
BMW Forschung und Technik GmbH, Munich, Germany

Claudia Eckert
Technische Universität München, Garching near Munich, Germany

Yibo Zhu, Zhong Zhou, Zheng Peng, Michael Zuba and Jun-Hong Cui
Department of Computer Science and Engineering, University of Connecticut, Storrs, CT 06269, USA

Surendar Chandra
FX Palo Alto Laboratory, Palo Alto, CA 94304, USA

Xuwen Yu
VMware Inc., Palo Alto, CA 94304, USA

Index

www.ingramcontent.com/pod-product-compliance
Lightning Source LLC
Chambersburg PA
CBHW080531200326
41458CB00012B/4398